# Genome Informatics 2010

# GENOME INFORMATICS SERIES (GIS)
## ISSN: 0919-9454

The Genome Informatics Series publishes peer-reviewed papers presented at the International Conference on Genome Informatics (GIW) and some conferences on bioinformatics. The Genome Informatics Series is indexed in MEDLINE.

| No. | Title | Year | ISBN Cl./Pa. |
|-----|-------|------|--------------|
| 1 | Genome Informatics Workshop I | 1990 | (in Japanese) |
| 2 | Genome Informatics Workshop II | 1991 | (in Japanese) |
| 3 | Genome Informatics Workshop III | 1992 | (in Japanese) |
| 4 | Genome Informatics Workshop IV | 1993 | 4-946443-20-7 |
| 5 | Genome Informatics Workshop 1994 | 1994 | 4-946443-24-X |
| 6 | Genome Informatics Workship 1995 | 1995 | 4-946443-33-9 |
| 7 | Genome Informatics 1996 | 1996 | 4-946443-37-1 |
| 8 | Genome Informatics 1997 | 1997 | 4-946443-47-9 |
| 9 | Genome Informatics 1998 | 1998 | 4-946443-52-5 |
| 10 | Genome Informatics 1999 | 1999 | 4-946443-59-2 |
| 11 | Genome Informatics 2000 | 2000 | 4-946443-65-7 |
| 12 | Genome Informatics 2001 | 2001 | 4-946443-72-X |
| 13 | Genome Informatics 2002 | 2002 | 4-946443-79-7 |
| 14 | Genome Informatics 2003 | 2003 | 4-946443-82-7 |
| 15 | Genome Informatics 2004 Vol. 15, No. 1 | 2004 | 4-946443-88-6 |
| 16 | Genome Informatics 2004 Vol. 15, No. 2 | 2004 | 4-946443-91-6 |
| 17 | Genome Informatics 2005 Vol. 16, No. 1 | 2005 | 4-946443-93-2 |
| 18 | Genome Informatics 2005 Vol. 16, No. 2 | 2005 | 4-946443-96-7 |
| 19 | Genome Informatics 2006 Vol. 17, No. 1 | 2006 | 4-946443-97-5 |
| 20 | Genome Informatics 2006 Vol. 17, No. 2 | 2006 | 4-946443-99-1 |
| 21 | Genome Informatics 2007 Vol. 18 | 2007 | 978-1-86094-991-3 |
| 22 | Genome Informatics 2007 Vol. 19 | 2007 | 978-1-86094-984-5 |
| 23 | Genome Informatics 2008 Vol. 20 | 2008 | 978-1-84816-299-0 |
| 24 | Genome Informatics 2008 Vol. 21 | 2008 | 978-1-84816-331-7 |
| 25 | Genome Informatics 2009 Vol. 22 | 2009 | 978-1-84816-569-4 |
| 26 | Genome Informatics 2009 Vol. 23 | 2009 | 978-1-84816-562-5 |

Genome Informatics Series Vol. 24            ISSN: 0919-9454

# Genome Informatics 2010

The 10th Annual International Workshop on Bioinformatics and Systems Biology (IBSB 2010)

Kyoto University, Japan            26 – 28 July 2010

Editors

## Tatsuya Akutsu • Minoru Kanehisa

Kyoto University, Japan

## Edda Klipp

Humboldt-Universität zu Berlin, Germany

## Satoru Miyano

University of Tokyo, Japan

## Scott Mohr • Thomas Tullius

Boston University, USA

## Iwona Wallach

Charité – Universitätsmedizin Berlin, Germany

Imperial College Press

*Published by*

Imperial College Press
57 Shelton Street
Covent Garden
London WC2H 9HE

*Distributed by*

World Scientific Publishing Co. Pte. Ltd.
5 Toh Tuck Link, Singapore 596224
*USA office:* 27 Warren Street, Suite 401-402, Hackensack, NJ 07601
*UK office:* 57 Shelton Street, Covent Garden, London WC2H 9HE

**British Library Cataloguing-in-Publication Data**
A catalogue record for this book is available from the British Library.

**GENOME INFORMATICS 2010**
**Genome Informatics Series Vol. 24**
**Proceedings of the 10th Annual International Workshop on Bioinformatics and Systems Biology**
**(IBSB 2010)**

ISBN-13 978-1-84816-657-8
ISBN-10 1-84816-657-5

# CONTENTS

# PREFACE

*Genome Informatics Vol. 24* contains the peer-reviewed papers presented at the Tenth Annual International Workshop on Bioinformatics and Systems Biology held on July 26-July 28 at Obaku Plaza, Uji Campus, Kyoto University. This workshop started in 2001, Berlin, as an event for doctoral students and young researchers to present and discuss their research results and approaches in Bioinformatics and Systems Biology.

Since 2001, the workshop has been held in Boston (2002), Berlin (2003), Kyoto (2004), Berlin (2005), Boston (2006), Tokyo (2007), Berlin (2008), and Boston (2009). In 2010, it was held in Kyoto as part of a collaborative educational program involving the leading institutions committing the following programs and partner institutions.

Programs

- Boston – Graduate Program in Bioinformatics, Boston University
- Berlin – The International Research Training Group (IRTG) "Genomics and Systems Biology of Molecular Networks"
- Kyoto – The JSPS International Training Program (ITP) "International Research and Training Program of Bioinformatics and Systems Biology"
- Tokyo – Global COE Program "Center of Education and Research for Advanced Genome-Based Medicine"

Partner Institutions

- Boston University
- Humboldt University Berlin
- Free University Berlin
- Max Planck Institute for Molecular Genetics
- Max-Delbrück Center for Molecular Medicine
- Bioinformatics Center, Institute for Chemical Research, Kyoto University
- Department of Bioinformatics and Chemical Genomics, Graduate School of Pharmaceutical Sciences, Kyoto University
- Human Genome Center, Institute of Medical Science, University of Tokyo

All papers included in this volume were reviewed by program committee members. These papers will be indexed in Medline, and their electronic versions are freely available from the website of Japanese Society for Bioinformatics as *Genome In-*

*formatics Online* (http://www.jsbi.org/modules/journal1/index.php/index.html). Former publications are also electronically available as Genome Informatics Vol. 15, No. 1 (2004), Vol. 16, No. 1 (2005), Vol. 17, No. 1 (2006), Vol. 18 (2007), Vol. 20 (2008), and Vol. 22 (2009).

We wish to thank all of those who submitted papers and helped with the reviewing process. We also wish to thank colleagues at Bioinformatics Center, Institute for Chemical Research, Kyoto University, and Human Genome Center, Institute of Medical Science, University of Tokyo, for their efforts in local arrangement, finance, and publication. In particular, Morihiro Hayashida, Emi Ikeda, Azumi Saito, Takeyuki Tamura, and Junko Yamamoto.

*Program Committee Chair*:
**Tatsuya Akutsu**


*Steering Committee*:
**Minoru Kanehisa**
**Edda Klipp**
**Satoru Miyano**
**Scott Mohr**
**Thomas Tullius**
**Iwona Wallach**

# PROGRAM COMMITTEE

# KINETIC MODELLING OF DNA REPLICATION INITIATION IN BUDDING YEAST

MATTEO BARBERIS
matteo.barberis@biologie.hu-berlin.de

THOMAS W. SPIESSER
thomas.spiesser@biologie.hu-berlin.de

EDDA KLIPP
edda.klipp@rz.hu-berlin.de

*Institute for Biology, Theoretical Biophysics, Humboldt University Berlin, Invalidenstraße 42, 10115 Berlin, Germany*

DNA replication is restricted to a specific time window of the cell cycle, called S phase. Successful progression through S phase requires replication to be properly regulated to ensure that the entire genome is duplicated exactly once, without errors, in a timely fashion. As a result, DNA replication has evolved into a tightly regulated process involving the coordinated action of numerous factors that function in all phases of the cell cycle. Biochemical mechanisms driving the eukaryotic cell division cycle have been the subject of a number of mathematical models. However, cell cycle networks reported in literature so far have not addressed the steps of DNA replication events. In particular, the assembly of the replication machinery is crucial for the timing of S phase. This event, called "initiation", which occurs in late M / early $G_1$ of the cell cycle, starts with the assembly of the pre-replicative complex (pre-RC) at the origins of replication on the DNA. Its activation depends on the availability of different kinase complexes, cyclin-dependent kinases (CDKs) and Dbf-dependent kinase (DDK), which phosphorylate specific components of the pre-RC to convert it into the pre-initiation complex (pre-IC). We have developed an ODE-based model of the network responsible for this process in budding yeast by using mass-action kinetics. We considered all steps from the assembly of the first components at the DNA replication origin up to the active replisome that recruits the polymerases and verified the computational dynamics with the available literature data. Our results highlighted the link between activation of CDK and DDK and the step-by-step formation of both     pre-RC and pre-IC, suggesting S-CDK (Cdk1-Clb5,6) to be the main regulator of the process.

*Keywords*: Cell cycle; budding yeast; CDK; DDK; kinetic modelling.

## 1.    Introduction

The maintenance of the eukaryotic genome integrity requires its precise and coordinated replication each time a cell divides. To achieve this coordination, eukaryotic cells employ an ordered sequence of events to form several protein complexes at specific sequences on the DNA, so called origins of replication. DNA replication begins at many origins scattered along each chromosome and origin activation or "firing" in budding yeast occurs throughout the S phase according to a temporal program [1, 44], being the origins activated continuously with the majority of the events in mid-S phase [55].

DNA replication has evolved into a tightly regulated process, involving the coordinated action of numerous factors. More than 20 different components have been identified to participate in DNA replication [8, 46] and the initiation of this process is considered the most critical event because all components have to assemble in an ordered fashion. First, in late mitosis and early $G_1$ phase, a large complex of initiator proteins, called pre-replicative complex (pre-RC), recognizes the origins and assembles to prepare them for firing [20]. This process is also referred to as "licensing". Second, in early S phase, the pre-RC is converted into an active pre-initiation complex (pre-IC) that facilitates unwinding of the origins and loading of the DNA synthesis machinery [8, 68, 75]. DNA replication is regulated by recruiting the replication machinery, or "replisome", to the origins on the chromosome. When a replication origin has fired, the replisome replicates the DNA bidirectionally – a process called "elongation" – as experimentally reported [50, 54]. The initiation and, hence, the recruitment process that forms pre-RC and pre-IC represents the central regulation of the entire process. In fact, when a change in replication rate is needed, adjustments are made to initiation. For example, the rate of DNA replication is almost two orders of magnitude faster in embryos than in somatic cells because more origins are used, resulting in a multiple activation of initiation complexes [60]. Once an origin has been activated, the replisome disassembles, and its reassembly is prevented until the next $G_1$ phase. As a result, each origin is used once, and only once, per cell cycle [41].

Cell cycle regulation by protein phosphorylation ensures that pre-RC assembly can only occur in $G_1$ phase, while pre-IC formation and activation of the replisome can only occur in S phase. These processes require the activity of two classes of protein kinases, CDKs (cyclin-dependent kinases: $G_1$-CDK, Cdk1-Cln1,2 and S-CDK, Cdk1-Clb5,6) and DDK (Dbf4-dependent kinase, Cdc7-Cbf4), both active at, and independently regulate, the beginning of S phase. These kinases trigger specific steps during the replication, particularly those leading to the association of Cdc45 – a pivotal factor in the transition to replication [2, 80] – with chromatin and the unwinding of the origins of DNA replication. The last step of the replisome formation is followed by Cdc45-dependent loading of the DNA polymerases $\alpha$, $\delta$ and $\varepsilon$ [3, 8, 36, 68]. As a consequence, both S-CDK and DDK control the timing of replication, acting at individual origins throughout S phase to activate early- and late-firing origins [12, 22, 23, 44]. The fundamental role of CDK in promoting origin activation and, thus, initiation of DNA replication has been recently elucidated [69, 77, 78], and many replication factors have been shown to be phosphorylated *in vitro* and *in vivo* (see [37] for a review). A bulk of evidence suggests that CDK and DDK regulate distinct steps in activation of replication origins [16, 19, 43, 47, 53, 62, 69, 77, 78, 80] but, although the essential mechanism of CDK activation of DNA replication has been established, exactly how DDK acts has been unclear. Genetic studies in budding yeast indicate that there is a sequential order to kinase action for replication, with S-CDK preceding DDK [53]. Dbf4 is absent in $G_1$ phase because it is targeted for degradation by the Anaphase-Promoting-Complex, APC [43, 61]. As cells enter into S phase, APC is inactivated by CDK phosphorylation, and Dbf4 is stabilized, thus activating DDK. Moreover, it has been suggested that S-CDK phosphorylation of target proteins might prime subsequent DDK phosphorylation of the same protein on adjacent residues. For instance, evidence show that DDK phosphorylation of Mcm2 is enhanced by prior CDK phosphorylation [16, 47], but it has also been reported that a proper balance between S-CDK and DDK activities on Mcm4 is necessary for efficient assembly or stabilization of the replisome [19, 62].

In this work, we present the first mathematical model of DNA replication initiation, defined according to the current knowledge of the key molecules of the process. To this end, we collected the available data concerning the assembly of pre-RC and pre-IC and highlighted the relation between activation of CDK and DDK and the step-by-step formation of the replisome. In our model, the precise temporal order in the activation of CDK and DDK during the replication process accounts for the correct initiation steps of DNA replication. Moreover, the computational dynamics simulated by the model have been fitted to the experimental data of the pre-RC formation. In summary, these data provide a picture of the global regulation of DNA replication initiation, where S-CDK exploits a key role in the fine tuning of the process. The findings support the fact that S-CDK is indeed crucial to trigger the temporal activation of the replication origins, as we have recently reported by developing both probabilistic and deterministic models of DNA replication [4-6, 66].

## 2. ODE Model of the DNA Replication Initiation

### 2.1. Model Construction

An ODE (Ordinary Differential Equation) model for the DNA replication initiation has been constructed according to the current biological knowledge (see Appendix for a detailed description). The model includes 47 metabolites (1 DNA element, 25 proteins, 16 DNA-protein complexes, 2 protein-protein complexes and 3 kinase complexes), 25 reactions and 25 kinetic parameters. The network described with the mathematical model is presented in Fig. 1. All reactions have been assigned with mass action kinetics and are listed in Table 1. DNA and proteins were specified with their respective names, and individual DNA-protein complexes formed in the course of DNA replication initiation have been numbered consecutively according to their successive appearance, from $c1$ to $c16$. The first complex $c1$ is formed by *ARS* (Autonomously Replicating Sequence) and Orc1-6, after the binding of Orc1-6 to the *ARS* sequence on the DNA. After this first step, the chain of reactions regards the formation of the complexes $c2$-$c15$, evolving from one to another through various association, phosphorylation and dissociation reactions, to form the last complex $c16$, the active replisome, loaded with all necessary components to start DNA replication. The protein-protein complexes (Mcm2-7-Cdt1, Sld2-P-Dpb11) and the kinase complexes (Cdk1-Cln1,2, Cdk1-Clb5,6 and Cdc7-Dbf4) were named according to their single components. Step-by-step formation of different complexes by the specific kinase complexes $G_1$-CDK (Cdk1-Cln1,2), S-CDK (Cdk1-Clb5,6) and DDK (Cdc7-Dbf4) is reported in detail in Fig. 1.

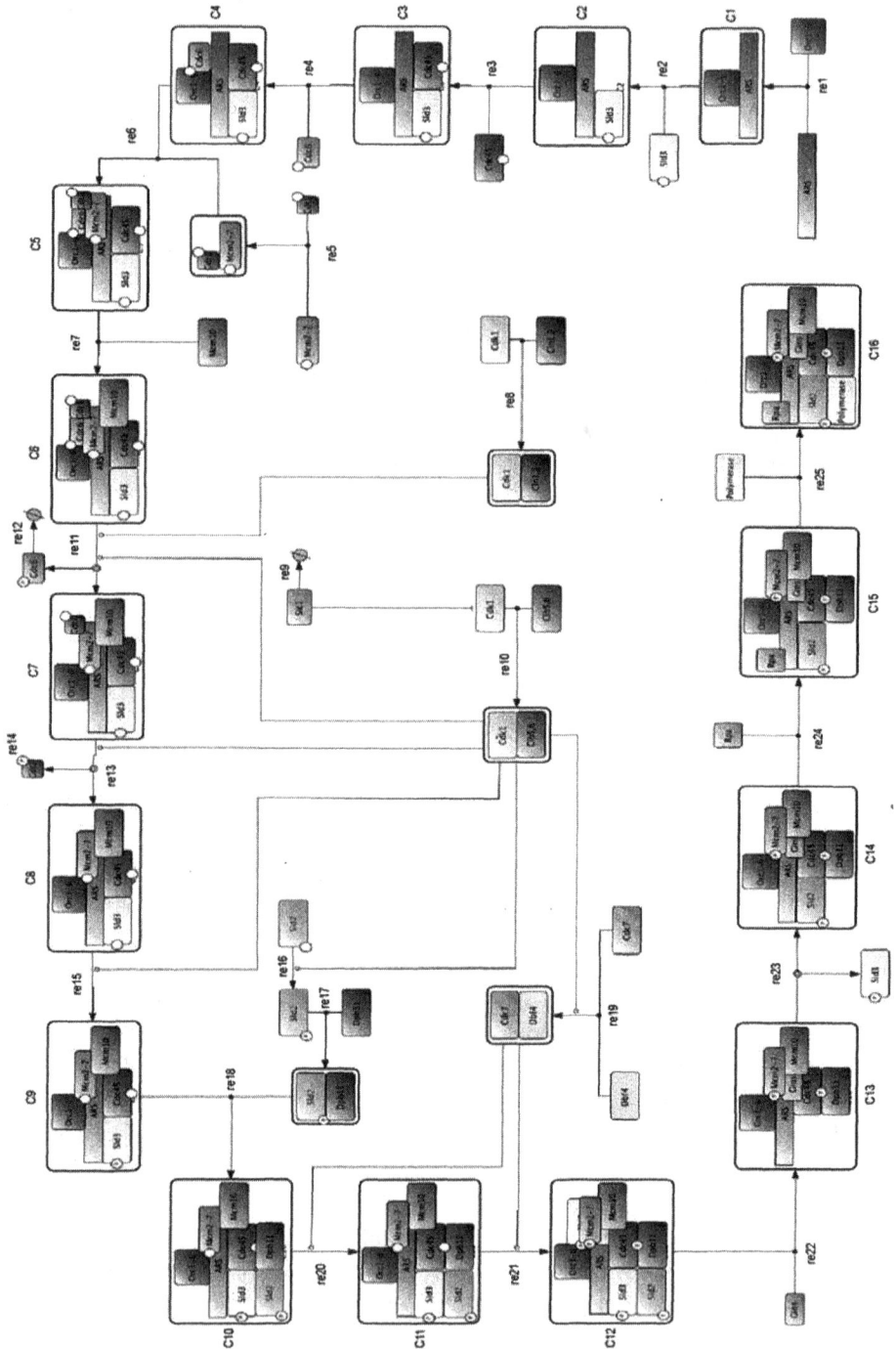

Fig. 1. Scheme of the ODE model representing the relevant steps of the DNA replication initiation. In each complex, phosphorylation events are indicated with white circles.

Table 1. Reactions implemented in the ODE model. Reaction modifiers are specified with algebraic signs: activator (+), inhibitor (−).

| Reaction ID | Reaction | Modifier |
|---|---|---|
| re1 | ARS + Orc1-6 → c1 | |
| re2 | c1 + Sld3 → c2 | |
| re3 | c2 + Cdc45 → c3 | |
| re4 | c3 + Cdc6 → c4 | |
| re5 | Mcm2-7 + Cdt1 → Mcm2-7-Cdt1 | |
| re6 | c4 + Mcm2-7-Cdt1 → c5 | |
| re7 | c5 + Mcm10 → c6 | |
| re8 | Cdk1 + Cln1,2 → Cdk1-Cln1,2 | |
| re9 | Sic1 → Sic1 degraded | |
| re10 | Cdk1 + Clb5,6 → Cdk1-Clb5,6 | − Sic1 |
| re11 | c6 → Cdc6-P + c7 | + Cdk1-Clb5,6; + Cdk1-Cln1,2 |
| re12 | Cdc6-P → Cdc6 degraded | |
| re13 | c7 → Cdt1-P + c8 | + Cdk1-Clb5,6 |
| re14 | Cdt1-P → Cdt1 exported | |
| re15 | c8 → c9 | + Cdk1-Clb5,6 |
| re16 | Sld2 → Sld2-P | + Cdk1-Clb5,6 |
| re17 | Sld2-P + Dpb11 → Sld2-P-Dpb11 | |
| re18 | c9 + Sld2-P-Dpb11 → c10 | |
| re19 | Cdc7 + Dbf4 → Cdc7-Dbf4 | + Cdk1-Clb5,6 |
| re20 | c10 → c11 | + Cdc7-Dbf4 |
| re21 | c11 → c12 | + Cdc7-Dbf4 |
| re22 | c12 + GINS → c13 | |
| re23 | c13 → Sld3-P + c14 | |
| re24 | c14 + RPA → c15 | |
| re25 | c15 + polymerase → c16 | |

Initial concentrations of the components considered in the model have been computed using the following equation:

$$[C] = \frac{N_C}{V_{cell} \times N_A} \tag{1}$$

where $[C]$ is the initial concentration of the component, $N_C$ the components abundance, $V_{cell}$ the volume of the cell, and $N_A$ the Avogadro constant (number of molecules in one mole). The volume of the cell has been considered to be $V_{cell} = 6.5 \times 10^{-14}$ *l* and the Avogadro constant was approximated with $N_A = 6.023 \times 10^{23}$ *mol*$^{-1}$. Protein abundances have been taken from the YeastGFP database, where a global analysis of protein expression and localization in the budding yeast is reported [81]. The database contains the measurements of absolute levels of a wide variety of proteins expressed during the log-phase growth, obtained by quantitative GFP labelling [29]. The complete list of the initial concentrations used is reported in Table 2. The following approximations have been made where protein abundances were ambiguous or not available:

1. The concentration of the DNA element, *ARS*, has been calculated using the relative abundance of replication origins (ORIs) in budding yeast, fixed to 450, in agreement with the previously reported data [55, 68, 76].

2. Protein abundances of the hexameric complexes Orc1-6 and Mcm2-7 have been approximated with the amounts found for Orc4 and Mcm4 subunits, respectively.
3. Protein abundance of the licensing factor Cdc6 has been chosen to be similar to the amount found for Cdt1.
4. Protein abundance of Sic1, cyclin-dependent kinase inhibitor (Cki) of the Cdk1-Clb complexes [45, 59] and homologous to the mammalian Cki p27$^{Kip1}$ [7], were arbitrarily chosen.
5. Protein abundances of the GINS (complex containing the four subunits Sld5, Psf1, Psf2 and Psf3, essential for both initiation and progression of DNA replication) and RPA complexes have been approximated with the amounts found for Psf1 and RPA1 subunits, respectively.
6. Protein abundance for the polymerase has been chosen as the mean of all values found for the different subunits.

Table 2. Initial concentrations used in the ODE model. Concentration units are expressed in *nmol/l* (nanomole/liter).

| Name | Initial concentration (*nmol/l*) | | Source |
|---|---|---|---|
| ARS | 11.5 | | approximated with 450 ORIs |
| Orc1-6 | 55.4 | (Orc4 subunit) | YeastGFP database |
| Sld3 | 3.2 | | YeastGFP database |
| Cdc45 | 44.2 | | YeastGFP database |
| Cdc6 | 55.0 | (similar to Cdt1) | arbitrarily chosen |
| Mcm2-7 | 224.8 | (Mcm4 subunit) | YeastGFP database |
| Cdt1 | 55.9 | | YeastGFP database |
| Mcm10 | 47.5 | | YeastGFP database |
| Sic1 | 2.0 | | arbitrarily chosen |
| Cdk1 | 170.3 | | YeastGFP database |
| Cln1,2 | 8.1 | | YeastGFP database |
| Clb5,6 | 13.3 | | YeastGFP database |
| Sld2 | 16.8 | | YeastGFP database |
| Dpb11 | 13.8 | | YeastGFP database |
| Cdc7 | 40.9 | | YeastGFP database |
| Dbf4 | 1.3 | | YeastGFP database |
| GINS | 36.5 | (Psf1 subunit) | YeastGFP database |
| RPA | 104.7 | (RPA1 subunit) | YeastGFP database |
| Polymerase | 40.0 | (mean of subunits) | YeastGFP database |

## 2.2. Deriving Parameters from Experimental Data

Three kinetic parameters of the ODE model have been derived from experimental time course data generated from Kawasaki and colleagues, which reconstituted *in vitro* the assembly of the budding yeast pre-RC [35]. They disclose the sequential loading of Orc1-6, Cdc6, Cdt1 and Mcm2-7 onto *ARS1* plasmid DNA in relative amounts. Hence, these data allowed us to derive kinetic parameters for the following reactions: (i) recruitment of Orc1-6 to the *ARS* binding site (*re1*), (ii) recruitment of Cdc6 into the complex c3 (*re4*), (iii) recruitment of the Cdt1-Mcm2-7 complex into the complex c4 (*re6*). These three reactions represent the formation of complex *C* from two reactants *A* and *B* and can, therefore, be described by bilinear rates

$$r = k \cdot [A] \cdot [B] \tag{2}$$

with $k$ being the kinetic constant. To calculate $k$, we applied a pseudo first-order approximation, which is appropriate if either $[A]$ or $[B]$ remains roughly constant throughout a reaction. In this case, the reaction effectively depends only on the concentration of one reactant. For example, the concentration of Orc1-6 is about 5 times higher than *ARS1* (Orc1-6 = 55.4 *nmol/l*, *ARS* = 11.5 *nmol/l*). Therefore

$$r = k \cdot [A] \cdot [B] = k' \cdot [A] \tag{3}$$

where $k' = k \cdot [B]_0$ and $[B]_0$ is the concentration of $[B]$ at time point 0. Making the assumption $C + A = \text{constant} = A_{total}$, we may write

$$\frac{dC}{dt} = \frac{d}{dt}(A_{total} - [A]_t) = -\frac{d}{dt}[A]_t = k' \cdot [A]_t \tag{4}$$

which can be integrated obtaining the following equation:

$$[A]_t = [A]_0 \cdot e^{-k't} \tag{5}$$

where $[A]_t$ is the concentration of $A$ at time point $t$ and $[A]_0$ the initial concentration of $A$. The initial concentrations used in the model are listed in Table 2.

The extracted values from the normalized experimental time course data for the reaction *re1*, the binding of Orc1-6 to *ARS1*, give the following: after $t = 60$ *sec*, 0.6 of the substrate (Orc1-6) is bound to the DNA (*ARS1*) and, therefore, 0.4 of the substrate is free (see Fig. 1B in [35]). Inserting this reasoning into the integrated Eq. (5), we obtain the following equation:

$$0.4 \cdot [A]_0 = [A]_0 \cdot e^{-k't} \tag{6}$$

The equation can be solved obtaining the value of $k'$

$$0.015 \frac{1}{s} = k' \tag{7}$$

Considering the previous assumption ($k' = k \cdot [B]_0$), $k'$ must be divided by the initial concentration of Orc1-6 (55.4 *nmol/l*):

$$\frac{0.015 \, s^{-1}}{55.4 \, (nmol/l)} = 0.000271 \cdot \frac{1}{(nmol \cdot s)} = k_1 \tag{8}$$

The result is an approximated kinetic parameter $k_1$ for the reaction *re1*. Accordingly, the kinetic parameters $k_4$ and $k_6$ have been derived for reactions *re4* and *re6*, respectively. The complete list of kinetic parameters is reported in Table 3.

Table 3. Kinetic parameter values used in the reactions of the ODE model.

| ID | Reaction ID | Parameter value in $l/(nmol \cdot s)$ | Source |
|---|---|---|---|
| $k_1$ | re1 | $2.71 \cdot 10^{-4}$ | this work |
| $k_2$ | re2 | $1.00$ | fitted to [Kawasaki et al., 2006] |
| $k_3$ | re3 | $1.00 \cdot 10^{-2}$ | fitted to [Kawasaki et al., 2006] |
| $k_4$ | re4 | $1.55 \cdot 10^{-4}$ | this work |
| $k_5$ | re5 | $1.93 \cdot 10^{-5}$ | fitted to [Kawasaki et al., 2006] |
| $k_6$ | re6 | $6.65 \cdot 10^{-4}$ | this work |
| $k_7$ | re7 | $3.82 \cdot 10^{-4}$ | fitted to [Kawasaki et al., 2006] |
| $k_8$ | re8 | $1.97 \cdot 10^{-6}$ | fitted to [Kawasaki et al., 2006] |
| $k_9$ | re9 | $1.00 \cdot 10^{-2}$ | fitted to [Kawasaki et al., 2006] |
| $k_{10}$ | re10 | $1.13 \cdot 10^{-6}$ | fitted to [Kawasaki et al., 2006] |
| $k_{11}$ | re11 | $4.28 \cdot 10^{-5}$ | fitted to [Kawasaki et al., 2006] |
| $k_{12}$ | re12 | $4.64 \cdot 10^{-1}$ | fitted to [Kawasaki et al., 2006] |
| $k_{13}$ | re13 | $2.17 \cdot 10^{-4}$ | fitted to [Kawasaki et al., 2006] |
| $k_{14}$ | re14 | $1.32 \cdot 10^{-5}$ | fitted to [Kawasaki et al., 2006] |
| $k_{15}$ | re15 | $6.63 \cdot 10^{-2}$ | fitted to [Kawasaki et al., 2006] |
| $k_{16}$ | re16 | $1.28 \cdot 10^{-5}$ | fitted to [Kawasaki et al., 2006] |
| $k_{17}$ | re17 | $2.68 \cdot 10^{-5}$ | fitted to [Kawasaki et al., 2006] |
| $k_{18}$ | re18 | $5.13 \cdot 10^{-3}$ | fitted to [Kawasaki et al., 2006] |
| $k_{19}$ | re19 | $1.94 \cdot 10^{-4}$ | fitted to [Kawasaki et al., 2006] |
| $k_{20}$ | re20 | $7.37 \cdot 10^{-3}$ | fitted to [Kawasaki et al., 2006] |
| $k_{21}$ | re21 | $5.18 \cdot 10^{-2}$ | fitted to [Kawasaki et al., 2006] |
| $k_{22}$ | re22 | $9.45 \cdot 10^{-4}$ | fitted to [Kawasaki et al., 2006] |
| $k_{23}$ | re23 | $8.32 \cdot 10^{-3}$ | fitted to [Kawasaki et al., 2006] |
| $k_{24}$ | re24 | $1.47 \cdot 10^{-3}$ | fitted to [Kawasaki et al., 2006] |
| $k_{25}$ | re25 | $1.61 \cdot 10^{-4}$ | fitted to [Kawasaki et al., 2006] |

## 2.3. Fitting Parameters to Experimental Data

In order to fix the remaining kinetic parameters, the ODE model has been fitted to the available experimental data concerning the assembly of the pre-RC in budding yeast [35]. The results revealed the sequential recruitment of ORC, Cdc6, Cdt1 and Mcm2-7 onto *ARS1*. When Mcm2-7 was maximally loaded, Cdc6 and Cdt1 were released, suggesting that these two proteins are co-ordinately regulated during pre-RC assembly [35].

We rescaled the experimental data points (see Figs. 1B and 1C in [35]) from relative amounts to concentrations in *nmol/l* using the abundance of the limiting reactant in each reaction of the ODE model, as shown in Fig. 2. For parameter estimation we used the Systems Biology Markup Language based Parameter Estimation Tool (SBML-PET) [79], which uses an evolutionary algorithm based on the Stochastic Ranking Evolution Strategy [33]. The three parameters $k_1$, $k_4$ and $k_6$ derived from the experimental data were considered fixed during the parameter estimation process. The remaining 22 parameters have been estimated during the fitting to the experimental time courses. The resulting parameters were fine tuned after the estimation process. Fig. 2 shows the time course for the concentrations of Orc1-6, Cdc6, Cdt1 and Mcm2-7. The ODE model of the DNA replication initiation, as sketched in Fig. 1, is able to reproduce the experimental time course data (Fig. 2). Since parameter values estimated with SMBL-PET reproduced the measured data points precisely, we used these results for further analysis.

## 3.    Results

### 3.1.    Simulation Results and Network Properties

The ODE model reported in Fig. 1 shows the ordered appearance and disappearance of the different complexes during DNA replication initiation (see Appendix for the detailed description). The first complex c1 is *ARS*-Orc1-6 and the last one is the active replisome c16. In between, all the other complexes c2-c15 define specific intermediate states on the highly regulated path that leads to the activation of the replication origins. The peak of pre-RC formation, which results in the complex c6, occurs at around 8 minutes after the initial recruitment of Orc1-6 to *ARS* (Fig. 3). The first intermediate complex of the pre-IC, c7, starts to appear at about the same time, which also represents the time when cells enter into S phase due to Cdk1-Cln1,2 and Cdk1-Clb5,6 activation. Complex c6 starts to be transformed into complex c7 when about the half of the final concentration of Cdk1-Clb5,6 is reached (Fig. 3). This is in agreement with our previous modelling studies, where firing of origins of replication was considered to occur at about 50 % of the maximum value of the Cdk1-Clb5,6 concentration [5, 6]. Moreover, Cdk1-Cln1,2 is active before Cdk1-Clb5,6. In the simulation, the accumulation of the complex c6 is rather high compared to the complexes c7 and c8, potentially indicating that there is a delay at the $G_1/S$ transition which allows for the correct completion of the pre-RC formation.

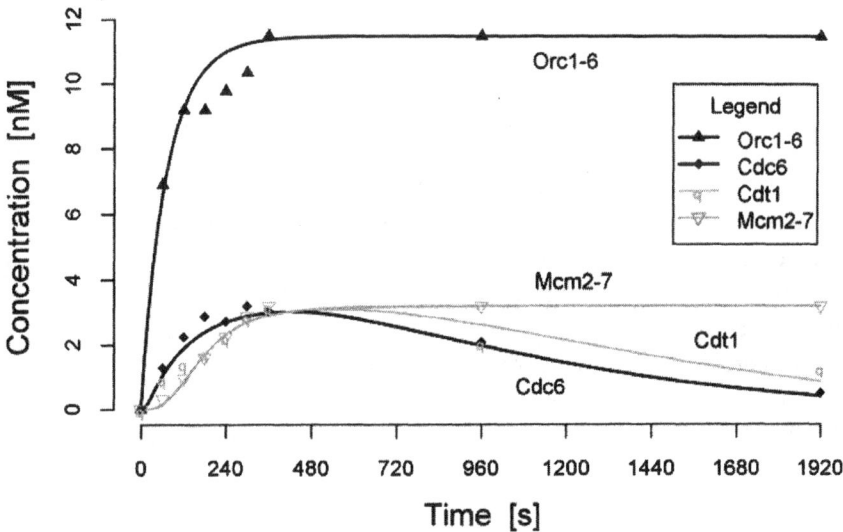

Fig. 2. Fitting of the ODE model to the experimental data of the pre-RC assembly. The fitting to the data has been performed with SBML-PET. Pre-RC assembly was performed from Kawasaki and colleagues for 6 *min* – measured every minute – and for 32 *min* – measured at 2, 4, 8, 16 and 32 *min* [41]. The experimental data points are shown as dots: Orc1-6 (solid triangles), Cdc6 (solid squares), Cdt1 (open circles) and Mcm2-7 (open triangles). The simulated curves are shown as lines.

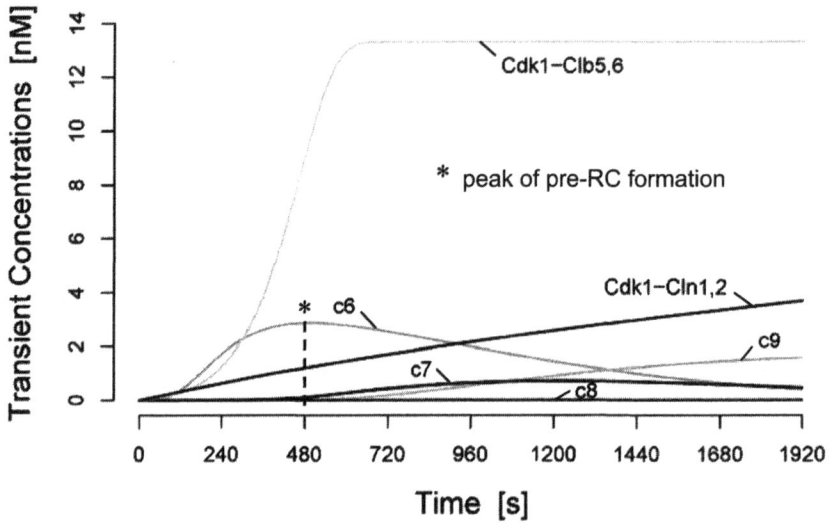

Fig. 3. Simulation of the complex formation during DNA replication initiation. The last complex c6 formed in G₁ phase, the pre-RC, and the complexes activated at the beginning of the S phase, c7, c8 and c9, are shown. Concentrations of the kinases active in S phase, Cdk1-Cln1,2 and Cdk1-Clb5,6, are also shown.

Indeed, it is possible to observe that, as soon as the cells enter into S phase and the Cdk1-Clb kinases have been activated, the complex transformation occurs rather fast, as indicated by low intermediate concentrations of both complexes c7 and c8. The complex c9, in contrast, accumulate again in bigger amounts compared to them (Figs. 3 and 4A).

The complex c9 appears after Cdc6 and Cdt1 being dislocated from the complexes and Cdk1-Clb5,6 having phosphorylated Sld3. The accumulation of the complex c9 mirrors a slow progression of DNA replication initiation, which is probably due to the gradual phosphorylation of Sld2 that has to take place [11]. In this way, the cell is thus able to assure that all preceding events are completed and that Cdk1-Clb activity has reached the necessary level to catalyze the specific complex activation. This process has been described as a switch for the S phase [11]. Figure 4B shows the simulation of the involved components by the model. The results of the simulations indicate that there seem to be more than one switch for the S phase. In fact, the progressive phosphorylation of Sld2 is the last halting point to check the appropriate course of the events that bring to the formation of the replisome [69, 78]. Nonetheless, a high accumulation of the complex c6 is denoted, which indicates that the activation of the Cdk1-Clb complexes and, therefore, the phosphorylation of both Cdc6 and Cdt1, act at a similar timing. The events that follow the phosphorylation of Sld2 occur fast, which is mirrored in their low transient concentrations in the simulation (data not shown).

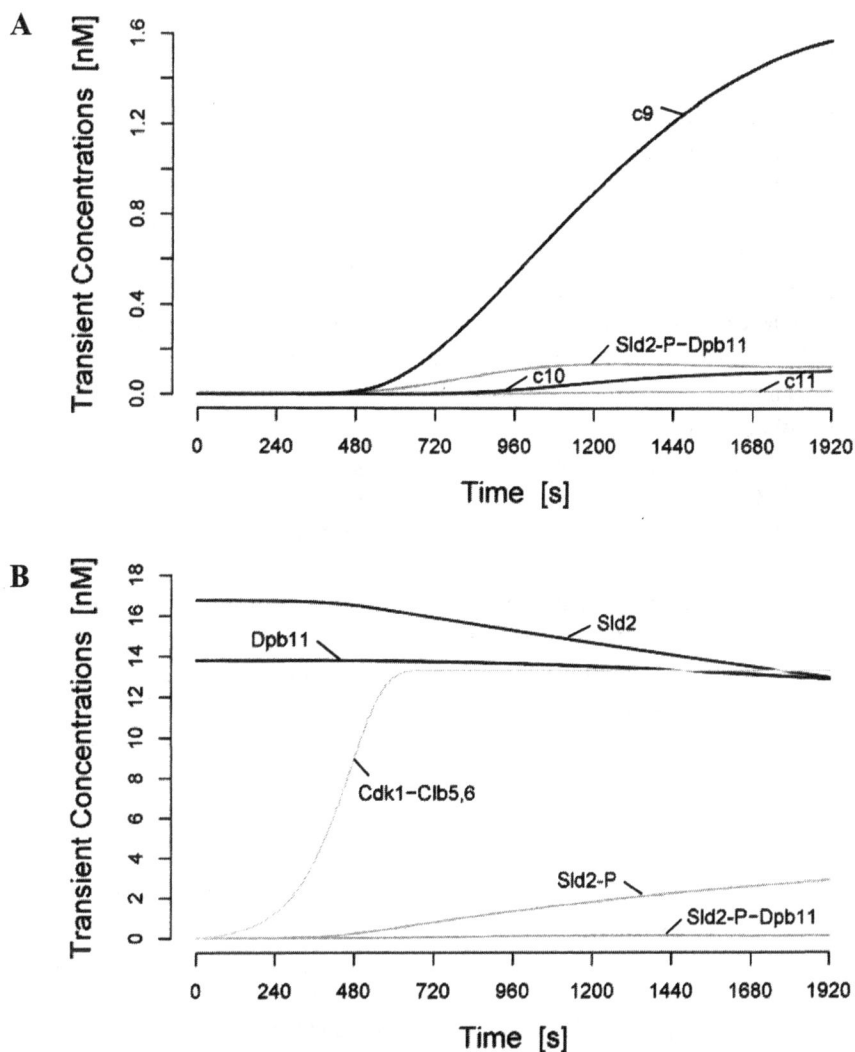

Fig. 4. Simulation of the switch for the S phase, according to Botchan, 2007 [46]. (A) The complex active prior to the switch, c9, is displayed, as well as the following complexes c10 and c11. The essential complex for the switch, Sld2-P-Dpb11, is also shown. (B) Concentrations of Sld2, Sld2-P, Dpb11, Sld2-P-Dpb11 and Cdk1-Clb5,6 are shown.

### 3.2. DNA Replication Mutants

Data about the single deletion mutants of the components involved in the assembly of the DNA replication machinery have been obtained from a large scale deletion study [30]. Nearly all single deletion mutants are lethal, with the only exception for the *clb5Δ clb6Δ* mutant, because other Cdk1-Clb complexes can substitute for Cdk1-Clb5,6 to a certain extent. Although there is considerable overlap in the function of Clb1-6 [48], the role of

Clb5 and Clb6 in DNA replication cannot be substituted by the mitotic Clb2 when all are expressed at physiological levels [18, 21]. In fact, the expression of Clb2 in $G_1$ phase instead of in late G2 phase and mitosis cannot correct the defect of the *clb5Δ* mutant, and the S phase takes longer because late origins do not fire [22]. Clearly, Clb2 has specificity distinct from that of Clb5 and Clb6 [21, 42]. However, despite that neither Clb2 nor Clb4 can completely substitute for Clb5 at late or even at early origins [21], in the *clb5Δ clb6Δ* mutant, initiation of DNA replication is delayed and late origins fire presumably due to other Cdk1-Clb activities. In support to it, the fact that a strain in which all six *CLB* genes have been deleted was viable after Clb1 overexpression, indicates that this specificity can be bypassed by overexpressing a single Clb protein [31].

The ODE model of the DNA replication initiation has been tested with the available deletion mutants for all components within the network. Due to its structure, where each component is necessary for the formation of the successive complex, simulations of single deletion mutants result in interruption of the signal transduction (data not shown). In the model, Cdk1-Cln1,2 can substitute for Cdk1-Clb5,6 in the reaction *re11*, but cannot substitute for it in all other reactions in which Cdk1-Clb5,6 is involved. For this reason, changes have to be applied in the model in the way that it could account for the viability of the *clb5Δ clb6Δ* mutant. In the future, the model could be further tested with depletion mutants (reduced availability of a specific component involved in the network) in order to investigate whether it is able to reproduce the altered phenotypes described in literature. In fact, this could be eventually provide a tool to study DNA replication initiation in the cell cycle regulation, where specific checkpoints due to deregulation of the components of the replisome formation can be observed.

## 4.   Discussion

One of the major features that the model of the DNA replication initiation shows is the role that CDK and DDK kinases exert in its fine regulation. We considered a specific order in which CDK and DDK act to regulate the DNA replication initiation in budding yeast, and the mathematical implementation of the network allowed us to highlight some essential features of the process. To begin with, we were able to earmark Cdk1-Clb5,6 as the key player and driving force of the process. Its crucial function in the cell cycle of budding yeast is widely known, which is the reason why the network has been structured around this kinase. The ODE model accounts for the fact that an essential function of Cdk1-Clb5,6 is to prevent the cell from re-replicating its DNA by phosphorylation of the licensing factors Cdc6 and Cdt1, and of the ORC and Mcm2-7 complexes as the cell enter into S phase (reviewed in [37, 51]). This preserves the essential switch for S phase, the progressive phosphorylation of Sld2 when Cdk1-Clb5,6 level rises, to ensure a temporally ordered complex formation at the $G_1$/S transition. Moreover, although DDK acts during the initiation process, Cdk1-Clb5,6 activity is required to activate DDK, which is as well implemented in the model.

In the simulation of the ODE model, we implemented a characteristic sigmoidal behavior of the Cdk1-Clb5,6 levels, in agreement with both the experimentally determined Clb5 cyclin levels [6]. The replication kinetics computed to study the temporal activation of the replication origins show the same sigmoidal behavior [5, 66]. This suggests that the replication efficiency might follow the dynamics of the Cdk1-Clb5,6 levels. Strikingly, it has been recently shown that efficiency of DNA replication in *Xenopus leavis* is indeed dependent on the nuclear activity of the homologous of Cdk1-Clb5,6, Cdk1- and Cdk2-cyclin E/A [38]. This strengthens our previous analysis, suggesting that Cdk1-Clb5,6 indeed plays a pivotal role in the DNA replication initiation and origin activation [5]. We have speculated that the dynamics in activation of replication origins could be possibly due to a different accessibility of the DNA to the components of the DNA replication initiation machinery [66], on the basis of conformational differences in the chromatin structure, histone acetylation or DNA methylation status at the origins of replication. Since CDK is well-known to be the driving force of the initial steps of the pre-RC and pre-IC formation [53], we speculate that origins can be activated dependent on a different accessibility to Cdk1-Clb5,6.

## 5. Conclusion

This work aimed at the understanding of the mathematically poorly elucidated DNA replication process in budding yeast. The ODE model describing the protein complexes assembly during DNA replication initiation is supported by the literature and its validity inferred by fitting to the available time course data. Yet, availability of experimental data is limited to the formation of the pre-RC. Therefore, biochemical and computational efforts are still required to further investigate the precise kinetics driving the replisome formation, possibly integrating them to the complexity of cell cycle regulation. However, our study highlights the role of Cdk1-Clb5,6 as key factor of DNA replication initiation, being the direct link between replication dynamics and cell cycle regulation.

## Acknowledgments

This work was supported by the Network of Excellence of the European Commission (Project ENFIN, contract number LSHG-CT-2005-518254) and the German Research Foundation (DFG) through IRTG 1360 (International Research Training Group "Genomics and Systems Biology of Molecular Networks").

## Appendix A.

### Appendix A.1.    Detailed Description of the Replisome Activation

*Appendix A.1.1.    Assembly of the Pre-RC*

The formation of the pre-RC begins in late M phase, when the origin recognition complex (Orc1-6), a six subunit and ATP-dependent DNA binding protein, binds to the replication origins [9, 10]. The DNA sequence of an origin is about 200 bp long and is called Autonomously Replicating Sequence (*ARS*) [49]. Within this region, an eleven base pair sequence, the so called *ARS* Consensus Sequence (ACS) is specifically recognized by the Orc1-6 complex [73]. A match to the ACS is essential, although the presence of this element alone does not define origin function per se [13, 52].

The next step in the pre-RC formation is the incorporation of the initiation factor Cdc6. Cdc6 interacts with Orc1-6, but only after Orc1-6 has bound to origin DNA [65]. The binding of Cdc6 increases the stability of Orc1-6 on the chromatin [32] and in turn the Orc1-6/Cdc6 complex shows an increased sequence specificity [64]. The loading of Cdc6 is directly followed by the recruitment of the minichromosome maintenance complex (Mcm2-7) to the origin of replication. At this juncture, Mcm2-7 has been suggested to be already in complex with the initiation factor Cdt1 [56]. ATP and Orc1-6/Cdc6 binding to the replication origin is required to load Cdt1/Mcm2-7 onto the origin [35], but the actual mode of recruitment is still unknown. It has been suggested that Cdc6 directs the loading via the interaction with Cdt1, although this has been shown so far only in mammalian cells [17]. Nevertheless, the fact that after the initial recruitment Mcm2-7 is still not fully and stably loaded on the chromatin, but simply in the close vicinity [56], supports this hypotheses. Mcm2-7 is considered to be the eukaryotic heterohexameric replicative DNA-helicase [28, 39] and can only be fully loaded onto the origin through Orc1-6's, Cdc6's and Cdt1's corporate action [56]. Once Mcm2-7 complexes are fully loaded, the other pre-RC components are dispensable [24, 32], indicating that the primary functions of Orc1-6, Cdc6 and Cdt1 in DNA replication are the origin recognition and the loading of the Mcm2-7 complex.

*Appendix A.1.2.    Assembly of the Pre-IC and Replisome Activation*

The transformation of the pre-RC into pre-IC occurs as cells enter into S phase and it is characterized by progressive activation of cyclin-dependent kinases ($G_1$-CDK, Cdk1-Cln1,2 and S-CDK, Cdk1-Clb5,6) and Dbf4-dependent kinase (DDK). DDK activation requires prior CDK activity [53]. $G_1$-CDK (Cdk1-Cln1,2) is the first kinase complex to become active and, by phosphorylation, targets Cdc6 for degradation in late $G_1$ phase [71]. As soon as S-CDK (Cdk1-Clb5,6) becomes active at the beginning of the S phase, both Cdc6 and Cdt1 are gradually phosphorylated. After phosphorylation, Cdc6 is degraded and Cdt1 is exported from the nucleus [25, 26, 70]. The nuclear export of Cdt1 and the degradation and Cdc6 are crucial steps during DNA replication initiation.

On the one hand, Cdc6 is a direct regulator of Cdk1 activity [14, 15, 27] and its degradation is necessary for rising the Cdk1 activity. On the other hand, phosphorylation of both Cdt1 and Cdc6 prevents *de novo* formation of pre-RC at this stage of the cell cycle and, therefore, prohibits chromosomal re-replication [68].

The next step of pre-IC formation is the loading of the initiation factors Sld3, Sld2 and the subunit Dpb11 of the DNA polymerase II $\varepsilon$ complex. Sld3 already associates with the replication origin during the $G_1$ phase of the cell cycle [34] and it is phophorylated by S-CDK (Cdk1-Clb5,6) after the cells enter into S phase. The phosphorylation triggers an essential interaction between ·Sld3 and Dpb11 [69, 78]. Furthermore, the phosphorylation of Sld2 by S-CDK triggers the formation of the Sld2/Dpb11 complex [67]. Interestingly, Sld2 presents multiple phosphorylation sites, which become phosphorylated when S-CDK activity rises. Only when a certain level of S-CDK activity is reached, Sld2 become hyperphosphorylated and changes its conformation to reveal another phosphorylation site, the threonine 84 residue (Thr84). This residue is the crucial target for S-CDK to form the complex between Sld2 and Dpb11 [69]. Both Sld2 and Sld3 bind to Dpb11, which in turn binds to the chromatin. Only at this stage the initiation factor Cdc45 can be stably incorporated into the nascent pre-IC [11]. By employing this control mechanism, the cell assures that a sufficient level of S-CDK activity is reached and that all previous events are completed at this stage of the initiation process. S-CDK phosphorylations bring Sld3 and Sld2 together with Dpb11 in progressively timed manner [11].

Cdc45 already associates with the origin during the $G_1$ phase of the cell cycle [34], but it is only stably bound to the pre-IC after phosphorylation of the Mcm4 subunit of the Mcm2-7 complex by DDK [62, 63]. The initiation factor Mcm10 directs the phosphorylation of Mcm4 by interaction with both Cdc45 and Mcm4 [40]. Mcm10 is found on the chromatin during both $G_1$ and S phases and is required for the stable loading of Cdc45 [58]. Mcm10 travels with the replication fork and maintains the DNA polymerase $\alpha$ on the chromatin [57]. The stable binding of Cdc45 requires DDK-dependent phosphorylation and is directly followed by the incorporation of the GINS complex. This complex is crucial for the maintenance of Cdc45 at the origin and both Cdc45 and the GINS complex move together with the replication fork. After the GINS complex is loaded, Sld3 is displaced from the pre-IC [34]. Ultimately, the formed complex initiates DNA replication by starting to uncoil the DNA strands. The single-stranded DNA (ssDNA) binding protein, RPA, binds to the DNA to stabilize the ssDNA and prevents it from rewinding [72]. In the final step, the polymerase is loaded onto the complex, which is called "replisome" and is able to initiate DNA replication [74].

## References

[1]  Alvino, G.M., Collingwood, D., Murphy, J.M., Delrow, J., Brewer, B.J., and Raghuraman, M.K., Replication in hydroxyurea: it's a matter of time, *Mol. Cell. Biol.*, 27(18):6396-6406, 2007.

[2]  Aparicio, O.M., Stout, A.M., and Bell, SP., Differential assembly of Cdc45p and DNA polymerases at early and late origins of DNA replication, *Proc. Natl. Acad. Sci. USA*, 96(16):9130-9135, 1999.

[3]  Aparicio, O.M., Weinstein, D.M., and Bell, S.P., Components and dynamics of DNA replication complexes in S. cerevisiae: Redistribution of MCM proteins and Cdc45p during S phase, *Cell*, 91(1):59-69, 1997.

[4]  Barberis, M., Spiesser, T.W., and Klipp, E., Replication origins and timing of temporal replication in budding yeast: How to solve the conundrum?, *Curr. Genomics*, 11(3):199-211, 2010.

[5]  Barberis, M. and Klipp, E., Insights into the network controlling the G1/S transition in budding yeast, *Genome Inform.*, 18:85-99, 2007.

[6]  Barberis, M., Klipp, E., Vanoni, M., and Alberghina, L., Cell size at S phase initiation: an emergent property of the G1/S network, *PloS Comput. Biol.*, 3(4):e64, 2007.

[7]  Barberis, M., De Gioia, L., Ruzzene, M., Sarno, S., Coccetti, P., Fantucci, P., Vanoni, M., and Alberghina, L., The yeast cyclin-dependent kinase inhibitor Sic1 and mammalian p27Kip1 are functional homologues with a structurally conserved inhibitory domain, *Biochem. J.*, 387(3):639-647, 2005.

[8]  Bell, S.P. and Dutta, A., DNA replication in eukaryotic cells, *Annu. Rev. Biochem.*, 71:333-374, 2002.

[9]  Bell, S.P., The origin recognition complex: from simple origins to complex functions, *Genes Dev.*, 16(6):659-672, 2002.

[10] Bell, S.P. and Stillman, B., ATP-dependent recognition of eukaryotic origins of DNA replication by a multiprotein complex, *Nature*, 357(6374):128-134, 1992.

[11] Botchan, M., Cell biology: a switch for S phase, *Nature*, 445(7125):272-274, 2007.

[12] Bousset, K. and Diffley, J.F., The Cdc7 protein kinase is required for origin firing during S phase, *Genes Dev.*, 12(4):480-490, 1998.

[13] Breier, A.M., Chatterji, S., and Cozzarelli, N.R. Prediction of Saccharomyces cerevisiae replication origins. *Genome Biol.*, 5(4):R22, 2004.

[14] Bueno, A. and Russell, P., Dual functions of Cdc6: a yeast protein required for DNA replication also inhibits nuclear division, *EMBO J.*, 11(6):2167-2176, 1992.

[15] Calzada, A., Sacristán, M., Sánchez, E., and Bueno, A., Cdc6 cooperates with Sic1 and Hct1 to inactivate mitotic cyclin-dependent kinases, *Nature*, 412(6844):355-358, 2001.

[16] Cho, W.H., Lee, Y.J., Kong, S.I., Hurwitz, J., and Lee, J.K., CDC7 kinase phosphorylates serine residues adjacent to acidic amino acids in the minichromosome maintenance 2 protein, *Proc. Natl. Acad. Sci.*, 103(31):11521-11526, 2006.

[17] Cook, J.G., Chasse, D.A.D., and Nevins, J.R., The regulated association of Cdt1 with minichromosome maintenance proteins and Cdc6 in mammalian cells, *J. Biol. Chem.*, 279(10):9625-9633, 2004.

[18] Cross, F.R., Yuste-Rojas, M., Gray, S., and Jacobson, M.D., Specialization and targeting of B-type cyclins, *Mol. Cell.*, 4(1):11-19, 1999.

[19] Devault, A., Gueydon, E., and Schwob, E., Interplay between S-cyclin-dependent kinase and Dbf4-dependent kinase in controlling DNA replication through phosphorylation of yeast Mcm4 N-terminal domain, *Mol. Biol. Cell*, 19(5):2267-2277, 2008.

[20] Diffley, J.F.X. and Labib, K., The chromosome replication cycle, *J. Cell Sci.*, 115(5):869-872, 2002.

[21] Donaldson, A.D., The yeast mitotic cyclin Clb2 cannot substitute for S phase cyclins in replication origin firing, *EMBO Rep.*, 1(6):507-512, 2000.

[22] Donaldson, A.D., Raghuraman, M.K., Friedman, K.L., Cross, F.R., Brewer, B.J., and Fangman, W.L., CLB5-dependent activation of late replication origins in S. cerevisiae, *Mol. Cell*, 2(2):173-182, 1998.

[23] Donaldson, A.D., Fangman, W.L., and Brewer, B.J., Cdc7 is required throughout the yeast S phase to activate replication origins, *Genes Dev.*, 12(4):491-501, 1998.

[24] Donovan, S., Harwood, J., Drury, L.S., and Diffley, J.F., Cdc6p-dependent loading of Mcm proteins onto pre-replicative chromatin in budding yeast, *Proc. Natl. Acad. Sci. USA*, 94(11):5611-5616, 1997.

[25] Drury, L.S., Perkins, G., and Diffley, J.F., The cyclin-dependent kinase Cdc28p regulates distinct modes of Cdc6p proteolysis during the budding yeast cell cycle, *Curr. Biol.*, 10(5):231-240, 2000.

[26] Elsasser, S., Chi, Y., Yang, P., and Campbell, J.L., Phosphorylation controls timing of Cdc6p destruction: A biochemical analysis, *Mol. Biol. Cell*, 10(10):3263-3277, 1999.

[27] Elsasser, S., Lou, F., Wang, B., Campbell, J. L., and Jong, A., Interaction between yeast Cdc6 protein and B-type cyclin/Cdc28 kinases, *Mol. Biol. Cell*, 7(11):1723-1735, 1996.

[28] Forsburg, S.L., Eukaryotic MCM proteins: beyond replication initiation, *Microbiol. Mol. Biol. Rev.*, 68(1):109-131, 2004.

[29] Ghaemmaghami, S., Huh, W.K., Bower, K., Howson, R.W., Belle, A., Dephoure, N., O'Shea, E.K., and Weissman, J.S., Global analysis of protein expression in yeast, *Nature*, 425(6959):737-741, 2003.

[30] Giaever, G. and Johnston, M., Functional profiling of the Saccharomyces cerevisiae genome, *Nature*, 418(6896):387-391, 2002.

[31] Haase, S.B. and Reed, S.I., Evidence that a free-running oscillator drives G1 events in the budding yeast cell cycle, *Nature*, 401(6751):394-397, 1999.

[32] Harvey, K.J. and Newport, J., Metazoan origin selection: origin recognition complex chromatin binding is regulated by Cdc6 recruitment and ATP hydrolysis, *J. Biol. Chem.*, 278(49):48524-48528, 2003.

[33] Ji, X. and Xu, Y., libSRES: a C library for stochastic ranking evolution strategy for parameter estimation, *Bioinformatics*, 22(1):124-126, 2006.

[34] Kanemaki, M. and Labib, K., Distinct roles for Sld3 and GINS during establishment and progression of eukaryotic DNA replication forks, *EMBO J.*, 25(8):1753-1763, 2006.

[35] Kawasaki, Y., Kim, H.D., Kojima, A., Seki, T., and Sugino, A., Reconstitution of Saccharomyces cerevisiae prereplicative complex assembly in vitro, *Genes Cells*, 11(7):745-756, 2006.

[36] Kawasaki, Y. and Sugino, A., Yeast replicative DNA polymerases and their role at the replication fork, *Mol. Cells*, 12(3):277-285, 2001.

[37] Kelly, T.J. and Brown, G.W., Regulation of chromosome replication, *Annu. Rev. Biochem.*, 69:829-880, 2000.

[38] Krasinska, L., Besnard, E., Cot, E., Dohet, C., Méchali, M., Lemaitre, J.M., and Fisher, D., Cdk1 and Cdk2 activity levels determine the efficiency of replication origin firing in Xenopus, *EMBO J.*, 27(5):758-769, 2008.

[39] Labib, K. and Diffley, J.F., Is the Mcm2-7 complex the eukaryotic DNA replication fork helicase?, *Curr. Opin. Genet. Dev.*, 11(1):64-70, 2001.

[40] Lee, J.-K., Seo, Y.-S., and Hurwitz, J., The Cdc23 (Mcm10) protein is required for the phosphorylation of minichromosome maintenance complex by the Dfp1-Hsk1 kinase, *Proc. Natl. Acad. Sci. USA*, 100(5):2334-2339, 2003.

[41] Li, J.J., DNA replication. Once, and only once, *Curr. Biol.*, 5(5):472-475, 1995.

[42] Loog, M. and Morgan, D.O., Cyclin specificity in the phosphorylation of cyclin-dependent kinase substrates, *Nature*, 434(7029):104-108, 2005.

[43] Masai, H. and Arai, K., Cdc7 kinase complex: a key regulator in the initiation of DNA replication, *J. Cell Physiol.*, 190(3):287-296, 2002.

[44] McCune, H.J., Danielson, L.S., Alvino, G.M., Collingwood, D., Delrow, J.J., Fangman W.L., Brewer, B.J., and Raghuraman, M.K., The temporal program of chromosome replication: genome-wide replication in clb5D Saccharomyces cerevisiae, *Genetics*, 180(4):1833-1847, 2008.

[45] Mendenhall, M.D., An inhibitor of p34CDC28 protein kinase activity from Saccharomyces cerevisiae, *Science*, 259(5092): 216-219, 1993.

[46] Méndez, J. and Stillman, B., Perpetuating the double helix: molecular machines at eukaryotic DNA replication origins, *Bioessays*, 25(12):1158-1167, 2003.

[47] Montagnoli, A., Valsasina, B., Brotherton, D., Troiani, S., Rainoldi, S., Tenca, P., Molinari, A., and Santocanale, C., Identification of Mcm2 phosphorylation sites by S-phase-regulating kinases. *J. Biol. Chem.*, 281(15):10281-10290, 2006.

[48] Nasmyth, K., At the heart of the budding yeast cell cycle, *Trends Genet.*, 12(10):405-412, 1996.

[49] Newlon, C.S. and Theis, J.F., The structure and function of yeast ARS elements, *Curr. Opin. Genet. Dev.*, 3(5):752-758, 1993.

[50] Newlon, C.S., Petes, T.D., Hereford, L.M., and Fangman, W.L., Replication of yeast chromosomal DNA, *Nature*, 247(5435):32-35, 1974.

[51] Nguyen, V.Q., Co, C., and Li, J.J., Cyclin-dependent kinases prevent DNA re-replication through multiple mechanisms, *Nature*, 411(6841):1068-1073, 2001.

[52] Nieduszynski, C.A., Knox, Y., and Donaldson, A.D., Genome-wide identification of replication origins in yeast by comparative genomics, *Genes Dev.*, 20(14):1874-1879, 2006.

[53] Nougarède, R., Della Seta, F., Zarzov, P., and Schwob, E., Hierarchy of S-phase-promoting factors: Yeast Dbf4–Cdc7 kinase requires prior S-phase cyclin-dependent kinase activation, *Mol. Cell. Biol.*, 20(11):3795-3806, 2000.

[54] Petes, T.D. and Williamson, D.H., Fiber autoradiography of replicating yeast DNA, *Exp. Cell Res.*, 95(1):103-110, 1975.

[55] Raghuraman, M.K., Winzeler, E.A., Collingwood, D., Hunt, S., Wodicka, L., Conway, A., Lockhart, D.J., Davis, R.W., Brewer, B.J., and Fangman, W.L., Replication dynamics of the yeast genome, *Science*, 294(5540):115-121, 2001.

[56] Randell, J.C.W., Bowers, J.L., Rodríguez, H.K., and Bell, S.P., Sequential ATP hydrolysis by Cdc6 and ORC directs loading of the Mcm2-7 helicase, *Mol. Cell*, 21(1):29-39, 2006.

[57] Ricke, R.M. and Bielinsky, A.-K., Mcm10 regulates the stability and chromatin association of DNA polymerase-alpha, *Mol. Cell*, 16(2):173-185, 2004.

[58] Sawyer, S.L., Cheng, I.H., Chai, W., and Tye, B.K., Mcm10 and Cdc45 cooperate in origin activation in Saccharomyces cerevisiae, *J. Mol. Biol.*, 340(2):195-202, 2004.

[59] Schwob, E., Bohm, T., Mendenhall, M.D., and Nasmyth, K., The B-type cyclin kinase inhibitor p40SIC1 controls the G1 to S transition in S. cerevisiae, *Cell*, 79(2): 233-244, 1994.

[60] Sclafani, R.A. and Holzen, T.M., Cell cycle regulation of DNA replication, *Annu. Rev. Genet.*, 41:237-280, 2007.

[61] Sclafani, R.A., Cdc7p-Dbf4p becomes famous in the cell cycle, *J. Cell Sci.*, 113(12):2111-2117, 2000.

[62] Sheu, Y.J. and Stillman, B., The Dbf4-Cdc7 kinase promotes S phase by alleviating an inhibitory activity in Mcm4, *Nature*, 463(7277):113-117, 2010.

[63] Sheu, Y.-J. and Stillman, B., Cdc7-Dbf4 phosphorylates MCM proteins via a docking site-mediated mechanism to promote s phase progression, *Mol. Cell*, 24(1):101-113, 2006.

[64] Speck, C. and Stillman, B., Cdc6 ATPase activity regulates ORC x Cdc6 stability and the selection of specific DNA sequences as origins of DNA replication, *J. Biol. Chem.*, 282(16):11705-11714, 2007.

[65] Speck, C., Chen, Z., Li, H., and Stillman, B., ATPase-dependent cooperative binding of ORC and Cdc6 to origin DNA, *Nat. Struct. Mol. Biol.*, 12(11):965-971, 2005.

[66] Spiesser, T.W., Klipp. E., and Barberis, M., A model for the spatiotemporal organization of DNA replication in Saccharomyces cerevisiae, *Mol. Genet. Genomics*, 282(1):25-35, 2009.

[67] Tak, Y.-S., Tanaka, Y., Endo, S., Kamimura, Y., and Araki, H., A Cdk-catalysed regulatory phosphorylation for formation of the DNA replication complex Sld2-Dpb11, *EMBO J.*, 25(9):1987-1996, 2006.

[68] Takeda, D.Y. and Dutta, A, DNA replication and progression through S phase, *Oncogene*, 24(17):2827-2843, 2005.

[69] Tanaka, S., Umemori, T., Hirai, K., Muramatsu, S., Kamimura, Y., and Araki, H., CDK-dependent phosphorylation of Sld2 and Sld3 initiates DNA replication in budding yeast, *Nature*, 445(7125):328-332, 2007.

[70] Tanaka, S. and Diffley, J.F.X., Interdependent nuclear accumulation of budding yeast Cdt1 and Mcm2-7 during G1 phase, *Nat. Cell Biol.*, 4(3):198-207, 2002.

[71] Tanaka, S. and Diffley, J.F., Deregulated G1-cyclin expression induces genomic instability by preventing efficient pre-RC formation, *Genes Dev.*, 16(20):2639-2649, 2002.

[72] Tanaka, T. and Nasmyth, K., Association of RPA with chromosomal replication origins requires an Mcm protein, and is regulated by Rad53, and cyclin- and Dbf4-dependent kinases, *EMBO J.*, 17(17):5182-5191, 1998.

[73] Theis, J.F. and Newlon, C.S., The ARS309 chromosomal replicator of Saccharomyces cerevisiae depends on an exceptional ARS consensus sequence, *Proc. Natl. Acad. Sci. USA*, 94(20):10786-10791, 1997.

[74] Walter, J. and Newport, J., Initiation of eukaryotic DNA replication: origin unwinding and sequential chromatin association of Cdc45, RPA, and DNA polymerase alpha, *Mol. Cell*, 5(4):617-627, 2000.

[75] Weinreich, M., DeBeer, M.A.P., and Fox, C.A., The activities of eukaryotic replication origins in chromatin, *Biochim. Biophys. Acta*, 1677(1-3):142-157, 2004.

[76] Wyrick, J.J., Aparicio, J.G., Chen, T., Barnett, J.D., Jennings, E.G., Young, R.A., Bell, S.P., and Aparicio, O.M., Genome-wide distribution of ORC and MCM proteins in S. cerevisiae: high-resolution mapping of replication origins, *Science*, 294(5550): 2357-2360, 2001.

[77] Yabuuchi, H., Yamada, Y., Uchida, T., Sunathvanichkul, T., Nakagawa, T., and Masukata, H., Ordered assembly of Sld3, GINS and Cdc45 is distinctly regulated by DDK and CDK for activation of replication origins, *EMBO J.*, 25(19):4663-4674, 2006.

[78] Zegerman, P. and Diffley, J.F., Phosphorylation of Sld2 and Sld3 by cyclin-dependent kinases promotes DNA replication in budding yeast, *Nature*, 445(7125):281-285, 2007.

[79] Zi, Z. and Klipp, E., SBML-PET: a Systems Biology Markup Language-based parameter estimation tool, *Bioinformatics*, 22(21):2704-2705, 2006.

[80] Zou, L. and Stillman, B., Assembly of a complex containing Cdc45p, replication protein A, and Mcm2p at replication origins controlled by S-phase cyclin-dependent kinases and Cdc7p-Dbf4p kinase, *Mol. Cell. Biol.*, 20(9):3086-3096, 2000.

[81] http://yeastgfp.yeastgenome.org/

# PREDICTING PROTEIN COMPLEX GEOMETRIES WITH LINEAR SCORING FUNCTIONS

OZGUR DEMIR-KAVUK[1]
odemir@chemie.fu-berlin.de

FLORIAN KRULL[1]
fkrull@chemie.fu-berlin.de

MYONG-HO CHAE[2]
pptayang@co.chesin.com

ERNST-WALTER KNAPP[1]
knapp@chemie.fu-berlin.de

[1]*Institute of Chemistry and Biochemistry, Freie Universität Berlin, Fabeckstrasse 36A, 14195 Berlin, Germany*

[2]*Department of Biology, University of Science, Unjong-District, Pyongyang, DPR Korea*

Protein-Protein interactions play an important role in many cellular processes. However, experimental determination of the protein complex structure is quite difficult and time consuming. Hence, there is need for fast and accurate *in silico* protein docking methods. These methods generally consist of two stages: (i) a sampling algorithm that generates a large number of candidate complex geometries (decoys), and (ii) a scoring function that ranks these decoys such that near-native decoys are higher ranked than other decoys. We have recently developed a neural network based scoring function that performed better than other state-of-the-art scoring functions on a benchmark of 65 protein complexes. Here, we use similar ideas to develop a method that is based on linear scoring functions. We compare the linear scoring function of the present study with other knowledge-based scoring functions such as ZDOCK 3.0, ZRANK and the previously developed neural network. Despite its simplicity the linear scoring function performs as good as the compared state-of-the-art methods and predictions are simple and rapid to compute.

*Keywords*: unbound structure docking; decoy; distance-dependent atom-pair potential; near-native decoys; protein-protein interaction; linear scoring function.

## 1. Introduction

Protein-protein interactions are known to have a vital role for many biological processes, such as the regulation of gene expression, signaling or recognition [4]. In order to understand the binding mode of two interacting proteins in detail, it is very helpful to know the geometry of the formed complex. Currently it is believed that a large number of protein complexes form transiently. Hence, they are often too unstable for crystallization to obtain their structure [2]. Many proteins are known to have multiple interaction partners. In particular in these cases prediction of the binding mode using the structures of the participating proteins is of great importance. Therefore, many protein docking algorithms have been developed [7, 21, 23], which try to predict the geometries of protein complexes, relying on the assumption that the free energy of the native complex structure is the lowest accessible state of the two interacting proteins [3].

During complex formation of two proteins, their structures may undergo conformational changes. Prediction of the complex geometry becomes particularly difficult, if these conformational changes involve also large backbone movements. Different strategies have been applied to tackle this problem. Flexible docking algorithms can in principle account for large structural differences between individual unbound proteins and their counterparts bound in a complex [6, 13, 15]. But usually the performance of such algorithms suffers from the large search spaces. Rigid-body docking, on the other hand, is computationally less expensive, but its ability to find acceptable solutions is limited, when the unbound protein structures undergo large conformational changes during their complex formation.

Generally rigid-body docking approaches start by generating complex geometries (decoys), considering primarily shape complementarity of the two individual proteins [3, 5, 20]. Slight penetrations of the surfaces are allowed, to account for moderate conformational changes upon complex formation. Since the number of decoys generated in this first step is quite large, a scoring function is applied, which discriminates approximately near-native decoys from decoys, which are far from the native complex geometry. Typically, such docking algorithms generate a subset of the initial decoys, containing decoys with high scores only. For these smaller decoy sets with enriched near-native complex structures, computationally more expensive refinement procedures can be applied.

Different methods to score protein complex geometries have been developed [8]. A prominent approach is to use scoring functions, which classify decoys according to characteristic quantities as for instance number of H-bonds or hydrophobic residues in the interface relative to native protein complexes. Alternatively, one can use empirical or physical energy functions. The latter describe the behavior of the participating proteins on an atomic level of description, considering for example electrostatic and van-der-Waals interactions. Scoring devices, which are based on knowledge based potentials, apply weights to atom or residue pairs found in a decoy. Those weights are optimized with respect to statistical data from native protein complexes.

Previously, we used a neural network to optimize atom pair potentials from a training set of 185 protein complexes with 2000 near-native decoys per complex.. We applied the resulting scoring function to decoys obtained by rigid-body docking, and could show that its scoring ability performs better than ZDOCK 3.0 [14] and comparable to ZRANK [16]. Details of the neural network can be found in [2]. In this work, we introduce a linear scoring function which is applied instead of the neural network to optimize the atom pair potentials. For comparison with the neural network approach, the linear scoring function uses exactly the same training and prediction data.

## 2.  Methods

### 2.1.  *Database of protein complexes and decoys*

Since we want to compare the discrimination power of the parameters resulting from the new learning device with the performance of the neural network approach, we took the same training and prediction data that we applied in previous work [2]. These are for training 191 protein complex structures, where 48 were taken from Benchmark 3.0 [9] and 143 of a protein complex set from Huang et. al. [8]. For prediction we used 65 protein complexes, which are all from Benchmark 3.0 [9]. For all protein complexes of the training set we generated near-native decoys with a maximum interface RMSD (*iRMSD*) of $d_{max} = 6.0$ Å, by applying random translations and rotations to one of the two proteins in the complex. The resulting decoys were sorted in ten distance classes of *iRMSD*s intervals [j*0.6 Å, (j+1)*0.6 Å] with j = 0, 1, 2 . . . 9. To control size and distribution of the decoys in these classes, we allowed 200 decoys per class only. Thus, in total we generated 370,000 decoys for the whole training set of protein complexes. These near-native decoys we call NN-$d_{max}$ (with $d_{max} = 6.0$) or NN-6.0 decoys. For the prediction set we took the 54.000 decoys that ZDOCK 3.0 [14] provides for each of these 65 protein complexes. They were generated by rigid-body sampling using the unbound structures (zlab.bu.edu/zdock/decoys.shtml).

### 2.2.  *Linear scoring function*

Each NN-6.0 decoy of the training and the prediction set can be characterized by features combined in a vector $\vec{x}_i \in \mathbb{R}^d$ in a *d* dimensional feature space. There are several techniques to extract features from decoys. From these atom-pair based or residue-pair based information are the most commonly used. It has been shown recently that atom-pair based information is superior to residue-pair based information to rank docking decoys [12]. Hence, the linear scoring function uses atom-pair based information which consists of 20 different heavy-atom types as defined in [8] and two polar hydrogen atom types (hydrogen atoms making H-bonds with sulfur and oxygen or alternatively with nitrogen) as defined in [2]. Non-polar hydrogen atoms are ignored. This results in a total of 253 = (22*23)/2 different types of atom-pairs.

Only decoys ranging from 0.0 to 6.0 Å *iRMSD* were considered in the training set. In the previous studies we divided the same *iRMSD* distance interval into 1-12 distance bins yielding an improved performance compared to using a single bin only. The neural network showed the best overall performance using eight equidistant distance bins. Therefore, in this study we also used the same distance bins for all decoys resulting in 253*8 = 2024 features. In the previous study using a neural network we furthermore showed that the reference points of the effective energy function have to be adjusted for the different protein complexes. This was necessary, since the different protein complexes are not directly comparable. Therefore, so called identity features were added. They label the protein complex to which the decoy refers to. Hence, there are as many identity features as there are native protein complexes in the training set. For each decoy in the training set the identity feature that labels the corresponding native protein complex is set to unity while all other identity features are set to zero. During learning the weights corresponding to the identity features are optimized such that the zero-point energy for the particular protein-pair is adjusted. During prediction the native protein complex structure is not known. Hence, for the prediction set all identity features are set to zero.

Prediction of the *iRMSD* of a new decoy $\vec{x}_i$ can be made with a linear scoring function defined by

$$g(\vec{x}_i) = \vec{w}' \cdot \vec{x}_i + b,$$ (1)

where $\vec{w} \in \mathbb{R}^d$ is the model parameter vector of the scoring function and $b \in \mathbb{R}$ the threshold or bias. In this scoring function the parameters correspond to the weights of the neural network and the function value to the value of the output neuron. Setting the linear scoring function to zero describes a hyperplane in the d-dimensional feature space $\mathbb{R}^d$. The orientation of the hyperplane is defined using the model parameter vector $\vec{w}$ as normal vector of the plane, while its distance from the origin is defined by the threshold value $b$ as $(b/|\vec{w}|)$. Note that for this type of scoring functions the number of model parameters is d+1, where d is the number of features. The d+1 model parameters are optimized during the so called training phase where data with known *iRMSD* are used to determine a hyperplane that is able to recall the learned *iRMSDs*.

To determine an optimal parameter vector $\vec{w}$ we use an objective function, which is minimized to yield a solution of the hyperplane. Given a set of *n* training decoys with their feature vectors $\vec{x}_i$ we define the objective function as a quadratic form in the parameter vector $\vec{w}$

$$L(\vec{w}, b) = \sum_{i=0}^{n} \left[ \mu_i \left\{ (\vec{w}' \cdot \vec{x}_i + b) - iRMSD_i \right\} \right]^2 + \lambda \vec{w}' \cdot \vec{w}.$$ (2)

Minimizing the objective function $L(\vec{w}, b)$ leads to a hyperplane where the distances of the data points to the hyperplane are proportional to their *iRMSD* values. The second term of the objective function is the so called Tikhonov regularization [22]. The Tikhonov regularization avoids over fitting by suppressing less important features. To optimize the regularization term the parameter $\lambda$ has to be chosen carefully: too small values will have no effect whereas too large values will yield to a model that is not able to learn the data. In this study $\lambda$ was set to 0.1, which yielded the best overall performance. Additional parameters $\mu_i > 0$ (i=1...n) can be used to weight each data point individually. Without the additional parameters $\mu_i$ all data points are weighted equally during learning. However, we are looking for a scoring function that is mainly able to score near native decoys accurately. Hence, to emphasize those data points that have a smaller *iRMSD* we used the weights $\mu_i = (d - iRMSD)^2$, where we set $d = 8$. Since the *iRMSD* values in the training set range from 0 to 6 the weights vary between 4 and 64.

Since the objective function is quadratic in the parameters $\vec{w}$ and b, the minimum of the objective function can be obtained analytically solving a corresponding set of linear equations [17] with the Cholesky decomposition [1]. However, in case of large data sets solving a linear equation system can be very time and memory consuming. In such a case the parameters of the linear equation system can be solved using a gradient descent procedure. In the present study we used the Limited Memory Broyden–Fletcher–Goldfarb–Shanno update (L-BFGS) algorithm [11, 24].

## 3. Results and Discussion

The linear scoring function, eq. (2), was trained on the total number of 370,000 NN-6.0 decoys for the 191 protein complexes (2000 decoys for each protein complex) of the training set. The trained model was tested by scoring 54,000 unbound decoys (formed by unbound protein structures) for each of the 65 protein complexes in the prediction set. These decoys were taken from the webpage (zlab.bu.edu/zdock/decoys.shtml) of Weng's Lab., where they were generated with ZDOCK 3.0 [10].

### 3.1. Performance comparison

The performance of the linear scoring function approach has been compared to ZRANK [16], ZDOCK 3.0 [14] and the previously developed neural network scoring function [2]. ZRANK uses an optimized combination of detailed electrostatics, van-der-Waals, and desolvation energy terms to rescore initial-stage docking predictions from ZDOCK. ZDOCK 3.0 is a new version of ZDOCK that uses a pair-wise atom-based statistical potential, electrostatic and shape-complementarity based scoring function. The neural network uses the same features as the linear scoring function as input and the back propagation algorithm [18, 19] is used for training. The neural network does not use hidden layers as these did not improve the prediction performance. Hence, the neural network essentially boils down to a linear predictor as well. The neural network is described in detail elsewhere [2].

The 54,000 decoy structures for each of the 65 test protein complexes obtained with ZDOCK 3.0 were ranked using the linear scoring function based on the distance-dependent atom-pair potentials with eight distance bins. The docking predictions for ZDOCK 3.0 and ZRANK were obtained from the webpage (zlab.bu.edu/zdock/decoys. shtml) of Weng's lab. Table 1 shows a comparison of the number of *HITs* in the top 2000 of all 54,000 ranked decoys of the prediction set and the rank of the highest scored NN–2.5 decoy. A *HIT* is defined as a predicted near native decoy with an *iRMSD* $\leq$ 2.5 Å relative to the corresponding native complex geometry. The linear scoring function has for 11 out of 65 protein complexes of the prediction set a NN-2.5 decoy at the first position, compared to 10 for ZDOCK 3.0 and 7 for ZRANK. Only the neural network performs as good as the linear scoring function. If the top 10 predictions are considered, the linear scoring function is successful for 17 protein complexes, compared to 16 for both ZDOCK 3.0 and ZRANK. However, the neural network performs better and is successful considering the top 10 predictions for 22 protein complexes. If the top 2000 predictions are considered both the linear scoring function and ZRANK have 51 *HITs* compared to 43 for ZDOCK 3.0 and 48 for the neural network. It can also be seen that both the linear scoring function and the neural network have more *HITs* in the top 2000 (on average 50 *HITs* per protein complex compared to 36 for ZDOCK 3.0 and 26 for ZRANK).

Plots of the success rates (fraction of protein complexes with at least one *HIT* within the given number of predictions) averaged over all 65 complexes of the prediction set are shown in Fig. 1. The success rate of the linear scoring function is comparable to the three other methods. For up to ten predictions the linear scoring function is slightly better than ZDOCK 3.0 and ZRANK but worse than the neural network. If more than 200 predictions are considered, the linear scoring function of the present study and ZRANK are the best performing models.

Table 1: Comparison of docking results for the prediction set of 65 protein complexes. ZDOCK 3.0 [14], ZRANK [16], neural network [2].

| PDB code | ZDOCK 3.0 | | | ZRANK | | neural network | | linear scoring function | |
|---|---|---|---|---|---|---|---|---|---|
| | NN–2.5 [a] | *HITs* [b] | rank [c] | *HITs* [b] | rank [c] | *HITs* [b] | rank [c] | *HITs* [b] | rank [c] |
| enzyme-inhibitor complexes | | | | | | | | | |
| 1AVX | 174 | 56 | 27 | 25 | 11 | 118 | 1 | 125 | 57 |
| 1AY7 | 74 | 0 | 5103 | 21 | 74 | 0 | 6655 | 0 | 2287 |
| 1BVN | 145 | 121 | 1 | 37 | 16 | 67 | 22 | 111 | 1 |
| 1CGI | 51 | 22 | 48 | 3 | 89 | 9 | 152 | 15 | 224 |
| 1DFJ | 89 | 84 | 1 | 15 | 2 | 75 | 1 | 87 | 1 |
| 1E6E | 115 | 70 | 16 | 82 | 5 | 63 | 70 | 85 | 21 |
| 1EAW | 122 | 16 | 328 | 23 | 42 | 115 | 1 | 114 | 18 |
| 1EWY | 77 | 41 | 10 | 12 | 21 | 63 | 7 | 60 | 20 |
| 1EZU | 55 | 1 | 1917 | 0 | 10154 | 1 | 1687 | 17 | 214 |
| 1F34 | 56 | 11 | 195 | 6 | 62 | 17 | 14 | 14 | 167 |
| 1KKL | 2 | 0 | 43815 | 1 | 1002 | 0 | 6407 | 0 | 3953 |
| 1MAH | 223 | 88 | 1 | 80 | 3 | 112 | 2 | 112 | 1 |
| 1PPE | 591 | 319 | 1 | 190 | 1 | 514 | 1 | 482 | 1 |
| 1TMO | 136 | 61 | 45 | 11 | 71 | 70 | 1 | 63 | 9 |
| 1UDI | 85 | 14 | 92 | 0 | 2741 | 58 | 1 | 52 | 1 |
| 2MTA | 120 | 5 | 1391 | 4 | 528 | 41 | 97 | 65 | 72 |
| 2PCC | 15 | 0 | 4682 | 2 | 920 | 11 | 8 | 7 | 394 |
| 2SIC | 248 | 106 | 13 | 83 | 1 | 138 | 3 | 159 | 1 |
| 2SNI | 81 | 1 | 1798 | 22 | 178 | 2 | 1074 | 19 | 433 |
| 7CEI | 299 | 147 | 1 | 145 | 3 | 210 | 1 | 259 | 1 |
| antigen-antibody complexes | | | | | | | | | |
| 1AHW | 107 | 0 | 2223 | 30 | 54 | 59 | 103 | 34 | 275 |
| 1BJ1 | 619 | 238 | 1 | 55 | 19 | 73 | 267 | 16 | 754 |
| 1BVK | 79 | 0 | 3953 | 0 | 3084 | 5 | 367 | 0 | 2979 |
| 1DOJ | 32 | 0 | 10150 | 0 | 3022 | 0 | 16187 | 0 | 9478 |
| 1E6J | 215 | 72 | 32 | 143 | 1 | 1 | 1981 | 23 | 302 |
| 1FSK | 255 | 110 | 5 | 88 | 1 | 185 | 1 | 174 | 2 |
| 1I9R | 42 | 12 | 139 | 10 | 40 | 4 | 1194 | 6 | 807 |
| 1IOD | 179 | 115 | 4 | 26 | 169 | 125 | 2 | 136 | 1 |

| PDB code | ZDOCK 3.0 | | | ZRANK | | neural network | | linear scoring function | |
|---|---|---|---|---|---|---|---|---|---|
| | NN–2.5 [a] | HITs [b] | rank [c] | HITs [b] | rank [c] | HITs [b] | rank [c] | HITs [b] | rank [c] |
| 1JPS | 115 | 0 | 2917 | 40 | 1 | 73 | 3 | 34 | 122 |
| 1K4C | 239 | 0 | 19538 | 30 | 197 | 50 | 153 | 0 | 2507 |
| 1KXO | 68 | 42 | 4 | 12 | 14 | 45 | 36 | 44 | 8 |
| 1MLC | 216 | 13 | 165 | 35 | 5 | 60 | 6 | 112 | 3 |
| 1NCA | 72 | 9 | 754 | 20 | 14 | 51 | 1 | 34 | 23 |
| 1NSN | 31 | 0 | 10627 | 2 | 473 | 1 | 1028 | 8 | 183 |
| 1OFW | 68 | 1 | 1619 | 8 | 192 | 0 | 6163 | 0 | 4938 |
| 1VFB | 50 | 0 | 2480 | 1 | 997 | 17 | 75 | 3 | 1060 |
| 1WEJ | 196 | 18 | 387 | 37 | 2 | 122 | 38 | 100 | 162 |
| 2JEL | 297 | 63 | 48 | 4 | 1239 | 108 | 1 | 206 | 1 |
| 2VIS | 252 | 5 | 1402 | 29 | 8 | 0 | 12181 | 0 | 11246 |
| | | | | other complexes | | | | | |
| 1A2K | 33 | 0 | 2695 | 0 | 17956 | 1 | 1101 | 5 | 526 |
| 1AK4 | 7 | 0 | 17326 | 0 | 6442 | 0 | 2239 | 0 | 3682 |
| 1AKJ | 81 | 0 | 2851 | 13 | 175 | 0 | 2566 | 27 | 142 |
| 1B6C | 52 | 34 | 1 | 18 | 2 | 39 | 14 | 45 | 3 |
| 1BUH | 129 | 55 | 38 | 6 | 353 | 0 | 2472 | 12 | 769 |
| 1E96 | 21 | 3 | 1026 | 18 | 24 | 0 | 2852 | 14 | 435 |
| 1F51 | 124 | 41 | 1 | 25 | 3 | 104 | 6 | 54 | 50 |
| 1FC2 | 1 | 0 | 33012 | 0 | 18629 | 0 | 15396 | 0 | 33410 |
| 1GCO | 11 | 0 | 21442 | 1 | 922 | 1 | 1689 | 0 | 4455 |
| 1GP2 | 15 | 11 | 353 | 0 | 3726 | 0 | 3471 | 0 | 11454 |
| 1GRN | 48 | 0 | 4078 | 1 | 1365 | 2 | 886 | 6 | 28 |
| 1HE1 | 87 | 31 | 2 | 18 | 43 | 10 | 179 | 19 | 130 |
| 1I2M | 7 | 0 | 2431 | 0 | 42277 | 0 | 3286 | 5 | 76 |
| 1I4D | 50 | 4 | 898 | 0 | 11099 | 0 | 3706 | 1 | 1671 |
| 1IJK | 51 | 1 | 1959 | 7 | 444 | 26 | 234 | 1 | 1932 |
| 1K5D | 28 | 3 | 321 | 2 | 1111 | 1 | 1205 | 2 | 1222 |
| 1KAC | 29 | 0 | 4202 | 10 | 11 | 17 | 8 | 3 | 849 |
| 1KLU | 14 | 0 | 27704 | 0 | 13333 | 0 | 5528 | 0 | 8883 |
| 1KTZ | 29 | 0 | 6972 | 6 | 397 | 0 | 2130 | 0 | 6196 |
| 1KXP | 51 | 51 | 1 | 10 | 40 | 51 | 3 | 47 | 1 |
| 1ML0 | 163 | 88 | 1 | 75 | 1 | 100 | 1 | 89 | 2 |
| 1OA9 | 15 | 0 | 41627 | 0 | 8818 | 0 | 2932 | 1 | 919 |
| 1RLB | 482 | 132 | 8 | 117 | 1 | 140 | 14 | 63 | 60 |
| 1SBB | 9 | 0 | 14481 | 0 | 8447 | 0 | 33427 | 0 | 28929 |
| 1WO1 | 22 | 10 | 21 | 0 | 7791 | 10 | 264 | 18 | 104 |
| 2BTF | 68 | 13 | 759 | 4 | 655 | 62 | 4 | 44 | 1 |
| Top 1 [d] | | | 10 | | 7 | | 11 | | 11 |
| Top 10 [d] | | | 16 | | 16 | | 22 | | 17 |
| Top 20 [d] | | | 18 | | 22 | | 25 | | 19 |
| Averages [e] | | 36 | | 26 | | 50 | | 50 | |

[a] Number of decoys for each individual protein pair with *iRMSD* ≤ 2.5 Å relative to the corresponding native complex geometry in the 54,000 decoy set generated by ZDOCK 3.0.

[b] Number of *HITs* (decoys with an *iRMSD* ≤ 2.5 Å relative to the corresponding native complex geometry) scored in the top 2000 of the 54,000 decoys generated by ZDOCK 3.0.

[c] Rank of the top scored NN–2.5 decoy.

[d] The number of successful cases when the top 1(top 10, top 20) prediction(s) were considered.

[e] Average number (over all 65 protein complexes of the prediction set) of *HITs* scored in the top 2000 from the 54,000 decoys generated by ZDOCK 3.0.

To test whether the performance differences are significant, we used a paired two-sample t-test.

Table 2 shows the results. Considering the first rank prediction only, the difference of the linear scoring function to the three other methods is not very

Table 2: T-test results of *iRMSD* prediction. The predicted decoys of the linear scoring function are compared to ZDOCK 3.0 [14], ZRANK [16] and the neural network [2] using a paired two-sample t-test.

| top[a] | ZDOCK 3.0 | | ZRANK | | neural network | |
|---|---|---|---|---|---|---|
| | difference[b] | p-value[c] | difference[b] | p-value[c] | difference[b] | p-value[c] |
| 1 | 1 | 0.74 | 4 | 0.29 | 0 | 1.00 |
| 5 | 1 | 0.74 | 0 | 1.00 | -2 | 0.48 |
| 10 | 1 | 0.77 | 1 | 0.80 | -5 | 0.13 |
| 50 | -2 | 0.60 | -5 | 0.23 | -5 | 0.10 |
| 100 | 1 | 0.78 | -6 | 0.18 | -4 | 0.16 |
| 500 | 8 | 0.06 | -1 | 0.82 | 2 | 0.60 |
| 1000 | 11 | 0.01 | 0 | 1.00 | 7 | 0.07 |

[a] number of predictions taken into account
[b] difference of the numbers of successful cases (*linear scoring function – other method*)
[c] probability value of the paired t-test

significant (p-values >= 0.29). Interestingly, the number of successful cases (i.e. if the top 1 predictions were considered) of the neural network and the linear scoring function are identical (difference of 0). This may be due to the fact that the neural network uses the same training points and features as the linear scoring function. If the top 100 predictions are considered, the performance of the linear scoring function seems to be comparable to ZDOCK (p-value = 0.78) and worse than ZRANK (p-value = 0.18) and the neural network (p-value = 0.16). For the top 1000 predictions the performance of the linear scoring function seems to be better than the neural network (p-value = 0.07) and significantly better than ZDOCK (p-value = 0.01).

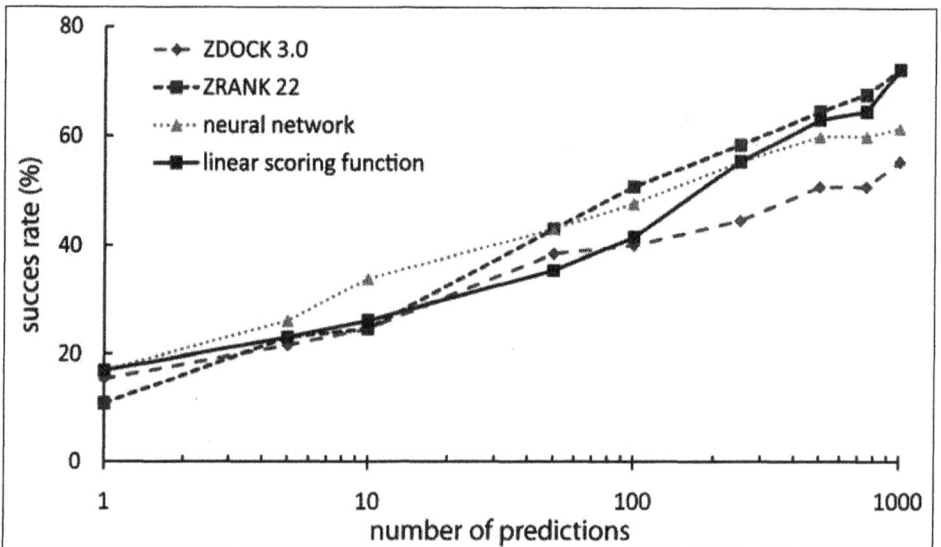

Fig. 1: Comparison of the success rate (defined in the method section) versus the number of highest ranked decoys (number of predictions per protein complex) for all 65 protein complexes of the prediction set. We compare ZDOCK 3.0 [14], ZRANK [16], the neural network [2] and the linear scoring function of the present study.

## 4. Conclusions

We developed a knowledge-based linear scoring function for protein docking prediction. It has been trained on distance dependent atom-pair based information. Due to its simple form ranking of new protein complexes can be performed most easily. We compared the performance of the linear scoring function with ZDOCK 3.0, ZRANK and the previously developed neural network on a benchmark of 65 protein pairs whose complex geometries were predicted. The linear scoring function of the present study has as many *HITs* at the top position as our previously developed neural network and shows in that respect superior performance to ZRANK. The neural network has more *HITs*, if the top 5, 10, 50 and 100 predictions are considered. However, applying a paired two-sample t-test it turns out that these differences between the neural network and the linear scoring function are statistically not significant. The neural network does not use hidden layers as these did not improve the prediction performance. Hence, the neural network essentially boils down to a linear predictor as well. The similarities of the two approaches can also be observed in the prediction behavior, since the top ranked results are identical for both methods. Nevertheless, the linear scoring function used in this study is the most basic one and therefore leaves space for further improvements.

## Acknowledgments

MHC is grateful to the Humboldt Foundation for financial support. This work was supported in the frame of the International Research Training Group (IRTG) on "Genomics and Systems Biology of Molecular Networks" (GRK1360) supported by the German Research Foundation (DFG)).

## References

[1] Bau, D. and Trefethen, L. N., Numerical linear algebra, *Philadelphia: Society for Industrial and Applied Mathematics*, 1997.
[2] Chae, M.-H., Krull, F., Lorenzen, S., and Knapp, E.-W., Predicting protein complex geometries with a neural network, *Proteins: Structure, Function, and Bioinformatics*, 78(4):1026-1039, 2010.
[3] Chen, R. and Weng, Z., Docking unbound proteins using shape complementarity, desolvation, and electrostatics, *Proteins*, 47(3):281-94, 2002.
[4] Gavin, A. C., Bosche, M., Krause, R., Grandi, P., Marzioch, M., Bauer, A., Schultz, J., Rick, J. M., Michon, A. M., Cruciat, C. M., Remor, M., Hofert, C., Schelder, M., Brajenovic, M., Ruffner, H., Merino, A., Klein, K., Hudak, M., Dickson, D., Rudi, T., Gnau, V., Bauch, A., Bastuck, S., Huhse, B., Leutwein, C., Heurtier, M. A., Copley, R. R., Edelmann, A., Querfurth, E., Rybin, V., Drewes, G., Raida, M., Bouwmeester, T., Bork, P., Seraphin, B., Kuster, B., Neubauer, G., and Superti-Furga, G., Functional organization of the yeast proteome by systematic analysis of protein complexes, *Nature*, 415(6868):141-7, 2002.

[5] Geppert, T., Proschak, E., and Schneider, G., Protein-protein docking by shape-complementarity and property matching, *Journal of Computational Chemistry*, 31(9):1919-1928, 2010.

[6] Gray, J. J., Moughon, S., Wang, C., Schueler-Furman, O., Kuhlman, B., Rohl, C. A., and Baker, D., Protein-protein docking with simultaneous optimization of rigid-body displacement and side-chain conformations, *J Mol Biol.*, 331(1):281-99, 2003

[7] Halperin, I., Ma, B., Wolfson, H., and Nussinov, R., Principles of docking: An overview of search algorithms and a guide to scoring functions, *Proteins*, 47(4):409-43, 2002.

[8] Huang, S. Y. and Zou, X., An iterative knowledge-based scoring function for protein-protein recognition, *Proteins*, 72(2):557-79, 2008.

[9] Hwang, H., Pierce, B., Mintseris, J., Janin, J., and Weng, Z., Protein-protein docking benchmark version 3.0, *Proteins*, 73(3):705-9, 2008.

[10]Jiang, L., Gao, Y., Mao, F., Liu, Z., and Lai, L., Potential of mean force for protein-protein interaction studies, *Proteins*, 46(2):190-6, 2002.

[11]Liu, D. C. and Nocedal, J., On the limited memory BFGS method for large scale optimization, *Mathematical Programming: Series A and B*, 45(3):503 - 528, 1989.

[12]Lu, H., Lu, L., and Skolnick, J., Development of unified statistical potentials describing protein-protein interactions, *Biophys J*, 84(3):1895-901, 2003.

[13]Mashiach, E., Nussinov, R., and Wolfson, H. J., FiberDock: Flexible induced-fit backbone refinement in molecular docking, *Proteins: Structure, Function, and Bioinformatics*, 78(6):1503-1519, 2009.

[14]Mintseris, J., Pierce, B., Wiehe, K., Anderson, R., Chen, R., and Weng, Z., Integrating statistical pair potentials into protein complex prediction, *Proteins*, 69(3):511-20, 2007.

[15]Noy, E. and Goldblum, A., Flexible protein-protein docking based on Best-First search algorithm, *Journal of Computational Chemistry*, 31(9):1929-1943, 2010.

[16]Pierce, B. and Weng, Z., ZRANK: reranking protein docking predictions with an optimized energy function, *Proteins*, 67(4):1078-86, 2007.

[17]Riedesel, H., Kolbeck, B., Schmetzer, O., and Knapp, E. W., Peptide binding at class I major histocompatibility complex scored with linear functions and support vector machines, *Genome Inform*, 15(1):198-212, 2004.

[18]Rojas, R., *Neural Networks: A Systematic Introduction*, Berlin, 1996.

[19]Rumelhart, D. E., Hinton, G. E., and Williams, R. J., Learning representations by back-propagating errors, *Nature*, 323:533-536, 1986.

[20]Schneidman-Duhovny, D., Inbar, Y., Nussinov, R., and Wolfson, H. J., Geometry-based flexible and symmetric protein docking, *Proteins*, 60(2):224-31, 2005.

[21]Smith, G. R. and Sternberg, M. J., Prediction of protein-protein interactions by docking methods, *Curr Opin Struct Biol*, 12(1):28-35, 2002.

[22]Tychonoff, A. N., On the stability of inverse problems, *Doklady Akademii Nauk SSSR*, 39(5):195-198, 1943.

[23]Vajda, S. and Kozakov, D., Convergence and combination of methods in protein-protein docking, *Curr Opin Struct Biol*, 19(2):164-70, 2009.

[24]Zhu, C., Byrd, R. H., Lu, P., and Nocedal, J., Algorithm 778: L-BFGS-B: Fortran subroutines for large-scale bound-constrained optimization, *ACM Transactions on Mathematical Software (TOMS)*, 23(4):550 - 560, 1997.

# CHARACTERIZING COMMON SUBSTRUCTURES OF LIGANDS FOR GPCR PROTEIN SUBFAMILIES

BEKIR ERGUNER[1]
bekir@kuicr.kyoto-u.ac.jp

MASAHIRO HATTORI[1]
hattori@kuicr.kyoto-u.ac.jp

SUSUMU GOTO[1]
goto@kuicr.kyoto-u.ac.jp

MINORU KANEHISA[1, 2]
kanehisa@kuicr.kyoto-u.ac.jp

[1] *Bioinformatics Center, Institute for Chemical Research, Kyoto University, Gokasho, Uji, Kyoto 611-0011, Japan*
[2] *Human Genome Center, Institute of Medical Science, University of Tokyo, 4-6-1 Shirokane-dai, Minato-ku, Tokyo 108-8639, Japan*

The G-protein coupled receptor (GPCR) superfamily is the largest class of proteins with therapeutic value. More than 40% of present prescription drugs are GPCR ligands. The high therapeutic value of GPCR proteins and recent advancements in virtual screening methods gave rise to many virtual screening studies for GPCR ligands. However, in spite of vast amounts of research studying their functions and characteristics, 3D structures of most GPCRs are still unknown. This makes target-based virtual screenings of GPCR ligands extremely difficult, and successful virtual screening techniques rely heavily on ligand information. These virtual screening methods focus on specific features of ligands on GPCR protein level, and common features of ligands on higher levels of GPCR classification are yet to be studied. Here we extracted common substructures of GPCR ligands of GPCR protein subfamilies. We used the SIMCOMP, a graph-based chemical structure comparison program, and hierarchical clustering to reveal common substructures. We applied our method to 850 GPCR ligands and we found 53 common substructures covering 439 ligands. These substructures contribute to deeper understanding of structural features of GPCR ligands which can be used in new drug discovery methods.

*Keywords*: G-protein coupled receptor; ligand; hierarchical clustering; chemical substructure.

## 1. Introduction

G-protein coupled receptor (GPCR) proteins are transmembrane receptor proteins which sense external signals and activate various signaling pathways inside the cell causing different cellular responses. They have a wide range of sensory functions including sensing light, olfaction, hormone reception and neurotransmission. They exist in every tissue of our body and play important roles in regulation of every major mammalian physiological system [1]. The GPCR superfamily contains the largest number of pharmaceutical target proteins with proven therapeutic value. It was reported that more than 40% of the prescription drugs are ligands of GPCR proteins [2]. Moreover, the fact that there are still hundreds of GPCR genes in the human genome with unknown function makes them promising targets for future drugs.

Up to date, there are more than 7000 GPCR sequences identified. However, high resolution crystal 3D structure is known for only 5 of these sequences [6]. Lack of structural information remains the biggest challenge for the *in silico* virtual screening researches for drug discovery. The best method to get 3D structure information is homology-based structure prediction; nonetheless, it is still very difficult to predict 3D structure accurately even with the most recent advancements in prediction methods [8]. Due to this reason, successful virtual screening is only possible with the help of ligand information. In fact, ligand-based virtual screening gives better results than target-based virtual screening [11]. Most recent virtual screening studies utilize ligand-based and target-based screening methods together to achieve the best results [15].

The goal of the virtual screening methods is to predict which compounds would bind to a specific protein. Therefore, specific features of proteins and ligands are more important and well studied rather than common features. To the best of our knowledge, common chemical features of ligands have yet to be studied. Even the specific features are not translated back to human knowledge because the learning part of virtual screening is done by machine learning. This is not much of a concern given the fact that the human mind is not capable of processing information of such high dimensionality. However, knowing common structural features of ligands can be a key element for people to understand GPCR-ligand interaction. This is especially important for manual selection of candidate compounds that will be used in further experimental procedures and trials.

Here we present a novel method for extraction of common chemical substructures by using chemical structure comparison aided by hierarchical clustering. Using hierarchical clustering as a guideline to find similar ligands greatly eased the manual examination of ligand groups as well as significantly reduced the amount of time spent. Distance values used in clustering were calculated using the SIMCOMP [3] program, which uses a graph based algorithm to calculate all possible common substructures and also the maximum common substructure. We applied our method to a set of 850 GPCR ligands which were initially separated into classes according to their target GPCR classification.

## 2.  Methods

### 2.1. *Dataset*

We used the GLIDA GPCR-ligand database (version 2.02) [10] for GPCR-ligand pair data. In this database there are 3738 GPCR entries from human, rat and mouse, but ligand data is available for only 400 of these entries. There are 24077 ligands paired with the GPCR entries yielding a total of 39140 GPCR-ligand pairs. The ligands vary greatly in terms of size; there are ions, small molecules as well as very large molecules like peptides. The ligand collection of the database is highly redundant and most of the GPCR-ligand pairs do not have any reference to an experiment or research probably because they were generated by using high-throughput screening. In order to get the most reliable data from the database, we used only the GPCR-ligand pairs with experimental evidence which is given via links to PubMed [12]. This selection reduced the number of ligands to 1028. Within these ligands we selected non-peptide ligands. Also we limited the size of ligands to be between 20 to 170 atoms. This size limitation was necessary because larger molecules would obstruct the structure comparison process and smaller ligands would not give satisfactory common substructures reflecting the ligand specificity of GPCRs. Finally we eliminated the ligands with 95% identity for a non-redundant dataset. After all the filtering, the number of ligands was reduced to 850 (~4%), the number of GPCR proteins with ligand binding data was reduced to 252 and the number of GPCR-ligand pairs was reduced to 3656 (~10%). We used the IUPHAR GPCR Families [5] to separate GPCRs for grouping functionally related GPCR-ligand pairs which would also allow us to gather structurally similar ligands corresponding to each GPCR family.

## 2.2. *Chemical structure comparison*

For calculating similarity scores of GPCR ligands we used the SIMCOMP software with default parameters. Given a list of compounds and their chemical structures in KEGG atom type [7] format, SIMCOMP can calculate all pairwise similarity scores for them. The software translates the chemical structure into a 2D graph form in which atoms are represented as vertices and the bonds are represented as edges connecting the vertices. Between these graphs, it uses a 2D graph matching algorithm to find the superimposing vertices and edges in the graphs then it calculates the ratio of superimposing vertices and edges to the total number of vertices and edges in the graphs to obtain the similarity score of the compounds. This way, the program can find the maximum connected common substructure and additional smaller common substructures between two compounds. It is an important feature of the software that it can output the smaller common substructures because in some cases there can be several separate common substructures comparable in size. Therefore, it is necessary to keep record of all common substructures so that it will be possible to evaluate which substructure is the most conserved and/or best represents the chemical properties of the ligands. However, the largest substructures were preferred in most cases.

## 2.3. *Extracting common substructures*

After calculating pairwise similarity scores for all ligands, we created distance matrices of ligands in each GPCR family. The distance values were calculated by subtracting the similarity scores from 1 (the highest score possible in SIMCOMP is 1). Next, we used these matrices to cluster ligands by using a hierarchical clustering function in the R Project [13]. We used 'average' as the method parameter in the clustering function because the average method would group similar ligands by maintaining the desired distance value among all of the ligands in subclusters under a cutoff value. This property eases manual substructure extraction in the further steps. It is also convenient because ligands gathered in these subclusters are also comparable in size as a result of normalization of similarity scores by SIMCOMP. Hence, it becomes easier to decide whether a common substructure in a subcluster of ligands is significant in terms of its size compared to the ligands.

The final and the most time-consuming step of extracting common substructures was the manual part in which similar ligands gathered in subclusters were examined one by one in order to reveal the common substructures shared among the ligands. Besides being time consuming, there was also the issue of defining subclusters of similar ligands. Further examinations showed that ligands in subclusters with a distance score of less than 0.6, which corresponds to ligand pairs with SIMCOMP similarity scores of more than 0.4, shared some significant substructures. Even though 0.4 seemed like a low similarity score, this could be explained by the distribution of all pairwise similarity scores. In Fig. 1, we could see that most of the similarity scores are less than 0.4. So, we decided that distance of 0.6 is significant enough (P-value < 0.047) to use as a cutoff value for defining subclusters of similar ligands.

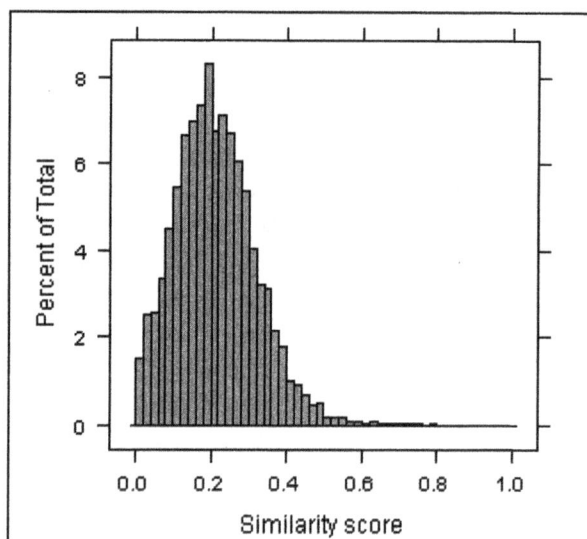

Fig. 1. Histogram of all pairwise similarity scores of 850 ligands in the filtered dataset.

The manual extraction process (also see Fig. 2) began with picking out the subclusters rooting from the cutoff value 0.6 and having at least 3 ligands. After that, we began comparing the most similar ligands with each other and extract their common substructures. Then we compared these substructures to ligands, which are less similar to the initial ligands, and extract new common substructures covering these later ligands. This process continued until we reached the threshold and as many ligands as possible were covered by substructures. For each subcluster, we tried to find a single substructure which covered every ligand, but for large subclusters we kept several substructures covering different combinations of ligands.

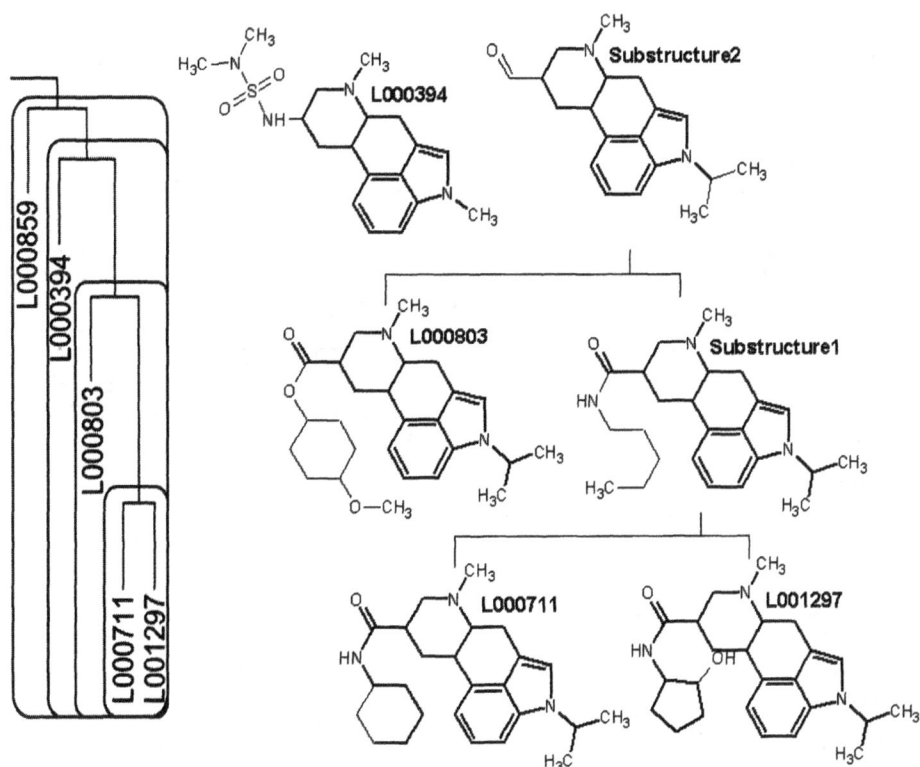

Fig. 2. Demonstration of manual substructure extraction. The tree structure on the left is a section from serotonin cluster dendrogram of similar ligands. The figure on the right is the 2D chemical structure of the ligands and their common substructures showed in bold. *Substructure1* is the common substructure extracted from *L000711* and *L001297*. *Substructure2* is the common substructure of *L000803* and *Substructure1*; therefore it is the common substructure of *L000711*, *L001297* and *L000803*.

## 3. Results

We successfully applied our substructure extraction method to the 12 largest GPCR families in terms of number of ligands. We were able to collect 130 substructures covering 439 ligands in our dataset. More detailed numbers are shown in Table 1. When looking at these numbers, the immediate fact attracting our attention is the irregularity of numbers of ligands per GPCR family. The serotonin family, for example, has 240 ligands which is more than twice its closest match dopamine, and more than 10 times thromboxane's number of ligands. The first explanation for this issue is the different levels of interest shown for different families because of their various biological functions. It is understandable that serotonin family proteins have attracted the most interest because they are known to influence neurological processes like anxiety, appetite, learning, and mood [9]. Hence, there are more studies on serotonin receptors than any other types of receptors. Another explanation would be that receptors of some families can be more selective in their ligand binding specificities, thus having fewer known ligands. The variance ($\mu$=134, $\sigma$=60) in numbers of ligands also affected the number of substructures defined for GPCR families. A large number of ligands yielded many subclusters of similar ligands with many common substructures.

Table 1: Number of ligands and GPCR proteins found in each GPCR family, number of common substructures found for these ligands, and coverage of ligands by the substructures. (M. Acetylcholine: Muscarinic acetylcholine; M. Glutamate: Metabotropic glutamate)

| GPCR family name | Number of GPCRs | Number of substructures | Number of ligands | Coverage |
|---|---|---|---|---|
| Serotonin | 31 | 53 | 240 | 181 (75%) |
| Dopamine | 14 | 17 | 113 | 76 (67%) |
| α - Adrenoceptors | 13 | 9 | 96 | 42 (44%) |
| β - Adrenoceptors | 8 | 4 | 44 | 33 (75%) |
| Histamine | 8 | 9 | 96 | 59 (51%) |
| M. Acetylcholine | 11 | 12 | 81 | 35 (43%) |
| Adenosine | 11 | 9 | 56 | 41 (73%) |
| M. Glutamate | 13 | 4 | 28 | 21 (75%) |
| Opioid | 10 | 6 | 56 | 32 (57%) |
| Prostacyclin | 3 | 2 | 20 | 12 (60%) |
| Prostaglandin E2 | 14 | 6 | 48 | 22 (46%) |
| Thromboxane | 3 | 1 | 23 | 8 (35%) |
| TOTAL | 138 | 130 | 850 | 439 (51%) |

Table 2: Some notable common substructures found in GPCR ligands. For each family, the structure of a natural ligand is also given for reference.

| GPCR family name | Natural ligand | Common substructures |
|---|---|---|
| Serotonin | | |
| Dopamine | | |

Table 2 Continued

| Alpha Adrenoceptors | |
| Beta Adrenoceptors | |
| Histamine | |
| Muscarinic Acetylcholine | |
| Adenosine | |

Table 2 Continued

| | | |
|---|---|---|
| Metabotropic Glutamate | | |
| Opioid | | |
| Prostacyclin | | |
| Prostaglandin E2 | | |
| Thromboxane | | |

In Table 2, we can see that there are a wide variety of common substructures with distinct sizes and shapes. It is interesting that many common substructures of the ligands of amine GPCR families (serotonin, dopamine, alpha & beta adrenoceptors, histamine and muscarinic acetylcholine) are much larger than their natural ligands. For example, natural ligand histamine itself is substructure of 7 ligands in histamine family. This shows that it is a common method to develop a new ligand by adding small groups onto the natural ligands. On the other hand, the large substructures show that it is also possible to discover new ligands that are very dissimilar in structure and size. By adding various chemical groups to these distinctive substructures it is possible to develop even more drugs and ligands targeting these receptors.

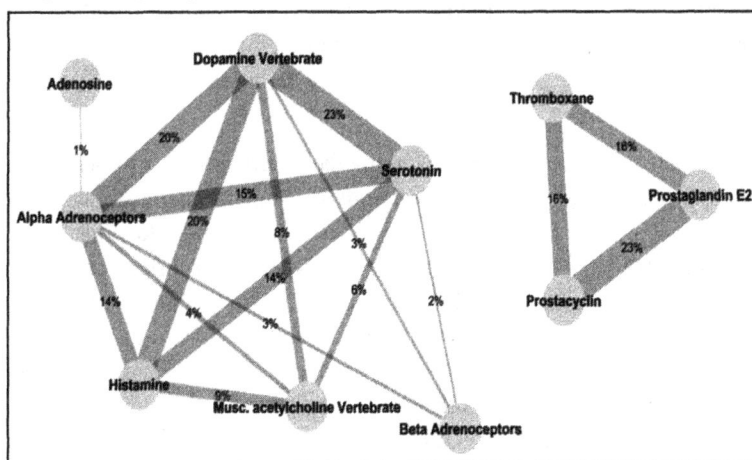

Fig. 3. Network diagram showing the connectivity of GPCR families in terms of shared ligands. Values on the edges give the percentage of shared ligands between two families.

By comparing common substructures of different families, we can see that some substructures are conserved in many families. For instance, the first substructures of serotonin, dopamine, alpha and beta adrenoceptors, are exactly the same. There are also other substructures which are actually a substructure of another family's common substructure, such as dopamine's 4th substructure which is a substructure of serotonin's and histamine's 4th common substructure. This can also be applied to the substructures of prostanoid families (prostacyclin, prostaglandin E2, and thromboxane). From our dataset, we know that ligands usually interact with more than one GPCR protein. However, we were assuming that most of the ligands interacted with proteins that are within the same family. The substructures common in several families put this assumption into question. To test if our assumption is reasonable, we evaluated amounts of ligands shared between different GPCR families. The results are illustrated in Fig. 3. There are two separate networks in Fig. 3; on the left, the amine receptor families plus adenosine are gathered; on the right, the prostanoid families are gathered. Metabotropic glutamate and opioid families are not shown since they didn't have any connection with other families. Overall, the network demonstrates that GPCR families are connected parallel to their sequence similarity because we know that primary parameter of GPCR classification is sequence similarity and, amine and opioid are two level 1 subfamilies in GPCR classes [4]. Nevertheless, it is difficult to explain why beta adrenoceptors have the weakest connections among amine families, especially with their close relative alpha adrenoceptors. Perhaps beta adrenoceptors have a key feature which makes them more selective but cannot be detected by sequence analysis. Obviously, further analysis is needed to have a satisfactory explanation for this matter. We further examined all of the shared ligands. We found that 88% of these ligands were covered by the defined common substructures, which is much higher than the overall coverage of 51%. So, our method is more successful in revealing common substructures of ligands shared among several GPCR families rather than ligands specific to a family.

## 4.   Discussion

In this research we presented a new method combining 2D chemical structure comparison and hierarchical clustering for finding common chemical substructures. By applying this method to 850 GPCR ligands we were able to find 130 common substructures in 439 ligands. Although most of the common substructures were specific to their GPCR families there were also some substructures common in multiple GPCR families. The substructures common in more than one GPCR family lead us to discover the correlation between sequence similarity of GPCRs and quantity of shared ligands. We also found out that our method is better in finding common substructures covering the ligands which interact with multiple GPCR family proteins, compared to GPCR-specific ligands. This indicates that our method was able to detect somewhat general chemical features of GPCR ligands. Evidently, the major goal of GPCR researches is to find out specific features of GPCR-ligand interaction; however, knowledge of common features can help us to see which features are more selective. For example, if we define the common substructures as core structures and the uncommon substructures as the modification patterns [14] added to the core structures, we can learn which modification pattern is specific to which target. Then we can use this information to discover new ligands for target proteins. This might even become an alternative strategy for drug discovery in the future.

## Acknowledgments

This work was supported by the Ministry of Education, Culture, Sports, Science and Technology of Japan, and the Japan Science and Technology Agency. Computational resources were provided by the Bioinformatics Center and the Supercomputer Laboratory at the Institute for Chemical Research of Kyoto University.
We would like to thank J.B. Brown and Masaaki Kotera for critical reading and overall improvement of our manuscript.

## References

[1]   Bockaert, J. and Pin, J. P., Molecular tinkering of G protein-coupled receptors: an evolutionary success, *EMBO J.,* 18(7):1723-1729, 1999.
[2]   Fredholm, B. B., Hökfelt, T., and Milligan G., G-protein-coupled receptors: an update, *Acta Physiol.,* 190:3-7, 2007.

[3] Hattori, M., Okuno, Y., Goto, S., and Kanehisa, M., Development of a chemical structure comparison method for integrated analysis of chemical and genomic information in the metabolic pathways, *J. Am. Chem. Soc.*, 125:11853-11865, 2003.

[4] Horn, F., Bettler, E., Oliveira, L., Campagne, F., Cohen, F. E., and Vriend, G.., GPCRDB information system for G protein-coupled receptors, *Nucleic Acids Res.*, 31:294-297, 2003.

[5] IUPHAR Database | GPCR Families, *http://www.iuphar-db.org/DATABASE/ReceptorFamiliesForward?type=GPCR*

[6] Katritch, V., Rueda M., Lam, P. C., Yeager, M., and Abagyan, R., GPCR 3D homology models for ligand screening: Lessons learned from blind predictions of adenosine A2a receptor complex, *Proteins*, 78(1):197-211, 2009.

[7] KEGG atom types, http://www.genome.jp/kegg/reaction/KCF.html

[8] Michino, M., Abola, E., Brooks, C.L., Dixon, J.S., Moult, J., and Stevens, R.C., Community-wide assessment of GPCR structure modelling and ligand docking: GPCR Dock 2008, *Nat. Rev. Drug Discov.*, 8:455–463, 2009.

[9] Nichols, D. E., and Nichols, C. D., Serotonin receptors, *Chem. Rev.*, 108 (5):1614–1641, 2008.

[10] Okuno, Y., Yang, J., Taneishi, K., Yabuuchi, H., and Tsujimoto, G., GLIDA: GPCR-ligand database for chemical genomic drug discovery, *Nucleic Acids Res.*, 36:D673-D677, 2006.

[11] Oprea, T. I. and Matte, H., Integrating virtual screening in lead discovery, *Curr. Opin. Chem. Biol.*, 8:349–358, 2004.

[12] PubMed, *http://www.ncbi.nlm.nih.gov/pubmed*

[13] R Development Core Team, R: A language and environment for statistical computing, *http://www.R-project.org*, Vienna, Austria, 2009.

[14] Shigemizu, D., Araki, M., Okuda, S., Goto, S., and Kanehisa, M., Extraction and analysis of chemical modification patterns in drug development, *J. Chem. Inf. Model.*, 49:1122-1129, 2009.

[15] Weill, N. and Rognan, D., Development and validation of a novel protein-ligand fingerprint to mine chemogenomic space: application to G protein-coupled receptors and their ligands, *J. Chem. Inf. Model.*, 49:1049-1062, 2009.

# A SYSTEMS BIOLOGY APPROACH: MODELLING OF AQUAPORIN-2 TRAFFICKING

MARTINA FRÖHLICH[1,3]
martina.froehlich@biologie.hu-berlin.de

PETER M. T. DEEN[2]
P.Deen@fysiol.umcn.nl

EDDA KLIPP[3]
edda.klipp@biologie.hu-berlin.de

[1] *Max Planck Institute for Molecular Genetics, Berlin, Germany*
[2] *Radboud University Nijmegen Medical Centre, Nijmegen, The Netherlands*
[3] *Humboldt-Universität zu Berlin, Berlin, Germany*

In healthy individuals, dehydration of the body leads to release of the hormone vasopressin from the pituitary. Via the bloodstream, vasopressin reaches the collecting duct cells in the kidney, where the water channel Aquaporin-2 (AQP2) is expressed. After stimulation of the vasopressin V2 receptor by vasopressin, intracellular AQP2-containing vesicles fuse with the apical plasma membrane of the collecting duct cells. This leads to increased water reabsorption from the pro-urine into the blood and therefore to enhanced retention of water within the body.

Using existing biological data we propose a mathematical model of AQP-2 trafficking and regulation in collecting duct cells. Our model includes the vasopressin receptor, adenylate cyclase, protein kinase A, and intracellular as well as membrane located AQP2. To model the chemical reactions we used ordinary differential equations (ODEs) based on mass action kinetics. We employ known protein concentrations and time series data to estimate the kinetic parameters of our model and demonstrate its validity.

Through generating, testing and ranking different versions of the model, we show that some model versions can describe the data well as soon as important regulatory parts such as the reduction of the signal by internalization of the vasopressin-receptor or the negative feedback loop representing phosphodiesterase activity are included.

We perform time-dependent sensitivity analysis to identify the reactions that have the greatest influence on the cAMP and membrane located AQP2 levels over time. We predict the time courses for membrane located AQP2 at different vasopressin concentrations, compare them with newly generated data and discuss the competencies of the model.

*Keywords*: mathematical model; systems biology; parameter estimation; time-dependent sensitivity analysis; aquaporin-2.

## 1. Introduction

### 1.1. *Biological background*

Aquaporin-2 (AQP2) is a water channel of the Aquaglyceroporin family [1]. It is located in the principal cells of the collecting duct in the kidney. In an unstimulated cell it is located mainly in intracellular vesicles. Upon a stimulus by the antidiuretic hormone vasopressin (AVP) an intracellular signaling cascade is activated including an increase in intracellular cAMP [16], activation of PKA and translocation of AQP2 into the apical plasma membrane. This leads to an osmotically driven water flow from the pro-urine back into the blood and is an important mechanism in the regulation of body water homeostasis.

### 1.2. *Existing models*

Knepper and Nielsen proposed the first mathematical model of the AQP2 trafficking [9]. They started with a 3-state model including activated, inactivated, and reserve AQP2 (also called by them transporters) and irreversible transition reactions between the states. Assuming that the transition between the inactivated and the reserve state is very fast, this resulted in a two state model including only activated (membrane located) and inactivated (intracellular) transporters as well as a transition from the inactivated to the activated state (exocytosis) and vice versa (endocytosis).

They fitted their model to data from osmotic water permeability (Pf) measurements in rat inner medulla collecting duct (IMCD) after AVP addition and washout. Thereby they proposed that the optimal fit results from a scenario where both the exocytosis as well as the endocytosis of AQP2 is regulated by vasopressin [9, 12].

### 1.3. *Aims*

Within the last years, a lot of knowledge has been gained concerning the regulation of AQP2. Our aim was to include this knowledge and build a more detailed and predictive mathematical model. With this model, questions should be answered through *in silico* experiments, for example which are the reactions that have the most influence on the output of the model? What happens if we use different concentrations of vasopressin and how is the response of the system achieved? In this paper, we propose a mathematical model of the AQP2 trafficking and use it to gain new insights into those questions.

## 2. Methods

### 2.1. *Time course simulation, parameter estimation and time-dependent sensitivity analysis*

Time course simulation and parameter estimation were performed with the systems biology tool COPASI [5]. Parameter estimation was performed by running evolutionary programming [4] in COPASI 4000 times with random initial values, random upper and lower bounds, 400 generations, a population size of 40, random number generator Mersenne Twister [11] and seed 1. For the optimal parameter set found the process was repeated once with boundaries from 1 to $\infty$ to finetune the parameters. The time course simulation was solved with the deterministic LSODA method [13].

Time dependent sensitivity analysis was performed using an algorithm proposed by Ingalls and Sauro [6] which was implemented in Mathematica7.0 [17].

## 2.2. *Cell surface biotinylation*

Madin-Darby Canine Kidney cells (MDCK-hAQP2-T269S) where seeded at a density of $2.7 \cdot 10^5$ cells/cm$^2$ on semipermeable 4.7 cm$^2$ filters (Transwell®, 0.4 $\mu$m pore size, Corning Costar, Cambridge, MA, USA). The cells grew at 37°C for two days, then the medium was changed and $5 * 10^{-5}$ M indomethacin was added for 1 more day to lower intracellular cAMP levels. After 3 days of seeding, the polarized cells were exposed to medium with (deamino-Cys1, D-arg8)-vasopressin (dDAVP, Sigma, St. Louis, MO, USA). Different concentrations of dDAVP were used ($10^{-6}$ M and $10^{-8}$ M) for durations of 0, 2, 5, 10, 20, 30 and 90 min. Apical cell surface biotinylation was performed as in [2]. The samples from total lysate were sonicated for 20 sec. After SDS-PAGE on a 12% acrylamide gel the proteins were immunoblotted as in [2]. Incubation with the primary antibody against the C-tail of AQP2 (750 K5007, kindly provided by Dr. M. Knepper, NIH, Bethesda) took place overnight at 4°C. The signal was amplified by 1:10,000-diluted biotinylated anti-rabbit IgGs and 1:8,000-diluted streptavidin-peroxidase (HRP; Sigma, St. Louis, MO, U.S.A.). Semi-quantification was performed with an Epson Expression 1640 XL (300dpi) using the software AIDA - Advanced Image Data Analyzer (V4.10.020; raytest Isotopemessgeräte GmbH). The experiment was done twice in duplicates.

## 3. Results

### 3.1. *The model*

We generated a mathematical model based on prior knowledge from Madin-Darby Canine Kidney (MDCK) cells to gain new information about AQP2 trafficking. The currently available data is mainly qualitative and rarely quantitative and time course data is not available for most model components. Therefore we choose a simplified model. The model consists of the species vasopressin, cyclic AMP, inactive and active PKA, and AQP2 in internal vesicles as well as in the apical membrane. The concentration dynamics of the involved components are represented as a system of ordinary differential equations using mass action kinetics. The complete model version contains nine reactions: increase of cAMP via vasopressin, activation and inactivation of PKA, and endo- and exocytosis of AQP2. cAMP can be degraded via negative feedback from active PKA on cAMP, which represents the decrease of cAMP by phosphodiesterases as well as via a PKA independent mechanism. Because it has been shown that treatment of MDCK cells by vasopressin leads to internalization of the vasopressin V2 receptor [14], we included the decrease of the stimulus into the model. Because Robben et al. [14] reported that the percentage of internalized receptors increases at a stronger stimulus of dDAVP we choose second order kinetics for that reaction. Further we assumed that, starting with a basal level of membrane located AQP2, this level does not change without any vasopressin stimulus. To achieve this, we added a constitutive exocytosis reaction to maintain a basal level of exocytosis and endocytosis.

Our model does not include the regulation of endocytosis via vasopressin as proposed by Knepper and Nielsen in rat terminal IMCD [12]. Using the model by Knepper and Nielsen and MDCK cell data [3] it could be shown that their model can reproduce the data reasonably well, when exclusively the exocytosis of AQP2 is regulated (Fig. 1, supplementary material).

A graphical representation of the model is shown in Fig. 1. The corresponding differential equations system is given in Table 1.

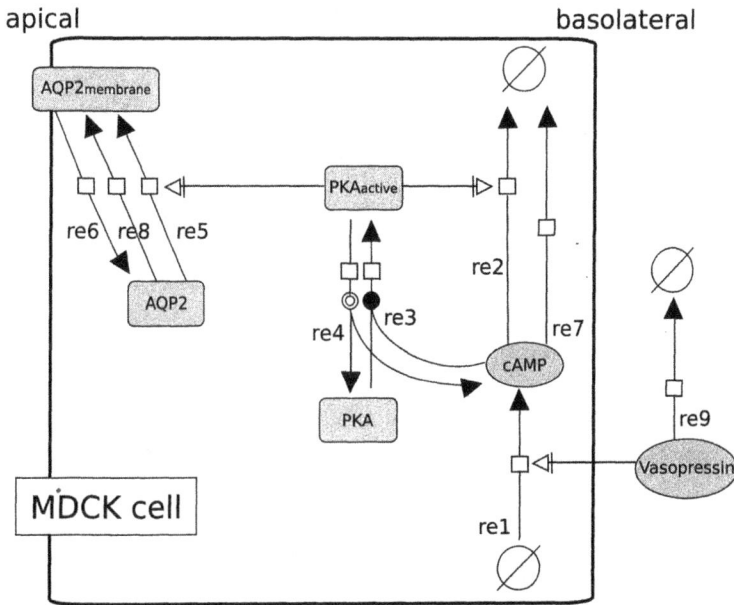

Fig. 1. The model, using the SBGN notation [7]. The model consists of 6 different species and 9 reactions. Vasopressin stimulates directly the increase of intracellular cAMP (re1). cAMP can form, together with PKA, in a reversible reaction PKA$_{active}$ (re3), (re4). cAMP can be degraded in a PKA$_{active}$ dependent (re2) and independent (re7) reaction. PKA$_{active}$ stimulates the integration of AQP2 into the apical membrane (re5) which is assumed to occur also via a PKA$_{active}$ independent reaction (re8). Endocytosis is assumed to depend only on the concentration of AQP2$_{membrane}$ (re6). The stimulus by Vasopressin can decrease over time, which represents the internalization and desensitization of the V2 receptor (re9).

Table 1.   Differential equations describing the model.

$$\frac{d[\text{PKA}]}{dt} = -(k3 \cdot [\text{PKA}] \cdot [\text{cAMP}]) + (k4 \cdot [\text{PKA}_{\text{active}}])$$

$$\frac{d[\text{PKA}_{\text{active}}]}{dt} = (k3 \cdot [\text{PKA}] \cdot [\text{cAMP}]) - (k4 \cdot [\text{PKA}_{\text{active}}])$$

$$\frac{d[\text{cAMP}]}{dt} = (k1 \cdot [\text{AVP}]) - (k2 \cdot [\text{cAMP}] \cdot [\text{PKA}_{\text{active}}]) - (k3 \cdot [\text{PKA}] \cdot [\text{cAMP}])$$
$$+ (k4 \cdot [\text{PKA}_{\text{active}}]) - k7 \cdot [\text{cAMP}]$$

$$\frac{d[\text{AQP2}]}{dt} = -(k5 \cdot [\text{AQP2}] \cdot [\text{PKA}_{\text{active}}]) + (k6 \cdot [\text{AQP2}_{\text{membrane}}]) - (k8 \cdot [\text{AQP2}])$$

$$\frac{d[\text{AQP2}_{\text{membrane}}]}{dt} = (k5 \cdot [\text{AQP2}] \cdot [\text{PKA}_{\text{active}}]) - (k6 \cdot [\text{AQP2}_{\text{membrane}}]) + (k8 \cdot [\text{AQP2}])$$

$$\frac{d[\text{AVP}]}{dt} = -(k9 \cdot [\text{AVP}] \cdot [\text{AVP}])$$

$$k8 = k6 \cdot \frac{[\text{AQP2}_{\text{membrane}}][0]}{[\text{AQP2}][0]}$$

$$k1, ..., k7, k9 = \text{estimated}$$

## 3.2.  *Time course simulation and parameter estimation*

We estimate the parameters to fit the experimental data taken from [3]. Deen et al. describe the increase in osmotic water permeability (Pf) and the peak in [cAMP] after a stimulus by dDAVP as well as the decrease in Pf after dDAVP washout. Furthermore, we used information taken from Xie et al. [18] to assign a range of the values for $[\text{AQP2}_{\text{membrane}}]$ between 114 $\mu$M and 255 $\mu$M with a total concentration of AQP2 of 1000 $\mu$M. We assumed that, while starting with a basal AQP2 level of 114 $\mu$M, this value should not change without any stimulus by vasopressin.

As concentration for total PKA we took 500 nM which is in consistency with published data [10, 15]. The absolute values for [cAMP] were taken from Deen et al. [3] and translated to nM.

Parameter estimation was performed as described in materials and methods. The estimated model parameters can be seen in Table 2 and the resulting fits in Fig. 2.

Table 2.   Estimated model parameters.

| Parameter name | Value | Units |
|:---:|:---:|:---|
| $k_1$ | $2.94e^{+5}$ | 1/min |
| $k_2$ | $1.02e^{-3}$ | 1/(nmol · min) |
| $k_3$ | $3.21e^{-4}$ | 1/(nmol · min) |
| $k_4$ | $4.04e^{-0}$ | 1/min |
| $k_5$ | $2.59e^{-5}$ | 1/(nmol · min) |
| $k_6$ | $2.15e^{-2}$ | 1/min |
| $k_7$ | $3.04e^{-2}$ | 1/min |
| $k_8$ | $2.76e^{-3}$ | 1/min |
| $k_9$ | $3.74e^{-0}$ | 1/(nmol · min) |

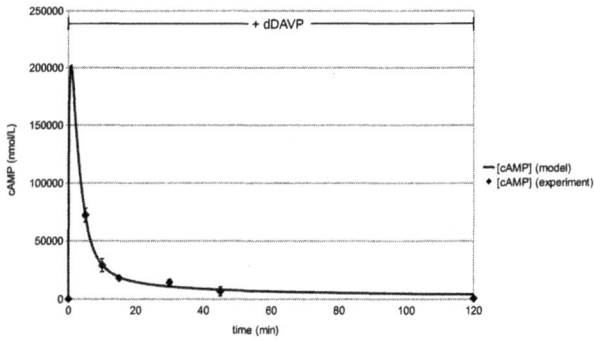

(a)

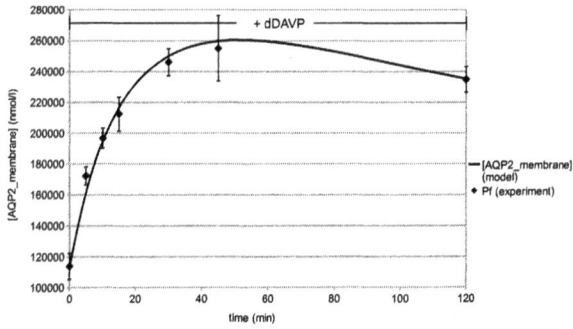

(b)

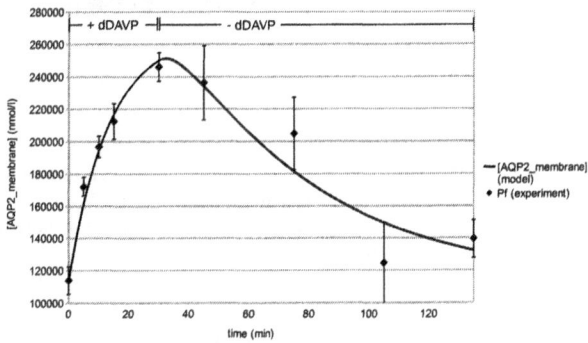

(c)

Fig. 2. Model fitting with COPASI. (a) Fitting of intracellular cAMP. (b) Fitting of membrane localized AQP2 (increase). (c) Fitting of membrane localized AQP2 (decrease).

### 3.3. *Model variations and ranking*

It was tested whether the dataset from Deen et al. [3] could also be reproduced with a simpler variant of the model to get an impression, which parts of the model are essential. Therefore different model variants were generated excluding either the negative feedback by phosphodiesterases (-re2), the reduction of the signal representing internalization of the receptor (-re9) or both (-re2 -re9). Parameter estimation was performed for all model variants as described in Sec. 3.2. They were ranked according to their error function (Fig. 3). It could be shown that models including the vasopressin receptor internalization or the negative feedback have a better fit than the minimal model, because they are needed to reproduce the peak of intracellular cAMP. The models including vasopressin receptor internalization could fit the data best.

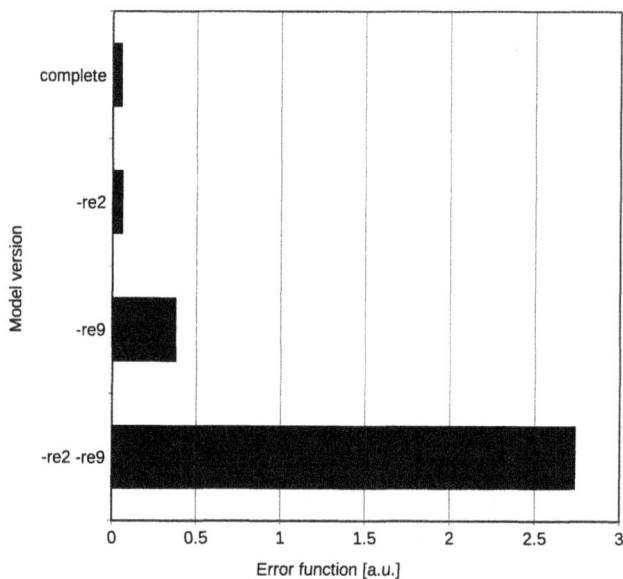

Fig. 3.   Ranking of different model versions. Complete: model version as shown in Fig. 1; -re2: complete model reduced by reaction re2 representing negative feedback via phosphodiesterases; -re9: complete model reduced by reaction re9, which represents the internalization and degradation of the receptor; -re2 -re9: reduced by both reactions re2 and re9

### 3.4. *Time-dependent sensitivity analysis*

We analyzed with time-dependent sensitivity analysis [6] how sensitive the model species cAMP and $AQP2_{membrane}$ react to minor changes in the model parameter values (Fig. 4). We applied the analysis on the complete model with the optimal parameter set found in Sec. 3.2.

It could be seen that cAMP is highly sensitive to the parameters $k1$, $k2$ and $k9$, which represent the activation of cAMP by vasopressin, the negative feedback loop, and decrease of vasopressin by reaction re9, which represents the internalization of the receptor. While looking at the sensitivities over time, we could observe that a change in $k1$ has the strongest (positive) effect when cAMP gets accumulated at the begin of the stimulation, while $k9$ and $k2$ affect cAMP mostly around the time when the peak starts and finishes decreasing, respectively.

$AQP2_{membrane}$ is highly sensitive to parameters $k5$ and $k6$, which characterize activation as well as deactivation of AQP2. The effect of a small change in $k5$ would not be as strong as a change in $k6$, but it would appear earlier after the begin of the stimulus.

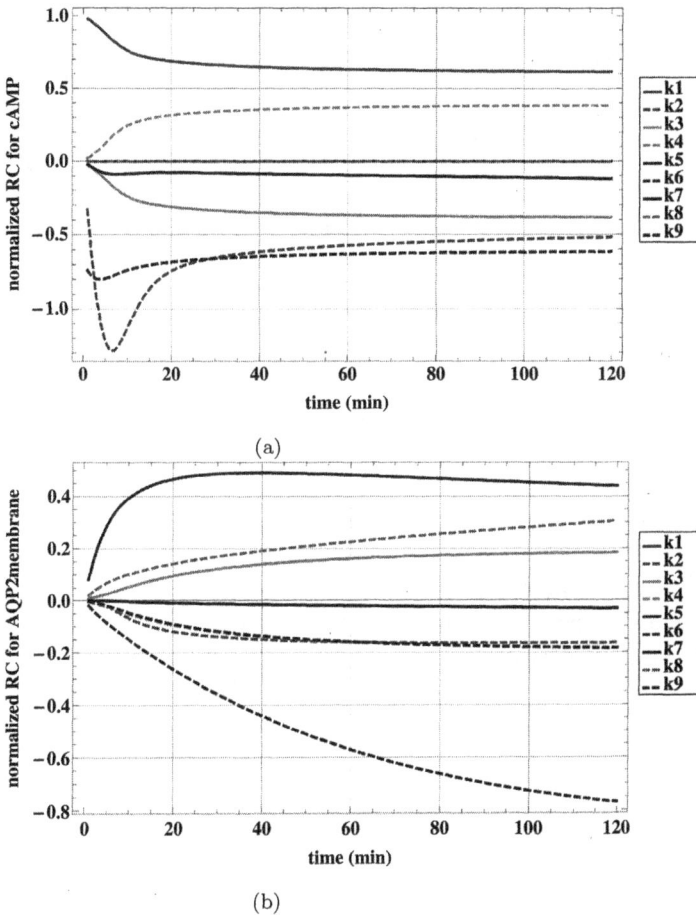

(a)

(b)

Fig. 4. Time-dependent sensitivity analysis: performed on complete model with the optimal parameterset found. (a) Time-dependent normalized respose coefficients for all parameters in respect to cAMP. (b) Time-dependent normalized respose coefficients for all parameters in respect to AQP2 in the membrane.

### 3.5. Experiment: AQP2 translocation after dDAVP stimulation

We as well as Knepper and Nielsen [9] used data from Pf measurements to estimate model parameters and assumed that $AQP2_{membrane}$ would be proportional to the Pf. We wanted to investigate, whether this assumption can be applied and furthermore, how the response of the system is changing at different concentrations of dDAVP. Therefore, we treated MDCK cells with $10^{-6}$ M and $10^{-8}$ M dDAVP and performed cell surface biotinylation to monitor the amount of AQP2 at the apical membrane over time.

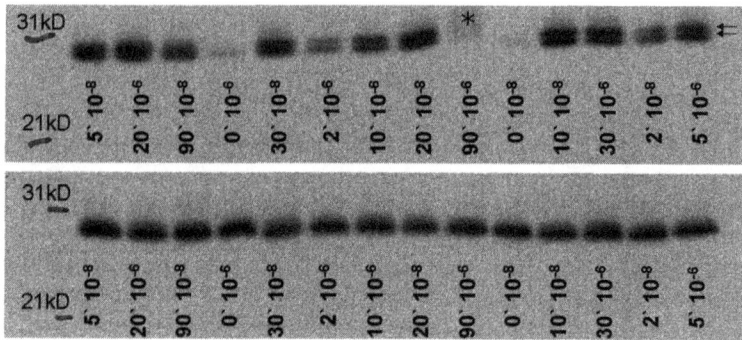

Fig. 5.    Cell surface biotinylation: membrane fraction (upper layer) and total lysate (lower layer) of cells treated with $10^{-6}M$ dDAVP and $10^{-8}M$ dDAVP for durations of 0, 2, 5, 10, 20, 30 and 90 min, loaded in random order. The sample labeled with * was left out from the calculation shown in Fig. 6.

Fig. 5 shows that in the membrane fraction two bands appeared at approximately 29 and 30 kD (indicated by arrows), which are AQP2 specific and are changing over time. In the total lysate, both bands were also detected, but the lower band was more intense. We could monitor an increase in AQP2 at the membrane over time, which was sustained after treatment with $10^{-6}$ M and transient after $10^{-8}$ M dDAVP (Fig. 6(a) and (b)). In Fig. 2 (supplementary material) $AQP2_{membrane}$ at both concentrations is compared with the Pf measured by Deen et al. [3] after $10^{-8}$ M dDAVP. The shape of the curve for $[AQP2_{membrane}]$ after $10^{-6}$ M dDAVP fits very well the curve for the Pf value. Whereas the transient curve at $10^{-8}$ M dDAVP looks rather different after 20 min.

To get an impression whether the amount of $AQP2_{membrane}$ is already saturated at a concentration of $10^{-8}$ M we compared the maximal value of $AQP2_{membrane}$ during the stimulation of both concentrations. The data shows that the maximal value of AQP2 at the membrane at $10^{-6}$ M is not higher than at $10^{-8}$ M dDAVP ($\frac{AQP2_{membrane}10^{-8}MdDAVP\cdot100\%}{AQP2_{membrane}10^{-6}MdDAVP} = 109.89\%$, stdev=15.27%, unnormalized data, n=2).

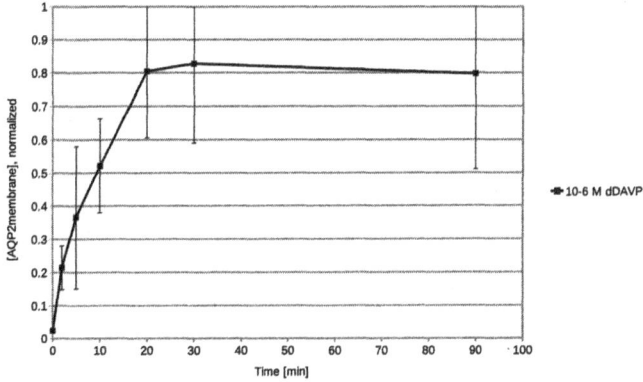

(a)

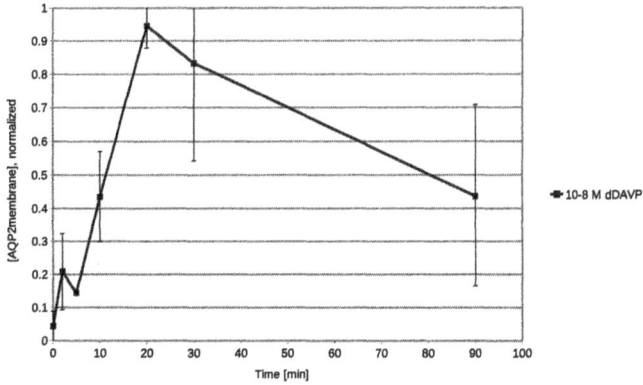

(b)

Fig. 6. Cell surface biotinylation: MDCK-hAQP2-T269S cells were treated with different concentrations of dDAVP. Biotinylation was performed and AQP2 was semiquantified. The amount of membrane located AQP2 was normalized to the amount of AQP2 in total lysate (see details in materials and methods). (a) Amount of AQP2 at the membrane after treatment with $10^{-6} M$ dDAVP (b) Amount of AQP2 at the membrane after treatment with $10^{-8} M$ dDAVP (done twice in duplicates)

### 3.6. *Model predictions, analysis, and comparison with recent data*

We used the complete model version with the optimal parameter set found in Sec. 3.2 to perform experiments *in silico*. The dDAVP concentration to which the model was fitted was $10^{-8}$ M. We simulated the change of $AQP2_{membrane}$ after a stimulus as well as a stimulus followed by washout for the dDAVP concentrations $10^{-6}$ to $10^{-14}$ (see Fig. 7).

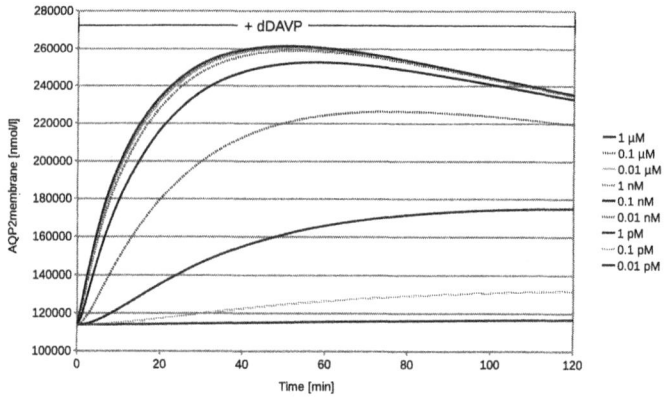

(a)

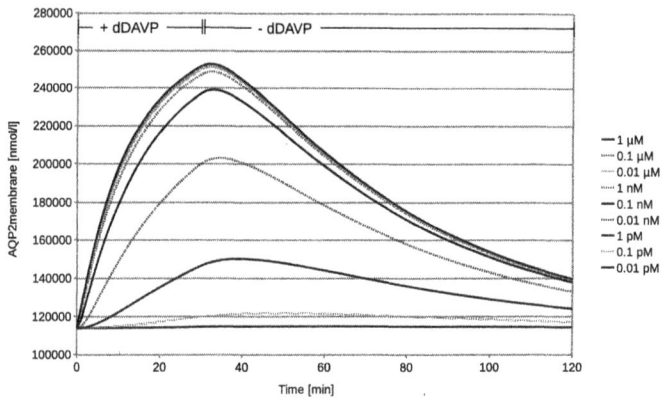

(b)

Fig. 7.   Model prediction: $AQP2_{membrane}$ over time at different dDAVP concentrations. dDAVP concentration in log scale from $10^{-6}$ M to $10^{-14}$ M. (a) Timecourse after stimulus with dDAVP at 0 min. (b) Timecourse after stimulus with dDAVP at 0 min and washout of dDAVP at 30 min.

Our model indicates that $AQP2_{membrane}$ is already saturated at a concentration of $10^{-9}$ M dDAVP, which explains why we see no higher value at $10^{-6}$ M than at $10^{-8}$ M dDAVP.

In our simulation we could not observe the difference between sustained and transient response of the system at a concentration of $10^{-6}$ M and $10^{-8}$ M dDAVP, respectively. At both concentrations $AQP2_{membrane}$ looks rather similar over time, reaches a maximum at around 50 min and decreases then slowly.

To investigate how it is possible to observe a high Pf over more than 90 min while the cAMP level is already low, we used the model to analyze the fluxes of AQP2 endocytosis and exocytosis over time. We showed that with our model structure and parameters there is a very rapid increase in AQP2 exocytosis directly after adding the stimulus while AQP2 endocytosis increases rather slowly. Around 50 min after adding the stimulus exocytosis becomes slightly lower than endocytosis so $AQP2_{membrane}$ starts decreasing (Fig. 7(a) and Fig. 3 (supplementary material)).

## 4. Discussion

We proposed a mathematical model of aquaporin-2 trafficking including the widely accepted pathway via intracellular cAMP and PKA. We further included a negative feedback representing phosphodiesterases as well as a reduction of the signal representing internalization of the receptors. In contrast to the proposed model by Knepper and Nielsen [12] based on work with isolated tubules, we did not include vasopressin dependent regulation of endocytosis.

To reproduce the experimental data for intracellular cAMP [3] we showed that either the negative feedback representing phosphodiesterases or the reduction of the vasopressin signal is required. Using mass action kinetics, the model version including the internalization of the receptor performed better than the model which includes only the negative feedback. This might be a hint that the internalization of the receptor is indeed an important mechanism for regulating the signal. We allowed the peak in cAMP to be higher than observed in the experimental data, because we lack information about the cAMP concentration between 0 and 5 min. The data we used [3] is based on a whole population of MDCK cells. It remains to be verified whether the peak of cAMP exists also in single cells. Also feasible would be an oscillatory behavior which sums up to a peak when looking at the whole population.

We used the model to simulate the time course for AQP2 at the membrane at different concentrations. Our model indicates that AQP2 at the membrane gets saturated already at a concentration of $10^{-9}$ M dDAVP. As stated before we used data for the Pf [3] for our model fitting and assumed that it would be proportional to AQP2 in the membrane. To investigate whether $AQP2_{membrane}$ is consistent with the Pf and to verify our predictions, we performed cell surface biotinylation on MDCK cells after a given stimulus and monitored the change of AQP2 at the membrane over time. We investigated that there was no difference in the maximal amount of AQP2 at the membrane at the concentrations $10^{-8}$ M and $10^{-6}$ M dDAVP, which supports the hypothesis that $AQP2_{membrane}$ is already saturated at those concentrations. At a concentration of $10^{-8}$ M dDAVP we could also see a transient response whereas at a concentration of $10^{-6}$ M the response was sustained. That slightly disagrees with our predictions, because there we did not observe a difference between the two concentrations.

By including the pathway prior of AQP2 in our model we are able to reproduce the increase and decrease of membrane located AQP2 very well even without including vasopressin regulated endocytosis of AQP2 into our model. Nevertheless, time dependent sensitivity analysis revealed that changing the parameter for the endocytosis reaction has the highest impact on AQP2 in the membrane. So even if a regulation of this reaction by vasopressin is not needed to fit the model to the data, it might be useful to integrate in future versions of the model and analyse it in more detail. One can include also information from Kamsteeg et al. [8] who proposed regulated endocytosis of AQP2 via short-chain ubiquitination after washout of a stimulus. Nevertheless, one has to be carefull with combining those datasets, because Kamsteeg et al. used forskolin to stimulate the cells instead of dDAVP. Furthermore they did not stimulate the cells with indomethacin prior to stimulation with forskolin, which might alter the cellular response significantly.

One of the major questions while looking at the pathway and the data from MDCK cells was, how it is possible that AQP2 is still in the membrane after two hours of stimulation while the stimulus in cAMP is already low. Our model demonstrates, that even with a simple pathway this behaviour can be observed.

Our results are a starting point to initiate the next iteration of quantitative experiments on AQP2 trafficking and subsequent model refinement.

## Acknowledgements

The authors thank Mark Knepper for providing the antibody against AQP2 C-tail and Guido Klingbeil for his helpful comments during the completion of this article. PMTD is a recipient of VICI grant 865.07.002 of the Netherlands Organization for Scientific research (NWO). This work was supported by grants of the Marie Curie Research Training Network Aquaglyceroporins LSHG-CT-2006-035995-2 to PMTD and EK, the Netherlands Organization for Scientific Research (NWO) 865.07.002 to PMTD, the International Research Training Group (IRTG) on "Genomics and Systems Biology of Molecular Networks" (GRK1360, German Research Council (DFG)) and the IMPRS for Computational Biology and Scientific Computing (Max Planck Society) to EK.

## References

[1] Agre, P., The aquaporin water channels, *Proc Am Thorac Soc.*, 3:5-13, 2006.

[2] Deen, P.M.T., Van Balkom, B.W., Savelkoul, P.J., Kamsteeg, E.J., Van Raak, M., Jennings, M.L., Muth, T.R., Rajendran, V., and Caplan, M.J., Aquaporin-2: COOH terminus is necessary but not sufficient for routing to the apical membrane, *Am J Physiol Renal Physiol.*, 282(2):F330–F340, 2002.

[3] Deen, P.M.T., Rijss, J.P.L., Mulders, S.M., Errington, R.J., Van Baal, J., and Van Os, C.H., Aquaporin-2 transfection of Madin-Darby canine kidney cells reconstitutes vasopressin-regulated transcellular osmotic water transport, *J. Am. Soc. Nephrol.*, 8(10):1493–1501, 1997.

[4] Fogel, D.B., Fogel, L.J., and Atmar, J.W., Meta-evolutionary programming, *25th Asiloma Conference on Signals, Systems and Computers. IEEE Computer Society*, 540–545, 1992.

 [5] Hoops, S., Sahle, S., Gauges, R., Lee, C., Pahle, J., Simus, N., Singhal, M., Xu, L., Mendes, P., and Kummer, U., COPASI - a COmplex PAthway SImulator, *Bioinformatics*, 22(24):3067–3074, 2006.

 [6] Ingalls, B.P. and Sauro, H.M., Sensitivity analysis of stoichiometric networks: an extension of metabolic control analysis to non-steady state trajectories, *Journal of Theoretical Biology*, 222(1):23–36, 2003.

 [7] Jansson, A. and Jirstrand, M., Biochemical modeling with systems biology graphical notation., *Drug Discovery Today*, 15(9-10):365–370, 2010.

 [8] Kamsteeg, E.J., Hendriks, G., Boone, M., Konings, I.B., Oorschot, V.,van der Sluijs, P., Klumperman, J., and Deen, P.M.T., Short-chain ubiquitination mediates the regulated endocytosis of the aquaporin-2 water channel, *PNAS*, 103(48):18344–18349, 2006.

 [9] Knepper, M.A. and Nielsen, S., Kinetic model of water and urea permeability regulation by vasopressin in collecting duct, *Am. J. Physiol.*, 265(2):F214–F224, 1993.

[10] Kopperud, R., Christensen, A.E., Kjarland, E., Viste, K., Kleivdal, H., and Doskeland, S.O., Formation of inactive cAMP-saturated holoenzyme of cAMPdependent protein kinase under physiological conditions, *J. Biol. Chem.*, 277(16):13443–13448, 2002.

[11] Matsumoto, M. and Nishimura, T., Mersenne twister: A 623-dimensionally equidistributed uniform pseudorandom number generator, *ACM Transactions on Modeling and Computer Simulation*, 8(1):3–30, 1998.

[12] Nielsen, S. and Knepper, M.A., Vasopressin activates collecting duct urea transporters and water channels by distinct physical processes, *Am. J. Physiol.*, 265(2):F204–F213, 1993.

[13] Petzold, L., Automatic selection of methods for solving stiff and nonstiff systems of ordinary differential equations, *SIAM Journal on Scientific and Statistical Computing*, 4(1):136-148, 1983.

[14] Robben, J.H., Knoers, N.V.A.M., and Deen, P.M.T., Regulation of the vasopressin V2 receptor by vasopressin in polarized renal collecting duct cells, *Molecular Biology of the Cell*, 15(12):5693–5699, 2004.

[15] Sette, C. and Conti, M., Phosphorylation and activation of a cAMP-specific phosphodiesterase by the cAMP-dependent protein kinase, *J. Biol. Chem.*, 271(28):16526–16534, 1996.

[16] Star, R.A., Nonoguchi, H., Balaban, R., and Knepper, M.A., Calcium and cyclic adenosine monophosphate as second messengers for vasopressin in the rat inner medullary collecting duct, *J Clin Invest.*, 81(6):1879–1888, 1988.

[17] Wolfram Research, Inc., *Mathematica, Version 7.0*, Champaign, IL, 2008.

[18] Xie, L., Hoffert, J.D., Chou, C.L., Yu, M.J., Pisitkun, T., Knepper, M.A., and Fenton, R.A., Quantitative analysis of aquaporin-2 phosphorylation, *Am J Physiol Renal Physiol.*, 298(4):F1018–F1023, 2010.

# COMPARISON OF GENE EXPRESSION PROFILES PRODUCED BY CAGE, ILLUMINA MICROARRAY AND REAL TIME RT-PCR

ANDRÉ FUJITA[1]
andrefujita@riken.jp

MASAO NAGASAKI[2]
masao@ims.u-tokyo.ac.jp

SEIYA IMOTO[2]
imoto@ims.u-tokyo.ac.jp

AYUMU SAITO[2]
s-ayumu@ims.u-tokyo.ac.jp

EMI IKEDA[2]
e-ikeda@ims.u-tokyo.ac.jp

TEPPEI SHIMAMURA[2]
shima@ims.u-tokyo.ac.jp

RUI YAMAGUCHI[2]
ruiy@ims.u-tokyo.ac.jp

YOSHIHIDE HAYASHIZAKI[3]
yoshihide@gsc.riken.jp

SATORU MIYANO[1,2]
miyano@ims.u-tokyo.ac.jp

[1] *Computational Science Research Program, RIKEN, Wako, Saitama, Japan*
[2] *Laboratory of DNA Information Analysis, Human Genome Center, Institute of Medical Science, University of Tokyo, Tokyo, Japan*
[3] *Genome LSA Technology Development Unit, Omics Science Center, RIKEN Yokohama Institute, Yokohama, Kanagawa, Japan*

Several technologies are currently used for gene expression profiling, such as Real Time RT-PCR, microarray and CAGE (Cap Analysis of Gene Expression). CAGE is a recently developed method for constructing transcriptome maps and it has been successfully applied to analyzing gene expressions in diverse biological studies. The principle of CAGE has been developed to address specific issues such as determination of transcriptional starting sites, the study of promoter regions and identification of new transcripts. Here, we present both quantitative and qualitative comparisons among three major gene expression quantification techniques, namely: CAGE, illumina microarray and Real Time RT-PCR, by showing that the quantitative values of each method are not interchangeable, however, each of them has unique characteristics which render all of them essential and complementary. Understanding the advantages and disadvantages of each technology will be useful in selecting the most appropriate technique for a determined purpose.

*Keywords*: CAGE; microarray; Real Time RT-PCR; transcriptome.

## 1. Introduction

The main goal of a transcriptome research is to comprehend the mechanisms underlying the information flow from genes to proteins and other functional molecules. This information flow is basically orchestrated by a set of mRNAs. The identification and quantification of these mRNAs are crucial to understanding the development of diverse diseases. In order to identify the transcriptome map, gene expression profiling has been systematically applied in the study of several diseases and in cells states under different physiological conditions or environmental stimuli. A variety of technologies for measuring gene expression are available, such as Real Time RT-PCR, microarrays and more recently, CAGE (Cap Analysis Gene Expression).

Large amounts of data are being produced from these different technologies, but it is not clear whether they are actually comparable and if they can be accurately combined for data analysis. qRT-PCR (quantitative Real Time RT-PCR) is a technique based on the PCR technology. The procedure is basically the same as PCR, i.e., the amplified cDNA is quantified as it accumulates in the reaction, in real time, after each amplification cycle. qRT-PCR is used to amplify and simultaneously quantify low abundance mRNA in a particular cell or tissue type. Microarray technology is a method based on hybridization, using fluorescent labels in order to study the expression of thousands of genes at the same time. This technology provides a high-throughput map of gene expression on a genome-wide scale, which greatly enhances the potential to understand the control of biological processes at the molecular level. CAGE was introduced in 2003 as a high-throughput sequencing-based method to identify transcription start sites by isolating 20bp tags derived from the 5' end of RNA transcripts [12]. In order to identify the transcription start site, the CAGE tag is mapped back to the genome. In the last few years, CAGE has proven to be a powerful and versatile tool for both, constructing transcriptome maps and analyzing promoter regions [1, 5].

To the best of our knowledge, it is the first comparative study among these three technologies. In order to determine the strength and weakness of each technology (CAGE, microarray and qRT-PCR), 12 datasets were quantitatively and qualitatively analyzed. Our results demonstrate that the quantitative expression levels obtained by them are not interchangeable, however, each technology has unique interesting characteristics which render all of them essential and complementary for a better comprehension of the transcriptome.

## 2. Materials and Methods

### 2.1. *CAGE data summary*

In order to compare CAGE, microarray and qRT-PCR gene expression profiling techniques, 12 CAGE libraries were constructed from a PMA treated THP-1 human cell line. All CAGE tag counts are after removal of duplicate dimers and linker sequences. The number of CAGE tags sequenced (before normalization, i.e., raw data) in each library is summarized in Table 1.

Fig. 1 describes the scheme used to analyze the three different platforms. For CAGE, a total of ~16 million tags were sequenced (in 12 CAGE libraries); each illumina microarray (Human-6 v2 expression beadchip) is composed of ~47,000 probes; and ~1,600 genes were quantified by qRT-PCR. The data was normalized and the CAGE tags were mapped to the genome. In order to compare the three technologies, only common transcripts which were identified by the three technologies were selected. And then, by using this set of commen transcripts, both quantitative and qualitative analyses were carried out.

A CAGE tag is matched to a gene if the tag is positioned between 500 bases upstream and 100 bases downstream the annotated transcription start site (Fig. 2). Other ranges of bases were also adjusted but no substantive differences were verified.

Table 1.   Tag counts of 12 CAGE libraries.

| Library | Library size | Unique tags |
|---|---|---|
| 1 | 1,542,163 | 434,123 |
| 2 | 1,626,454 | 455,282 |
| 3 | 448,020 | 160,121 |
| 4 | 1,362,747 | 392,224 |
| 5 | 1,110,267 | 282,627 |
| 6 | 797,753 | 209,117 |
| 7 | 1,073,970 | 292,322 |
| 8 | 1,277,701 | 359,958 |
| 9 | 939,076 | 303,106 |
| 10 | 767,000 | 241,967 |
| 11 | 955,133 | 279,021 |
| 12 | 1,532,887 | 376,505 |

Microarray probes which matched at least one base to a qRT-PCR fragment or to a CAGE tag were considered in our analysis. The curated transcription start sites information were retrieved from DBTSS (Data Base of Transcriptional Start Sites) version 6.0 [15].

Fig. 3 illustrates the distribution of unique CAGE tags after normalizing by one million $(1,000,000 \times \frac{\text{number of tags}}{\text{total number of tags in the library}})$. Error bars represent one standard error.

Fig. 4 presents the scatter plot between the total number of sequenced tags (library size) versus the number of unique tags sequenced in each CAGE library. The relationship between them is fitted by a linear regression with $R^2 = 0.94$.

We examined the mapping information for the 20bp tags obtained with counts greater than zero. Based on the hg18 version of the human genome assembly, around ~81.6% of tags were assigned to the genome and ~9.8% of unique tags have been mapped to already known transcription start sites (described in DBTSS) of genes classified as "validated" by EntrezGene database.

Fig. 5 illustrates the distribution of the quantity of unique tags which matched to the genome. The higher is the number of matches in different regions of the genome, the lower is the number of unique tags. Around 80% of unique tags have matched only once, ~10% have matched in two different regions and ~3% have matched in three different regions.

Comparing the transcription profiles obtained by CAGE and microarray, among the total of 2,276,658 unique CAGE tags (total of 16,449,323 CAGE tags) sequenced in 12 libraries, 499,928 unique tags have matched to 9,527 microarray probes, i.e., ~22% of unique CAGE tags have matched to 19.6% of the illumina microarray probes (Fig. 6).

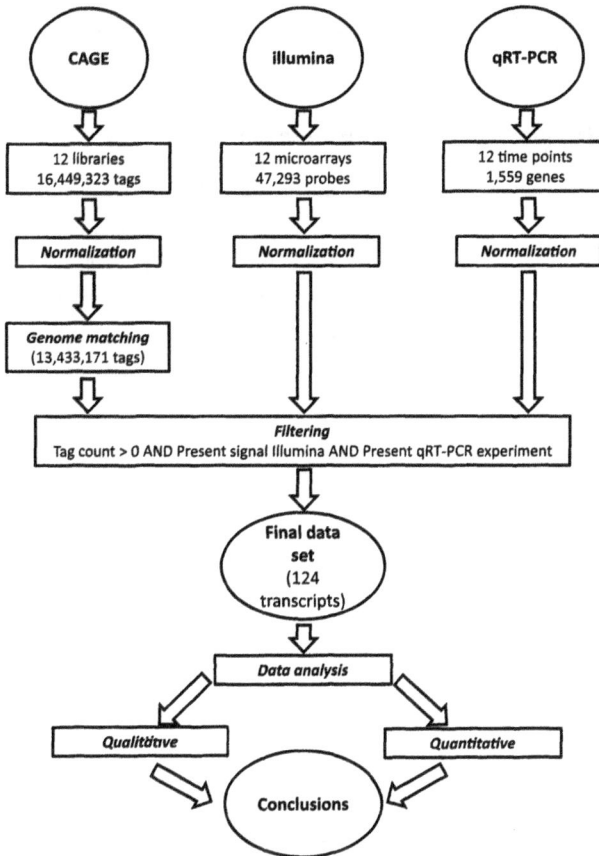

Fig. 1. Flow chart for methods comparison. CAGE tags were matched into genome using BLAST/Vmatch. Platforms are matched according to their sequences. Data are filtered for tag counts > 0 for CAGE and present calls on microarray platform. The pairwise comparisons (qRT-PCR versus CAGE, qRT-PCR versus microarray and CAGE versus microarray) were performed using only common transcripts.

## 2.2. *Qualitative comparison between CAGE, illumina microarray and qRT-PCR*

As discussed by Wang (2006) [16] in a comparison between SAGE and microarrays, similar factors must be considered when comparing CAGE, microarray data and Real Time RT-PCR:

(1) *The different transcription sites detected by CAGE, microarray and qRT-PCR.* CAGE specifically targets the 5' region of cap-purified cDNAs, and the presence of the restriction site for releasing the CAGE tag from the template is the determining factor. Microarray targets different regions of the transcript and the hybridization's specificity is closely related to probe base composition and its length. The detection and specificity of qRT-PCR depends essentially on the

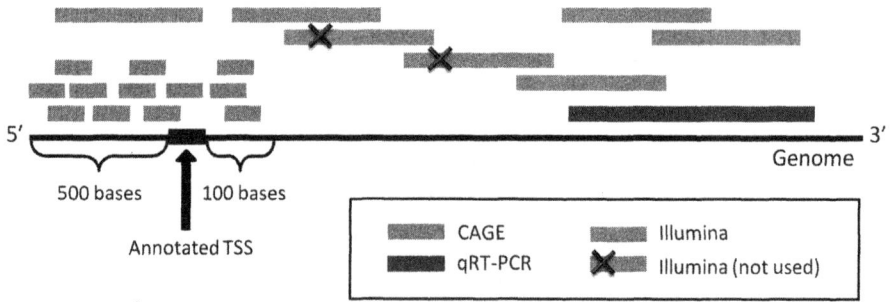

Fig. 2.    Scheme for matching CAGE, microarray and qRT-PCR datasets. TSS: Transcription Start Site.

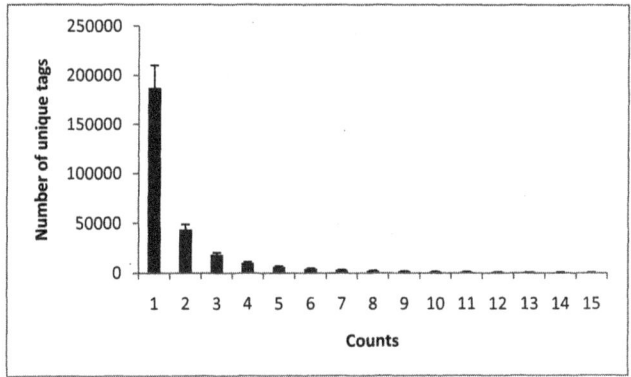

Fig. 3.    Mean distribution of unique tags in 12 CAGE libraries. Error bars represent one standard error.

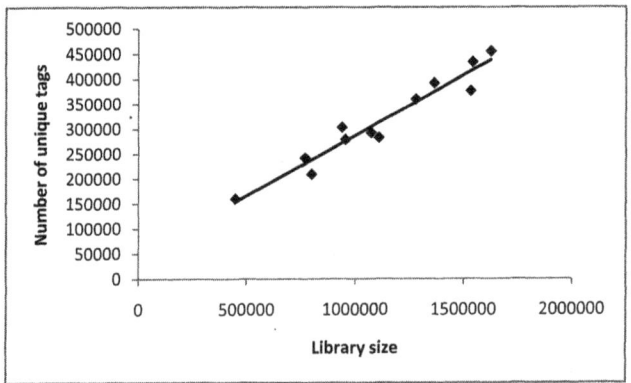

Fig. 4.    A scatter plot of the total number of sequenced tags (library size) versus the number of unique tags identified for each CAGE library. The line represents the linear regression with $R2=0.94$ indicating a linear increasing correlation.

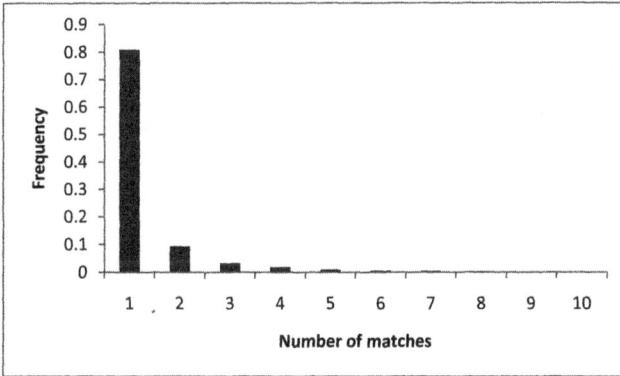

Fig. 5.  Distribution of the number of "single-mapped" (number of matches = 1) and "multi-mapped" (number of matches > 1) CAGE tags in the genome.

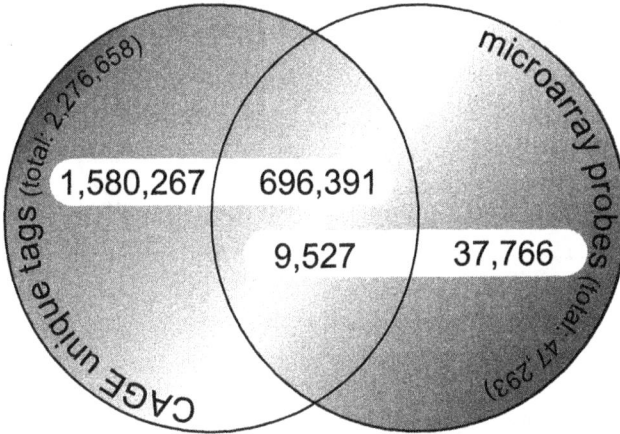

Fig. 6.  Number of CAGE tags that match to illumina microarray proves. 696,391 unique CAGE tags matched to 9,527 illumina microarray probes.

design of primers.

(2) *Transcript detection characteristics.* CAGE collects sequence information for the detected transcripts and quantification is obtained from the CAGE tag copy number; microarray identifies the transcript based on the presence or absence of fluorescent signals and quantification is calculated using the fluorescent intensity; qRT-PCR depends on fluorescent signals detected per RT-PCR cycle.

(3) *The bias present in each platform.* CAGE tag may contain sequencing error, presence of artifacts and quantification bias; microarray can contain labeling bias and noise signals from nonspecific hybridization; qRT-PCR may contain quantification bias due to sample initial quantification and primer's efficiency.

In Table 2, several other characteristics related to CAGE, microarray and qRT-PCR are summarized. Notice that all the three methods are able to identify and quantify expression levels of known transcripts. Moreover, all of them may detect alternatively spliced transcripts and also antisense transcripts. CAGE and qRT-PCR are the most sensitive techniques, i.e., they may identify transcripts with very low expression level while microarray has a sensitivity lower than the other two approaches. CAGE is the least specific, i.e., one CAGE tag may match to different transcripts (see Fig. 5). qRT-PCR combines both advantages, sensitivity and specificity, however, differently from microarray and CAGE, it is not a high-throughput approach, rendering difficult analysis of thousands of genes. Usually, qRT-PCR are used to quantify accurately the expression level of a few specific genes of interest.

Table 2.   Comparison of CAGE, illumina microarray and qRT-PCR methods for gene expression quantifications.

| Features | CAGE | microarray | qRT-PCR |
|---|---|---|---|
| Detects known transcripts | Yes | Yes | Yes |
| Detects unkown transcripts | Yes | No | No |
| Detects alternatively spliced transcripts | Yes | Yes | Yes |
| Detects antisense transcripts | Yes | Yes | Yes |
| Quantification | Absolute measure | Relative measure | Relative measure |
| Sensitivity | High | Moderate | High |
| Specificity | Moderate | High | High |
| High-Throughput method | Yes | Yes | No |

## 2.3. Correlation analysis

Since illumina microarray is a closed system that detects only known transcripts, and qRT-PCR requires the design of primers, i.e., the target sequence must be known, a quantitative comparison among CAGE, illumina microarray and qRT-PCR is restricted only to known transcripts.

In order to certify both robustness and reproducibility in our statistical analysis, two different correlation methods were applied, namely, Hoeffding's D measure and Spearman's correlation. Since Hoeffding's D measure is bounded in the interval [-1/60; 1/30] and Spearman's correlation in the interval [-1;1], comparison between correlation coefficients are difficult. Therefore, corresponding Z-values (the p-value corresponding quantile of the standard normal distribution of mean of zero and variance of one) were calculated in order to compare the results.

We observed a high and statistically significant correlation (high Z-values, $p < 0.0001$) between qRT-PCR and microarray, and a modest correlation (moderate Z-values, $p < 0.001$) between qRT-PCR and CAGE (Table 3). On the other hand, correlation between microarray versus CAGE resulted in a low correlation coefficient (not statistically significant, low Z-values, $p > 0.05$).

Calculating local correlations, i.e., correlations for each gene expression value

Table 3. Spearman's rank order and Hoeffding's D measure correlations of gene expression measurements from three quantification techniques (qRT-PCR, microarray and CAGE). Z-values are represented in average absolute values.

| Comparison | Platform A | Platform B | Spearman's mean Z-value | Spearman's SD Z-value | Hoeffding's mean Z-value | Hoeffding's SD Z-value |
|---|---|---|---|---|---|---|
| 1 | qRT-PCR | microarray | 5.713 | 0.580 | 5.612 | 0.001 |
| 2 | qRT-PCR | CAGE | 3.267 | 0.705 | 3.392 | 0.733 |
| 3 | microarray | CAGE | 1.764 | 0.890 | 1.549 | 0.950 |

*Note*: SD: Standard Deviation.

between qRT-PCR and microarray and also between qRT-PCR and CAGE, the results are as expected. Correlation increases slightly as higher is the gene expression value (Fig. 7).

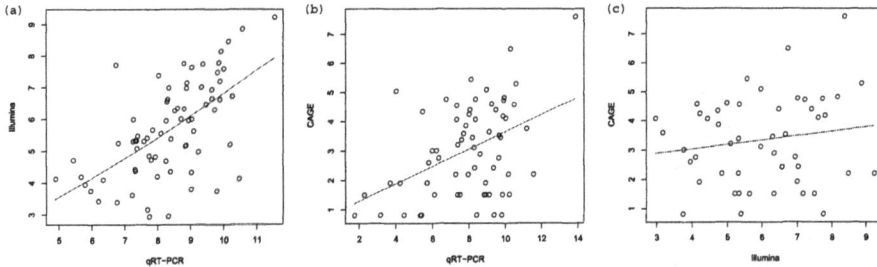

Fig. 7. Illustrative examples of scatter plots between (a) qRT-PCR versus microarray; (b) qRT-PCR versus CAGE and (c) microarray versus CAGE. The x-axis and y-axis represent the gene expression levels in the log scale. The red curve represents the regression obtained by splines smoothing. The number of knots used in splines was determined by the number which minimizes the Generalized Cross-Validation.

## 3. Results and Discussions

### 3.1. *Qualitative analysis*

Each technological platform currently used to study the transcriptome map of several organisms (CAGE, illumina microarray and qRT-PCR) has its own strengths and weakness.

Hybridization-based illumina microarrays provide high-throughput capacity to quantify expression levels for known but not for unknown transcripts. Amplification-based Real Time RT-PCR provides high sensitivity but low-throughput quantification and are limited to known transcripts. For sequencing-based approaches, such as CAGE, two main factors influence the efficiency in transcript detection. One is the sequenced tags' length for each transcript; the second one is the total number of sequenced tags. CAGE is based on the high-throughput sequencing of concatemers of short (20 bp) sequence tags originated from the 5' ends cap-purified cDNA. Therefore, CAGE obtains higher sensitivity to identify rare transcripts than full-length cDNA and ESTs (Expressed Sequence Tags) which are composed of hundreds of bases [8]. However, the cost to be paid is a lower specificity since one CAGE tag may map to several parts of the genome ($\sim$19% of unique tags have matched more than one place in the genome) (Fig. 5).

Since CAGE is based on a cap-trapper system, it may capture only full-length mRNAs (rRNA and tRNA transcripts are avoided). In contrast to microarrays, CAGE estimates the abundances (expression levels) of tens of thousands of transcripts without prior knowledge about their sequence. The proportion of the number of a given tag within the total number of tags sequenced in the library gives an estimate of the abundance of the transcript within this sample. The advantage of the sequence-based CAGE technique is that it performs a random sampling from the pool of all expressed transcripts allowing the discovery of new transcripts. Moreover, the nature of this kind of data enables exchange of CAGE libraries among different centers, allowing the creation of large public CAGE data bases. Another advantage is that CAGE can be used to identify transcriptional initiation sites and also to locate gene promoters in the genome. Disadvantages of CAGE are that it is expensive, labor-intensive and sensitive to sequencing errors. In addition, the annotation of short 20 bp sequence tags may identify more than one transcript, i.e., a CAGE tag may match to different regions of the genome.

Illumina microarrays are used to measure relative expression levels between samples of thousands of known transcripts. They are suitable for high-throughput analysis of multiple samples, and data can easily be shared and used for comparisons with other researchers using the same chips. Disadvantages of illumina microarrays are that they are only commercially available, require expensive specialized equipment and are inflexible in design. Furthermore, microarrays only measure the expression of genes represented on the chip in contrast to CAGE, in which the expression profile of the complete transcriptome can be mapped. By mapping the CAGE tags on the illumina microarray, only $\sim$22% of the transcripts (unique tags) identified by CAGE were represented in the illumina microarray platform, i.e., there are a lot of transcripts which cannot be captured by illumina microarrays.

Real Time RT-PCR is a low-throughput technique which is characterized by high specificity and sensitivity. The target transcript's expression level is measured using a pair of primers composed of approximately 20 bases which guarantees a high specificity if they are designed adequately. Since it is based on PCR reaction, even low expressed transcripts may be accurately quantified. Disadvantages are that this technique needs a pair of primers for each transcript and the target transcript's sequence must be known. Moreover, for normalization, housekeeping genes are necessary. Selection of an "actual" housekeeping gene is speculative and derived several discussions [14].

## 3.2. *Quantitative analysis*

By analyzing CAGE data, it was possible to note that the number of singletons (unique tags with one count) is very high (Fig. 3), probably because the majority of transcripts has low expression levels. However, it is necessary to point out that ~19% of the tags did not match to the genome, i.e., part of these singletons may be derived from sequencing errors or artifacts. Therefore, it is necessity to confirm the existence of these singletons by other accurate techniques such as qRT-PCR.

Although CAGE is the most sensitive method among these three platforms for detecting new transcripts, the 12 libraries show no saturation. By plotting the number of sequenced tags versus the number of unique tags (Fig. 4), it is not possible to identify a plateau in this graph (if it is saturated, it is expected to find an upper bound plateau in the graph). The unsaturated detection of transcripts maybe due to the limited number of tags sequenced or to sequencing errors. Unfortunately, it is not possible to differentiate "true" tags from "false" tags (resulted from sequencing error). By analyzing Fig. 4, notice that the number of unique tags grow linearly in relation to the number of tags sequenced. Therefore, by performing deeper CAGE sequencing, more promoter regions may be identified, contributing to the construction of the complete transcriptome map. At the moment, there are both, time and financial limitations to sequence larger libraries using the actual sequencing systems. With the advances in sequencing technologies, it will be possible to obtain more tags with higher accuracy. When larger CAGE libraries become available, it will be possible to verify how much far or closer we are from the detection of the whole transcriptome.

Only around ~9.8% of the CAGE tags are matched to known genes transcription start sites and ~81% have matched in only one region of the genome, supporting the hypothesis that the majority of them are not derived from sequencing errors (Fig. 5). From these results, it is possible to construct at least two hypotheses: (i) there are a lot of unknown genes that probably have very low expression levels (Fig. 3) and consequently they were not captured yet and/or (ii) there are a lot of unknown transcription start sites.

Around 22% of the unique CAGE tags have matched to ~19.6% illumina microarray probes (Fig. 6), showing at least that: (i) there are a lot of CAGE tags sequence information which may be used to design new microarray probes; (ii) the sensitivity of illumina microarray is lower than CAGE; (iii) microarray captures a wide range of RNAs' expressions while CAGE only identifies mRNA expression due to the necessity of cap structure; and (iv) the low number of matches may be due to the DBTSS limitation since only ~9.8% of the CAGE tags have matched to already known transcription start sites.

Two different correlation methods namely, Spearman's and Hoeffding's D correlations were applied in the comparison between different platforms in order to verify whether the quantitative values obtained by each technology are interchangeable or not. Standard Pearson's correlation was not applied since it is known that it is highly sensitive to outliers. Spearman's and Hoeffding's measures are less sensitive to outliers than Pearson since they are non-parametric methods based on ranks.

Although significant correlations ($p < 0.05$) were observed between (CAGE and qRT-PCR) and between (microarray and qRT-PCR), Spearman's correlation coefficients were lower than 0.7, indicating poor quantitative relations (Fig. 7). Even in high gene expression levels where it is expected to obtain higher correlations due to lower variance, the association is not satisfactory ($\sim$0.6). Therefore, these comparisons showed a low association between different platforms. Correlations between illumina microarray and qRT-PCR were higher than between CAGE and qRT-PCR. The lowest correlation was obtained between CAGE and illumina microarray (Fig. 7). Since similar results were obtained using Hoeffding's D measure, it is possible to say that these results are robust. However, it is necessary to point out that since the transcripts used in this comparison are basically derived from transcription factor genes (limited to genes quantified by qRT-PCR in Genome Network Project), this correlation may increase if others are considered. Notice that transcription factor genes have usually low expression, and consequently, high variance. Therefore, these correlations may be considered as a lower bound to the correlation coefficient.

Lower correlation was identified in low values of gene expression using local correlation. This result is probably due to the high variance in low gene expression values and low variance in high expression levels (Fig. 7). Notice that small perturbations in low expression levels may cause high relative quantification. With the advances in sequencing technology, where more bases may be sequenced by time, by increasing the number of sequenced tags may improve the resolution and consequently decreasing the variance of genes with low expression.

All these three technologies (CAGE, illumina microarray and qRT-PCR) have strengths and weakness. Notice that for transcript discovery using CAGE, qualitative rather than quantitative issues are more suitable since it is able to identify new transcripts, however, with low quantification accuracy. Microarray is indicated when one wants to quantify in a high-throughput manner, the expression values, however, it is limited to known transcripts. qRT-PCR is useful to quantify the transcripts with high accuracy although this technique is limited to a small number of genes per time.

Although qRT-PCR versus microarray and qRT-PCR versus CAGE comparisons showed a significant Spearman's and Hoeffding's correlations using the common set of reliable transcripts, the correlation were low, showing that although they are correlated (statistically significant), quantitative results are not interchangeable.

The number of CAGE tags sequenced nowadays is not enough to estimate the total number of transcription start sites (Fig. 4) nor the number of tags to be sequenced to obtain the same accuracy of illumina microarray. When both, larger CAGE libraries and technical replicates become available, it will be possible to analyze the intrinsic variance of CAGE and also estimate the optimum CAGE library size.

## 4. Conclusions

In summary, a qualitative and quantitative comparison study of CAGE, illumina microarray and qRT-PCR were conducted in order to highlight strengths and weakness for each technology, and also to determine whether data are interchangeable among them.

CAGE and qRT-PCR demonstrated to be the most sensitive methods, however, the false positive transcipts identified by CAGE is still quite high, probably due to sequencing errors. CAGE and microarray allow the identification and quantification of thousands of transcripts while qRT-PCR is a low-thoughtput method. In terms of precision in quantification, qRT-PCR is the best one followed by microarray, but both requires that the transcript is known. CAGE can quantify yet unknown transcripts.

The quantitative results indicate that the gene expression quantifications obtained by each technology is not interchangeable, however, the strength of one technology may complement the weakness of the others.

### Acknowledgements

Computational analyses were carried out using the Super Computer System, Human Genome Center, Institute of Medical Science, University of Tokyo. This work was supported by grants of the Genome Network Project from the Ministry of Education, Culture, Sports, Science and Technology, Japan.

## References

[1] Bajic, V.B., Tan, S.L., Christoffeis, A., Schonbach, C., Lipovich, L., Yang, L., Hofmann, O., Kruger, A., Hide, W., Kai, C., Kawai, J., Hume, D.A., Carninci, P., and Hayashizaki, Y., Mice and men: their promoter properties, *PLoS Genetics*, 2:e54, 2006.

[2] Bjerve, S. and Doksum, K., Correlation curves: measures of association as functions of covariate values, *Annals of Statistics*, 21:890-902, 1993.

[3] Bustin, S.A., Benes, V., Nolan, T., and Pfaffl, M.W., Quantitative real-time RT-PCR - a perspective, *Journal of Molecular Endocrinology*, 34:597-601, 2005.

[4] Doksum, K., Blyth, S., Bradlow, E., Meng, X., and Zhao, H., Correlation curves as local measures of variance explained by regression, *Journal of the American Statistical Association*, 89:571-582, 1994.

[5] Gustincich, S., Sandelin, A., Plessy, C., Katayama, S., Simone, R., Lazarevic, D., Hayashizaki, Y., and Carninci, P., The complexity of the mammalian transcriptome, *J Physiol.*, 575:321-32, 2006.

[6] Hoeffding, W. A non-parametric test of independence, *The Annals of Mathematical Statistics*, 19:546-557, 1948.

[7] Hollander, D.J. and Wolfe, D.A., *Nonparametric statistical methods*, John Wiley, New York, 1973.

[8] de Hoon M and Hayashizaki Y., Deep cap analysis gene expression (CAGE) genome-wide identification of promoters, quantification of their expression, and network inference, *BioTechniques*, 44:627-632, 2008.

[9] Hume, D.A. and Gordon, S., Optimal conditions for proliferation of bone marrow-

derived mouse macrophages in culture: the roles of CSF-1, serum, Ca2+, and adherence, *J Cell Physiol.*, 117:189-194, 1983.

[10] Lu, J., Lal, A., Merriman, B., Nelson, S., and Riggins, G., A comparison of gene expression profiles produced by SAGE, long SAGE, and oligonucleotide chips, *Genomics*, 84:631-636, 2004.

[11] Rozen, S. and Skaletsky, H., Primer 3 on the WWW for general users and for biologist programmers, *Methods Mol Biol.*, 132:365-386, 2000.

[12] Shiraki, T., Kondo, S., Katayama, S., Waki, K., Kasukawa, T., Kawaji, H., Kodzius, R., Watahiki, A., Nakamura, M., Arakawa, T., Fukuda, S., Sasaki, D., Podhajska, A., Harbers, M., Kawai, J., Carcini, P., and Hayashizaki, Y., Cap analysis gene expression for high-throughput analysis of transcriptional starting point and identification of promoter usage, *Proc. Natl. Acad. Sci. USA.*, 100:15776-15781, 2003.

[13] Tukey, J.W., *Exploratory Data Analysis*, Addison-Wesley, 1977.

[14] Vandesompele, J., De Preter, K., Pattyn, F., Poppe, B., Van Roy, N., De Paepe, A., and Speleman, F., Accurate normalization of real-time quantitative RT-PCR data by geometric averaging of multiple internal control genes, *Genome Biology*, 3:research0034.1-0034.11, 2002.

[15] Wakaguri, H., Yamashita, R., Suzuki, Y., Sugano, S., and Nakai, K., DBTSS: database of transcription start sites, progress report 2008, *Nucleic Acids Research*, 2007.

[16] Wang, S.M., Understanding SAGE data, *TRENDS in Genetics*, 23:42-50, 2006.

[17] Wang, T. and Brown, M.J., mRNA quantification by real-time TaqMan polymerase chain reaction: validation and comparison with RNase protection, *Analytical Biochemistry*, 269:198-201, 1999.

# ON THE PERFORMANCE OF METHODS FOR FINDING A SWITCHING MECHANISM IN GENE EXPRESSION

MITSUNORI KAYANO[1,3]     ICHIGAKU TAKIGAWA[1,3]     MOTOKI SHIGA[1,3]
kayano@kuicr.kyoto-u.ac.jp takigawa@kuicr.kyoto-u.ac.jp shiga@kuicr.kyoto-u.ac.jp

KOJI TSUDA[2,3]     HIROSHI MAMITSUKA[1,3]
koji.tsuda@aist.go.jp     mami@kuicr.kyoto-u.ac.jp

[1] *Bioinformatics Center, Institute for Chemical Research, Kyoto University, Gokasho, Uji 611-0011, Japan*
[2] *Computational Biology Research Center (CBRC), National Institute of Advanced Industrial Science and Technology (AIST), 2-42 Aomi, Koto-ku, Tokyo 135-0064, Japan*
[3] *Institute for Bioinformatics Research and Development (BIRD), Japan Science and Technology Agency (JST), Japan*

We address an issue of detecting a switching mechanism in gene expression, where two genes are positively correlated for one experimental condition while they are negatively correlated for another. We compare the performance of existing methods for this issue, roughly divided into two types: interaction test (IT) and the difference of correlation coefficients. Interaction test, currently a standard approach for detecting epistasis in genetics, is the log-likelihood ratio test between two logistic regressions with/without an interaction term, resulting in checking the strength of interaction between two genes. On the other hand, two correlation coefficients can be computed for two experimental conditions and the difference of them shows the alteration of expression trends in a more straightforward manner. In our experiments, we tested three different types of correlation coefficients: Pearson, Spearman and a midcorrelation (biweight midcorrelation). The experiment was performed by using $\sim 2.3 \times 10^9$ combinations selected out of the GEO (Gene Expression Omnibus) database. We sorted all combinations according to the p-values of IT or by the absolute values of the difference of correlation coefficients and then visually evaluated the top ranked combinations in terms of the switching mechanism. The result showed that 1) combinations detected by IT included non-switching combinations and 2) Pearson was affected by outliers easily while Spearman and the midcorrelation seemed likely to avoid them.

*Keywords*: differential/diverse correlation; likelihood ratio test; correlation coefficient.

## 1. Introduction

Expression analysis on genes with microarrays is an important and basic technique in biology. A typical but simple approach is to examine the difference in expression for a single gene between different experimental conditions, such as case and control, which is generally called differential expression [18]. Another approach is slightly more complex and to focus on a combination of multiple genes, and it is checked whether they are over- or under-expressed simultaneously, which is called co-expression [20].

In this paper, we address an issue of finding interaction between two genes in expression. This issue is related with differential co-expression, in which co-expression patterns differ depending upon experimental conditions [4, 5, 10, 13, 14, 23]. More concretely our issue is a switching mechanism in expression of two genes, being controlled by experimental conditions, where two genes are positively correlated for one experimental condition while they are negatively correlated for another [3, 7, 10, 16, 21, 24]. A simple, well-known example is Max, a transcription factor, which plays a role of an activator or a suppressor, depending on whether it binds to Myc (i.e. Myc-Max) or Mad (i.e. Mad-Max) [1]. Finding this type of interaction would be a first key step to elucidating complex biological systems.

Existing methods for detecting the switching mechanism can be divided into two types: interaction test (IT) and the difference of correlation coefficients. Interaction test, currently a standard approach for detecting epistasis in genetics [6], is the log-likelihood ratio test between two logistic regressions with/without an interaction term, resulting in checking the strength of interaction between two genes. A serious disadvantage of IT is its high computational burden, making hard to apply IT to a large number of gene combinations practically. Recently this problem is relaxed by a method, which allows IT to run around ten times faster by using the idea of removing combinations to which IT does not have to be applied [12]. This method is called FTGI (Fast finding Three-way Gene Interaction), which can prune unnecessary combinations by using a randomness test (mean-covariance test) and linear discriminant analysis (LDA). FTGI is originally for among three genes (i.e. three-way gene interaction), considering expression (numerical) values of two genes and genotypes (discrete values) of one gene. We can apply this method by regarding genotypes of one gene as classes of experimental conditions. FTGI can be applied to our issue of finding a switching mechanism in expression. Thus instead of applying IT directly to our dataset, we used FTGI in our experiments, with slightly modifying it by removing the part of LDA. In fact, the LDA part affects the performance only if there are so many cases for which expression values are easily separated by experimental conditions, but practically it would be hard to find these cases. On the other hand, the difference of two correlation coefficients, each corresponding to one of two experimental conditions, shows the alteration of expression trends more straightforwardly [3, 7, 10, 21]. We can consider several measures for correlations, and evaluated three measures, Pearson, Spearman and midcorrelation, in our experiments. Both these approaches, FTGI and the difference of correlation coefficients, can rank given gene combinations according to $p$-values or by the absolute values of the difference.

In this paper we explore the best approach for finding the switching mechanism. For this purpose, we sorted a huge number of gene combinations by using each of the existing approaches and evaluated the performance of each approach by visually examining the top ranked combinations in terms of outliers, computation speed and sample size.

## 2. Method

### 2.1. *Notations and preliminaries*

Let $\mathcal{X}$ be an input matrix, in which each row is an experimental condition and each column is a numerical vector of gene expression. Let $E$ be the set of genes for which expressions are measured in $\mathcal{X}$. To detect the switching mechanism in gene expression, we choose one *combination*, i.e. two genes $(e_1, e_2)$ out of $E$, and we write $\mathcal{X}(e_1, e_2)$ which has only two columns of $\mathcal{X}$, corresponding to $e_1$ and $e_2$. Hereafter until Sec. 2.3.3, we assume that we already choose one combination, which is the only input.

The input has two genes, and so we have two numerical variables $X_1$ and $X_2$, each corresponding to one of two genes. Variable $X_3$ corresponds to the interaction between these two genes. Let $n$ be the size of rows of $\mathcal{X}$, resulting in $\mathcal{X} = (x_1, \ldots, x_n)'$, where $x_i$ is expression values of the $i$-the experimental condition (row). In our switching mechanism, the number of groups (or classes) is two. We denote two classes by $G_1$ and $G_2$, and out of $n$ inputs, $n_1$ and $n_2$ inputs are for $G_1$ and $G_2$, respectively. Let $Y$ be the class variable, taking values $y = 1$ if $x \in G_1$ and $y = 0$ if $x \in G_2$.

The average expression values can be defined for each class $c$ and all classes: $\bar{x}_c = n_c^{-1} \sum_{i=1|i \in G_c}^{n} x_i$ and $\bar{x} = n^{-1} \sum_{i=1}^{n} x_i = n^{-1} \sum_{c=1}^{2} n_c \bar{x}_c$, respectively. We can define covariance matrix $S_c$ for class $c$ and total covariance matrices $S_T$ and $S$ as follows:

$$S_c = \frac{1}{n_c} \sum_{i=1|i \in G_c}^{n_c} (x_i - \bar{x}_c)(x_i - \bar{x}_c)',$$

$$S_T = \frac{1}{n} \sum_{i=1}^{n} (x_i - \bar{x})(x_i - \bar{x})',$$

$$S = \frac{1}{n} \sum_{c=1}^{2} \sum_{i=1|i \in G_c}^{n_c} (x_i - \bar{x}_c)(x_i - \bar{x}_c)' \quad \left( = \frac{1}{n} \sum_{c=1}^{2} n_c S_c \right).$$

In the next section, we assume that the inputs in class $c$ ($= 1, 2$) follow the bivariate normal distribution $N_2(\mu_c, \Sigma_c)$ with mean (vector) $\mu_c$ and covariance (matrix) $\Sigma_c$.

### 2.2. *Fast finding three-way gene interaction (FTGI)*

FTGI is a fast method to obtain combinations with small $p$-values by IT, keeping the consistency with the outputs of IT (i.e. the log-likelihood ratio test between two logistic regressions with/without an interaction term) with a high probability. The idea behind FTGI is to remove combinations not to be applied to IT by using mean-covariance (MC) test and LDA. In this paper we use FTGI, skipping the part of LDA, and so in this section, we describe logistic regression, log-likelihood ratio test and MC test. Finally we summarize the entire procedure of FTGI used in this paper.

### 2.2.1. *Logistic regression*

We first denote the probability that $x$ is in $G_1$ by $p_1(x)$, and similarly the probability that $x$ is in $G_2$ by $p_2(x)$ $(= 1 - p_1(x))$. We use logistic regression to link these probabilities to input $x$ by using weight parameters (or coefficients) $w$ with the same dimension as $x$ in the following:

$$\begin{cases} p_1(x) = \dfrac{\exp(w'x)}{1 + \exp(w'x)} \\ p_2(x) = \dfrac{1}{1 + \exp(w'x)} \end{cases} \Leftrightarrow \log \frac{p_1(x)}{p_2(x)} = w'x$$

Here we denote $p_1(x)$ and $p_2(x)$ by $p_1(x; w)$ and $p_2(x; w)$ $(= 1 - p_1(x; w))$, respectively, because they are functions of $w$. We can then write the likelihood of logistic regression for given $n$ examples and parameters $w$, as follows:

$$L(w) = \prod_{i=1}^{n} p_1(x_i; w)^{y_i} p_2(x_i; w)^{1-y_i}.$$

We can obtain the maximum likelihood estimator $\hat{w}$ for $w$ by maximizing the log-likelihood $l(w) = \log L(w)$. A standard approach for this purpose is the Newton-Raphson method, which is an iterative gradient descent, having the following updating rule by which we can have $\hat{w}^{(t+1)}$ at the $(t + 1)$-th iteration, using $\hat{w}^{(t)}$ of the $t$-th iteration:

$$\hat{w}^{(t+1)} = \hat{w}^{(t)} - \left( H(w)|_{w=\hat{w}^{(t)}} \right)^{-1} U(w)|_{w=\hat{w}^{(t)}}, \tag{1}$$

where $H(w)$ $(= \partial^2 l/\partial w \partial w')$ is the Hessian matrix and $U(w)$ $(= \partial l/\partial w)$ is the gradient vector. In practice, we start with some initial values $\hat{w}^{(0)}$ and update $\hat{w}^{(t+1)}$ according to Eq.(1) until convergence.

### 2.2.2. *Log-likelihood test*

We then examine the significance of the interaction in expression between two genes in terms of classes. Let $x_{i1}$ and $x_{i2}$ be expression values of the corresponding two genes for input $i$. The interaction term is $x_{i1}x_{i2}$, meaning that our purpose is to find the case that the logistic model is well fitted to the data when this term is added. We then let $x_i = (1, x_{i1}, x_{i2}, x_{i1}x_{i2})'$ and $w = (w_0, w_1, w_2, w_3)'$, and the logistic model with the interaction term is given as follows:

$$\log \frac{p_1(x; w)}{p_2(x; w)} = w_0 + w_1 x_{i1} + w_2 x_{i2} + w_3 x_{i1}x_{i2} .$$

If $w_3 = 0$, the model does not have the interaction term, meaning that the null hypothesis and $w_0$ are given as follows:

$$H_0 : w_3 = 0$$
$$w_0 = (w_0, w_1, w_2, 0)'. \tag{2}$$

**Require:** $\mathcal{X}(e_1, e_2)$: Input two vectors of genes $e_1, e_2$.
　　$\alpha_i$: Significance level for interaction test
**Ensure:** One if $e_1$ and $e_2$ are interacting with each other under class; otherwise zero.
Interaction_test($e_1$, $e_2$, $\alpha_i$)
1: $\boldsymbol{w}_0 \leftarrow$ some initial value.
2: **repeat**
3:　　Update $\boldsymbol{w}_0$, according to the iterative rule of Eq.(1)
4: **until** convergence
5: $\boldsymbol{w} \leftarrow$ some initial value.
6: **repeat**
7:　　Update $\boldsymbol{w}$, according to the iterative rule of Eq.(1)
8: **until** convergence
9: **if** $-2\{l(\hat{\boldsymbol{w}}) - l(\hat{\boldsymbol{w}}_0)\} > \chi_1^2(\alpha_i)$ **then**
10:　　**return** 1
11: **else**
12:　　**return** 0
13: **end if**

Fig. 1.　Pseudocode of interaction test.

Then the test statistic by the log-likelihood ratio and its asymptotic distribution can be given:

$$-2\log\lambda = 2\{l(\hat{\boldsymbol{w}}) - l(\hat{\boldsymbol{w}}_0)\} \sim \chi_1^2(\alpha_i),$$

where $\chi_1^2(\alpha_i)$ is the $\chi^2$ distribution with $df$ (the degree of freedom) of one, meaning that interacting genes can be obtained as those which have lower $p$-values under this distribution than the input significance level $\alpha_i$. Fig. 1 shows a pseudocode of IT.

### 2.2.3. Mean-Covariance (MC) test

We consider the following hypotheses over both the means and the covariances:

$H_0: \boldsymbol{\mu}_1 = \boldsymbol{\mu}_2$ and $\Sigma_1 = \Sigma_2$
$H_1: \boldsymbol{\mu}_c \neq \boldsymbol{\mu}_{c^*}$ or $\Sigma_c \neq \Sigma_{c^*}$ for some pair of $c$ and $c^*$

The statistic $-2\log\lambda$ of this test is given as follows.

$$-2\log\lambda = \sum_{c=1}^{C} n_c \log\det(S_c^{-1} S_T) . \tag{3}$$

The above test statistic can be easily obtained by maximum log-likelihoods in Box's M test and multivariate analysis of variance (MANOVA) and follows approximately a $\chi^2$ distribution with $df$ of five (since in our case the number of classes is two and the number of genes is two). MC test can be thought as a combination of Box's M test and MANOVA which examines the equality of two means $\boldsymbol{\mu}_1$ and $\boldsymbol{\mu}_2$ and is a multivariate extension of the univariate analysis of variance (ANOVA). Fig. 2 shows a pseudocode of MC test. See [12, 17] for detail.

**Require:** $\mathcal{X}(e_1, e_2)$: Input two vectors of genes $e_1, e_2$.
    $\alpha_m$: Significance level for MC test
**Ensure:** One if two genes $e_1$ and $e_2$ are randomly generated in terms of class; otherwise zero.
MC_test$(e_1, e_2, \alpha_m)$
1: Compute $-2 \log \lambda$ according to Eq. (3).
2: **if** $-2 \log \lambda < \chi_5^2(\alpha_m)$ **then**
3:    **return** 1
4: **else**
5:    **return** 0
6: **end if**

Fig. 2.   Pseudocode of Mean-covariance (MC) test.

**Require:** $\mathcal{X}$: Input dataset
    $\alpha_i$: Significance level for interaction test,
    $\alpha_m$: Significance level for MC test
**Ensure:** $\mathcal{I}$: Interacting gene pairs
Proposed_procedure$(\mathcal{X}, \alpha_i, \alpha_m)$
1: **for** each combination of genes $e_1 \in E$ and $e_2 \in E$ **do**
2:    // Pruning by MC test
3:    **if** MC_test$(e_1, e_2, q, \alpha_m) == 1$ **then**
4:        Expressions of this combination should be randomly distributed. **go to** Pruned
5:    **end if**
6:    // Interaction test for unpruned combinations
7:    **if** Interaction_test$(e_1, e_2, \alpha_i) == 1$ **then**
8:        $\mathcal{I} \leftarrow \mathcal{I} \cup (e1, e2)$
9:    **end if**
10:    Pruned
11: **end for**

Fig. 3.   Pseudocode of our entire procedure of FTGI

### 2.2.4. *Procedure of FTGI*

Fig. 3 shows a pseudocode of FTGI, skipping LDA. We first generate all possible combinations of two genes out of given data. For each of gene combinations, MC test is first run, and this combination is removed if expression values of this combination is randomly distributed (or consistent in means and covariances with between two classes); otherwise, interaction test is applied to this combination.

### 2.3. *The difference of correlation coefficients*

The difference of correlation coefficients is simply given by

$$|r_1 - r_2|,$$

where $r_1$ and $r_2$ are correlation coefficients of class 1 and class 2, respectively. We describe the manner of computing three types of correlation coefficients. Fig. 4 shows a pseudocode of the method by correlations.

**Require:** $\mathcal{X}(e_1, e_2)$: Input two vectors of genes $e_1, e_2$.
    $t$: Threshold value for the absolute value of the difference of the correlations
**Ensure:** One if $e_1$ and $e_2$ are differentially coexpressed under class; otherwise zero.
Difference_correlation($e_1$, $e_2$, $t$)
1: Compute the correlation coefficient $r_1$ for one class.
2: Compute the correlation coefficient $r_2$ for another.
3: Compute the difference of the correlations $|r_1 - r_2|$.
4: **if** $|r_1 - r_2| > t$ **then**
5:     **return** 1
6: **else**
7:     **return** 0
8: **end if**

Fig. 4.　Pseudocode of the method by correlation coefficients.

### 2.3.1. *Pearson correlation coefficient*

Pearson correlation coefficient $r_P$ is a measure of the linear dependence between $X_1$ and $X_2$ as follows:

$$r_P = \frac{\sum_i (x_{i1} - \bar{x}_1)(x_{i2} - \bar{x}_2)}{\sqrt{\left\{\sum_i (x_{i1} - \bar{x}_1)^2\right\}\left\{\sum_i (x_{i2} - \bar{x}_2)^2\right\}}}, \tag{4}$$

where $\bar{x}_1$ and $\bar{x}_2$ are the means of $X_1$ and $Y_2$ respectively. The correlation coefficient ranges from $-1$ to $1$. When each $x_{i1}$ is exactly the same as the corresponding $x_{i2}$, then $r_P$ is 1. In contrast, when each $x_{i1}$ is $-x_{i2}$, then $r_P$ is $-1$.

### 2.3.2. *Spearman correlation coefficient*

Let $R_i$ be the rank of $x_{i1}$ among $\{x_{11}, \ldots, x_{n1}\}$. Similarly let $Q_i$ be the rank of $x_{i2}$ among $\{x_{12}, \ldots, x_{n2}\}$. Spearman correlation coefficient $r_S$ can be defined by using $(R_i, Q_i)$ as follows:

$$r_S = \frac{\sum_i (R_i - \bar{R})(Q_i - \bar{Q})}{\sqrt{\left\{\sum_i (R_i - \bar{R})^2\right\}\left\{\sum_i (Q_i - \bar{Q})^2\right\}}}, \tag{5}$$

where $\bar{R}$ and $\bar{Q}$ are the means over $\{R_1, \ldots, R_n\}$ and $\{Q_1, \ldots, Q_n\}$ respectively. If there are examples ranked in a tie, we assign the mean of those ranks, called midrank, to these examples. If there are no such examples, $r_S = 1 - (6\sum_i d_i^2)/(n^3 - n)$ with $d_i = R_i - Q_i$. Spearman correlation coefficient can be robust or cannot be affected by outliers, because of using ranks, instead of raw values.

Table 1.   List of eight GDS used in our experiments (#ex.≥20).

| GDS | #genes | #ex. in a class | | | annotation |
|---|---|---|---|---|---|
| GDS531 | 12555 | 36 | 137 | | Multiple myeloma and bone lesions |
| GDS806 | 21939 | 32 | 28 | | Estrogen positive breast cancer recurrence during ta- moxifen therapy: whole tissue tumor |
| GDS807 | 21939 | 32 | 28 | | Estrogen positive breast cancer recurrence during ta- moxifen therapy: microdissected tumor |
| GDS1412 | 2759 | 47 | 42 | | Hormone replacement therapy effect on whole blood |
| GDS1615 | 22215 | 42 | 26 | 59 | Ulcerative colitis and Crohn's disease comparison: peripheral blood mononuclear cells |
| GDS2190 | 22215 | 31 | 30 | | Bipolar disorder: dorsolateral prefrontal cortex |
| GDS2519 | 22215 | 22 | 33 | 50 | Early-stage Parkinson's disease: whole blood |
| GDS2960 | 4132 | 41 | 60 | | Marfan syndrome: cultured skin fibroblasts |

### 2.3.3.  *Biweight midcorrelation*

Biweight midcorrelation $r_B$ [9, 19, 22], is a weighted correlation coefficient given by

$$r_B = \frac{\sum_i w(u_i)w(v_i)(x_{i1} - m_1)(x_{i2} - m_2)}{\sqrt{\left\{ \sum_i w^2(u_i)(x_{i1} - m_1)^2 \right\} \left\{ \sum_i w^2(v_i)(x_{i2} - m_2)^2 \right\}}}, \quad (6)$$

where

$$w(x) = (1 - x^2)^2 \quad \text{if } |x| \le 1$$
$$= 0 \qquad \qquad \text{otherwise,}$$

$u_i = (x_{i1} - m_1)/(K \text{ mad}_1)$, $v_i = (x_{i2} - m_2)/(K \text{ mad}_2)$, where $m_1$ and $m_2$ are medians of the $X_1$ and $X_2$ and $\text{mad}_1, \text{mad}_2$ are median absolute deviations (MAD) of the $X_1$ and $X_2$, which are robust measures of the variability defined as the sample medians for $\{|x_{i1} - m_1|\}$ and $\{|x_{i2} - m_2|\}$ respectively. In our experiments, $K$, a tuning constant, was set to 9, according to [11, 15, 22] (although this value is not sensitive). Here $w(x)$ is called the biweight function or Tukey's bisquare [8].

If weights $w(u)$ and $w(v)$ are both 1 and $m_1$ and $m_2$ are means of $X_1$ and $X_2$, then $r_B$ is exactly the same as Pearson correlation coefficient $r_P$, which is given by Eq. (4). Biweight midcorrelation can be a robust measure of the linear dependence of two variables, due to the biweight function $w(x)$.

## 3. Experimental Result

### 3.1.  *Data*

We retrieved datasets from the GEO (Gene Expression Omnibus) database [2]. Out of the 2089 gene datasets (called GDS in GEO) in GEO, we selected eight gene datasets by using the following two conditions: (1) experimental conditions can be divided into two or more classes; (2) each class has 20 or more experiments. We

Table 2. Comparison of real computational time [hours].

| Pearson | Spearman | Biweight | FTGI (MC + IT) |
|---------|----------|----------|----------------|
| 8.8 | 8.8 | 14.4 | 50.0 (36.4 + 13.6) |

used only genes which have no missing values, and from eight GDSs, we generated totally $\sim 2.3 \times 10^9$ gene combinations for which each of competing methods, FTGI and three methods based on the difference of correlation coefficients were applied to rank. Table 1 shows a list of eight GDSs, where for GDSs with more than two classes, we considered all possible pairwise combinations of classes, meaning that we generated multiple datasets, such as GDS1615_1, GDS1615_2 and GDS1615_3 for GDS1615.

All experiments were run on machines with Dual-Core AMD Opteron Processor 2222 SE (3.0 GHz), 48 GB RAM and Intel XEON CPU 3.33 GHz, 32 GB RAM.

## 3.2. *Computation time*

For FTGI, we chose $\alpha_m$ (the significance level of MC test) of $5.0 \times 10^{-10}$, resulting in that 99% of all around $2.3 \times 10^9$ gene combinations were pruned. Table 2 shows the computation time of each method. FTGI needed around 50 hours for ranking all combinations according to p-values, despite of a relatively conservative choice of $\alpha_m$. The computation time of FTGI was able to be divided into two: around 36 hours of the MC test part (which pruned 99% combinations out of around all 2.3 billions) and around 14 hours of the IT part (which was applied to the rest 1% combinations). On the other hand, Pearson and Spearman needed around 9 hours, and the biweight needed around 14 hours, respectively. This means that IT was 100 times slower than the difference of correlation coefficients.

## 3.3. *Top ranked gene combinations*

Tables 3 to 6 show lists of top ten gene combinations ranked by four methods, i.e. FTGI, Pearson, Spearman and biweight midcorrelation. Figs. 5 to 8 are the distributions of gene expressions of these corresponding top ten ranked combinations.

**FTGI**: Table 3 and Fig. 5 show a list and distributions of expression values for the top 10 ranked gene combinations by FTGI. Fig. 5 shows that expression values are, for at least one class, distributed from the upper left to the lower right (negatively correlated) or the upper right to the lower left (positively correlated). In some cases, this can be seen clearly for both two classes. For example, combinations ranked at the 7th and 10th showed these two distributions, meaning that both two classes contributed the significance of IT and that each combination has a clear switching mechanism. However, there are cases that this distribution can be found for one class only. For example, combinations ranked at the 6th, 8th and 9th showed that expression values colored black (and plotted by •) contributed the significance of IT (by distributions from the upper left to the lower right), while expression

Table 3.   Top 10 ranked pairs by FTGI.

| | p-value | test stat. | GDS | $n_1$ | $n_2$ | classes |
|---|---|---|---|---|---|---|
| 1 | $1.19 \times 10^{-17}$ | 73.2 | GDS1615_2 | 42 | 59 | normal / Crohn's disease |
| 2 | $2.99 \times 10^{-17}$ | 71.3 | GDS1615_2 | 42 | 59 | normal / Crohn's disease |
| 3 | $2.00 \times 10^{-16}$ | 67.6 | GDS1615_2 | 42 | 59 | normal / Crohn's disease |
| 4 | $1.11 \times 10^{-15}$ | 64.2 | GDS1615_2 | 42 | 59 | normal / Crohn's disease |
| 5 | $1.44 \times 10^{-15}$ | 63.7 | GDS1615_2 | 42 | 59 | normal / Crohn's disease |
| 6 | $1.68 \times 10^{-15}$ | 63.4 | GDS1615_2 | 42 | 59 | normal / Crohn's disease |
| 7 | $1.96 \times 10^{-15}$ | 63.1 | GDS1615_2 | 42 | 59 | normal / Crohn's disease |
| 8 | $2.09 \times 10^{-15}$ | 63.0 | GDS1615_2 | 42 | 59 | normal / Crohn's disease |
| 9 | $2.68 \times 10^{-15}$ | 62.5 | GDS1615_2 | 42 | 59 | normal / Crohn's disease |
| 10 | $3.82 \times 10^{-15}$ | 61.8 | GDS1615_2 | 42 | 59 | normal / Crohn's disease |

*Note*: "test stat." means the value of test statistic by IT

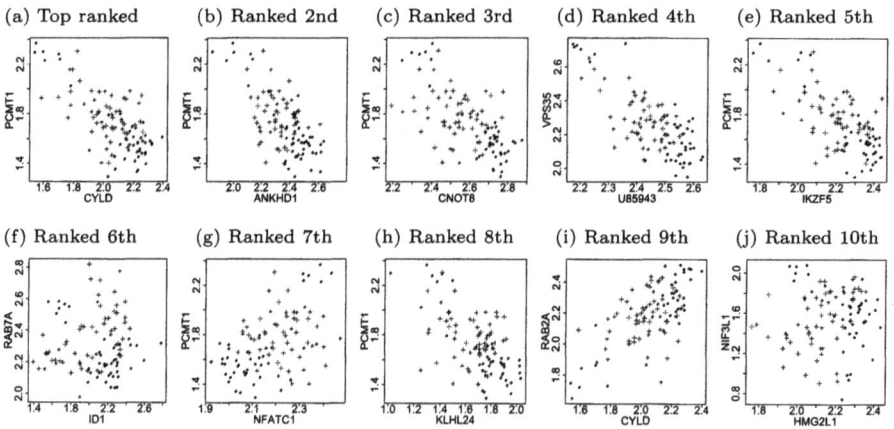

Fig. 5.   The expression value distribution of the resultant top 10 ranked combinations by FTGI.

values colored red (and plotted by +) seemed not affect the significance of IT since the distribution was rather center-focused. Another point of note is that the number of points (examples) in Fig. 5 was larger than those in other figures, since hypothesis testing is used.

**Pearson correlation**: Table 4 and Fig. 6 show a list and distributions of expression values for the top 10 ranked gene combinations by Pearson correlation coefficients. Fig. 6 shows that most distributions were heavily affected by outliers, each being shown in the corner of each two-dimensional figure. This directly draw a serious problem that there are some cases (the 6,7,8 and 10th ranked combinations), where their scores will be lower if we remove outliers in these combinations. Another point is that the number of examples shown in Table 4 are relatively small.

**Spearman correlation**: Table 5 and Fig. 7 show a list and distributions of expression values for the top 10 ranked gene combinations by Spearman correlation coefficients. Fig. 7 shows that the outlier problem was overcome by using Spearman correlation, which is robust against outliers. In fact, the top ranked combination

Table 4.  Top 10 ranked combinations by the difference of Pearson correlation coefficients (shown by diff.).

|   | diff. | $r_P$(class1) | $r_P$(class2) | GDS | $n_1$ | $n_2$ | classes |
|---|-------|---------------|---------------|-----|-------|-------|---------|
| 1 | 1.617 | -0.821 | 0.796 | GDS2519_1 | 22 | 33 | healthy control/ neurodegenerative disease control |
| 2 | 1.560 | 0.850 | -0.710 | GDS2519_1 | 22 | 33 | healthy control/ neurodegenerative disease control |
| 3 | 1.556 | 0.920 | -0.636 | GDS2519_1 | 22 | 33 | healthy control/ neurodegenerative disease control |
| 4 | 1.533 | -0.854 | 0.679 | GDS2519_1 | 22 | 33 | healthy control/ neurodegenerative disease control |
| 5 | 1.530 | -0.740 | 0.790 | GDS1615_1 | 42 | 26 | normal/ulcerative colitis |
| 6 | 1.524 | 0.729 | -0.795 | GDS2519_1 | 22 | 33 | healthy control/ neurodegenerative disease control |
| 7 | 1.519 | 0.865 | -0.654 | GDS2519_1 | 22 | 33 | healthy control/ neurodegenerative disease control |
| 8 | 1.517 | 0.862 | -0.655 | GDS2519_1 | 22 | 33 | healthy control/ neurodegenerative disease control |
| 9 | 1.514 | 0.676 | -0.838 | GDS1615_1 | 42 | 26 | normal/ulcerative colitis |
| 10 | 1.512 | 0.921 | -0.591 | GDS2519_2 | 22 | 50 | healthy control/ Parkinson's disease |

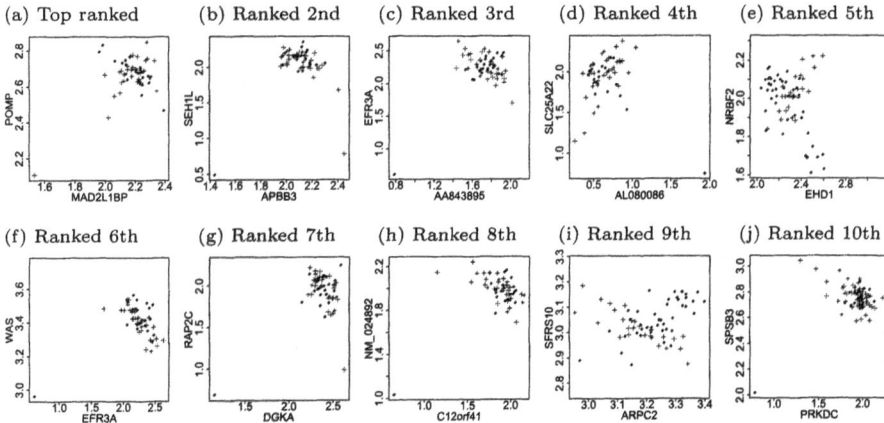

Fig. 6.  The distribution of the resultant top 10 ranked combinations by the difference of Pearson correlations.

shows a clear switching mechanism, being not affected by an outlier in the class colored red (+) which can be found on the bottom of the figure. Particularly in the 4th ranked combination, it seems that this effect of Spearman correlation can be shown in both two classes.

**Biweight midcorrelation**: Table 6 and Fig. 8 show a list and distributions of expression values for the top 10 ranked gene combinations by biweight midcorrelation coefficients. Similar to the results by Spearman correlation, Fig. 8 shows that top-ranked distributions reveal relatively clear switching forms, because outliers are

Table 5.   Top 10 ranked pairs by the difference of Spearman's rank correlation coefficients (shown by diff.).

|    | diff. | $r_S$(class 1) | $r_S$(class 2) | GDS | $n_1$ | $n_2$ | classes |
|----|-------|----------------|----------------|-----|-------|-------|---------|
| 1  | 1.552 | -0.855 | 0.697  | GDS1615_1 | 42 | 26 | normal/ulcerative colitis |
| 2  | 1.545 | 0.782  | -0.763 | GDS1615_1 | 42 | 26 | normal/ulcerative colitis |
| 3  | 1.536 | 0.773  | -0.763 | GDS1615_1 | 42 | 26 | normal/ulcerative colitis |
| 4  | 1.529 | 0.807  | -0.722 | GDS1615_1 | 42 | 26 | normal/ulcerative colitis |
| 5  | 1.506 | 0.651  | -0.856 | GDS1615_1 | 42 | 26 | normal/ulcerative colitis |
| 6  | 1.499 | -0.676 | 0.823  | GDS1615_1 | 42 | 26 | normal/ulcerative colitis |
| 7  | 1.495 | 0.721  | -0.774 | GDS1615_1 | 42 | 26 | normal/ulcerative colitis |
| 8  | 1.490 | -0.778 | 0.712  | GDS1615_1 | 42 | 26 | normal/ulcerative colitis |
| 9  | 1.485 | 0.761  | -0.724 | GDS1615_1 | 42 | 26 | normal/ulcerative colitis |
| 10 | 1.484 | 0.762  | -0.722 | GDS1615_1 | 42 | 26 | normal/ulcerative colitis |

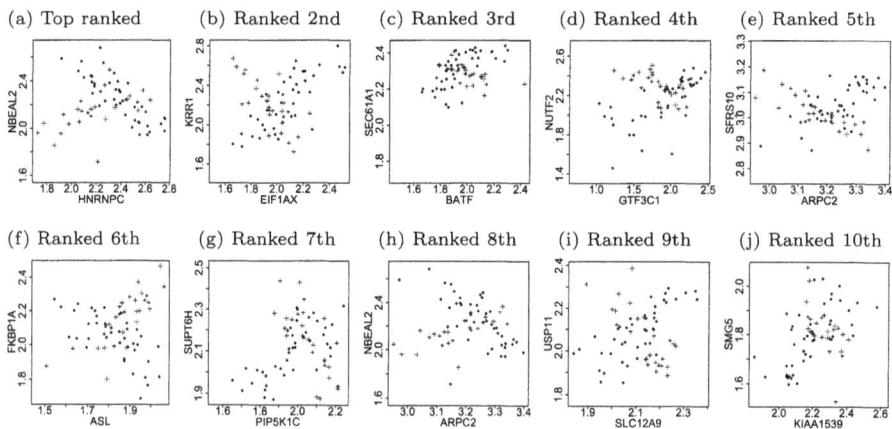

Fig. 7.   The distribution of the resultant top 10 ranked combinations by the difference of Spearman's rank correlation coefficients.

Table 6.   Top 10 ranked pairs by the difference of biweight midcorrelation coefficients (shown by diff.).

|    | diff. | $r_B$(class 1) | $r_B$(class 2) | GDS | $n_1$ | $n_2$ | classes |
|----|-------|----------------|----------------|-----|-------|-------|---------|
| 1  | 1.549 | 0.780  | -0.769 | GDS1615_1 | 42 | 26 | normal/ulcerative colitis |
| 2  | 1.516 | -0.745 | 0.771  | GDS1615_1 | 42 | 26 | normal/ulcerative colitis |
| 3  | 1.511 | 0.775  | -0.736 | GDS1615_1 | 42 | 26 | normal/ulcerative colitis |
| 4  | 1.510 | -0.842 | 0.668  | GDS1615_1 | 42 | 26 | normal/ulcerative colitis |
| 5  | 1.497 | 0.792  | -0.705 | GDS1615_1 | 42 | 26 | normal/ulcerative colitis |
| 6  | 1.493 | -0.678 | 0.815  | GDS1615_1 | 42 | 26 | normal/ulcerative colitis |
| 7  | 1.492 | 0.652  | -0.840 | GDS1615_1 | 42 | 26 | normal/ulcerative colitis |
| 8  | 1.491 | 0.797  | -0.694 | GDS1615_1 | 42 | 26 | normal/ulcerative colitis |
| 9  | 1.486 | -0.747 | 0.739  | GDS2519_1 | 22 | 33 | healthy control/ neurodegenerative disease control |
| 10 | 1.483 | 0.651  | -0.832 | GDS1615_1 | 42 | 26 | normal/ulcerative colitis |

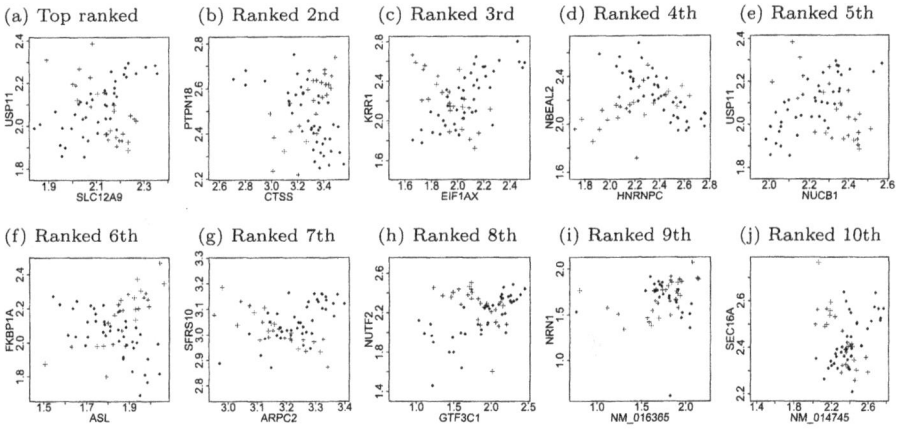

Fig. 8.   The distribution of the resultant top 10 ranked combinations by the difference of biweight midcorrelation coefficients.

avoided by using biweight midcorrelation, which can ignore outliers being far from the center of given data. In fact, in the 2nd and 3rd ranked combinations, the effect of outliers in the class colored black (•) were removed. Similarly, in the 4th and 6th ranked combinations, a few outliers in the class colored red (+) were also removed. Particularly, both classes has outliers in the 8th and 9th ranked gene combinations but these outliers did not affect to find these switching combinations.

## 4.  Discussion

**Comparison**: Our result showed pros and cons of the methods we used. First, a serious problem of FTGI is that top ranked combinations included non-switching combinations. This problem can be found when we used Pearson correlation, which was affected by outliers heavily, while this issue was overcome by Spearmen correlation and biweight midcorrelation. Another problem of FTGI is computation speed, which is slower than all three methods using the idea of the difference of correlation coefficients. However, FTGI uses IT in the final step, by which the number of examples in each combination is larger, making the results more reliable, while the sample size issue is not considered in the approach by using the difference of correlation coefficients. This is an advantage of FTGI.

   **Future work**: There are three points to be considered: outliers, speed and sample size. That is, we first need a robust approach to avoid finding combinations including outliers or non-switching mechanisms. The method we need then has to be a fast and efficient method. Finally, it would be better for the method to allow to consider the sample size by using $p$-values based on some statistical testing.

## Acknowledgements

This work has been supported in part by JSPS KAKENHI 20700134, 20700269, 21680025.

## References

[1] Ayer, D. E. and Eisenman, R. N., A switch from Myc:Max to Mad:Max heterocomplexes accompanies monocyte/macrophage differentiation, *Genes Dev.*, 7: 2110–2119, 1993.

[2] Barrett, T., Troup, D. B., Wilhite, S. E., *et al.*, NCBI GEO: mining tens of millions of expression profiles.database and tools update, *NAR*, 35: D760-D765, 2007.

[3] Braun, R., Cope, L., and Parmigiani, G., Identifying differential correlation in gene/pathway combinations, *BMC Bioinformatics*, 9:488 doi:10.1186/1471-2105-9-488, 2008.

[4] Cho, S. B., Kim, J., and Kim, J. H., Identifying set-wise differential co-expression in gene expression microarray data, *BMC Bioinformatics*, 10:109, doi:10.1186/1471-2105-10-109, 2009.

[5] Choi, J. K., Yu, U., Yoo, O. J., and Kim, S., Differential coexpression analysis using microarray data and its application to human cancer, *Bioinformatics*, 21(24):4348-4355; doi:10.1093/bioinformatics/bti722, 2005.

[6] Cordell, H. J., Epistasis: what it means, what it doesn't mean, and statistical methods to detect it in humans, *Human Molecular Genetics*, 11(20):2463–2468, 2002.

[7] Dettling, M., Gabrielson, E., and Parmigiani, G., Searching for differentially expressed gene combinations, *Genome Biology*, 6:R88 (doi:10.1186/gb-2005-6-10-r88), 2004.

[8] Hampel, F. R., Ronchetti, E. M., Rousseeuw, P. J., and Stahel, W. A., *Robust Statistics*: the approach based on influence functions, Wiley, 1986.

[9] Hardin,J., Mitani, A., Hicks, L., and VanKoten, B., A robust measure of correlation between two genes on a microarray, *BMC Bioinformatics*, 8:220, doi:10.1186/1471-2105-8-220, 2007.

[10] Ho, Y. Y., Cope, L., Dettling, M., and Parmigiani, G., Statistical methods for identifying differentially expressed gene combinations, In: Ochs MF, ed. *Methods in Molecular Biology: Gene Function Analysis*, 408, New Jersey, NJ: Humana; 171–191, 2007.

[11] Hoaglin, D. C., Mosteller, F., and Tukey, J. W. (ed.), *Understanding Robust and Exploratory Data Analysis*, John Wiley & Sons Inc, 1983.

[12] Kayano, M, Takigawa, I, Shiga, M., Tsuda, K., and Mamitsuka, H, Efficiently finding genome-wide three-way gene interactions from transcript- and genotype-data, *Bioinformatics*, 25(21):2735–43, 2009.

[13] Kostka, D. and Spang, R., Finding disease specific alterations in the co-expression of genes, *Bioinformatics*, 20(Suppl. 1):i194–i199, 2004.

[14] Lai, Y., Wu, B., Chen, L., and Zhao, H., A statistical method for identifying differential gene-gene co-expression patterns, *Bioinformatics*, 20(17):3146-3155; doi:10.1093/bioinformatics/bth379, 2004.

[15] Lax, D. A., Robust estimators of scale: finite-sample performance in long-tailed symmetric distributions, *Journal of the American Statistical Association*, 80(391):736-741, 1985.

[16] Li, K.-C., Genome-wide coexpression dynamics: theory and application, *PNAS*, 99(26):16875–80, 2002.

[17] Mardia, K. V., Kent J. T., and Bibby, J. M., *Multivariate Analysis*, Academic Press, New York, 1979.

[18] Pan, W., A comparative review of statistical methods for discovering differentially

expressed genes in replicated microarray experiments, *Bioinformatics*, 18:546–554, 2002.

[19] Shedden,K. and Taylor,J., Differential correlation detects complex associations between gene expression and clinical outcomes in lung adenocarcinomas, In *Methods of Microarray Data Analysis IV*, Springer, New York, US. 121–131, 2005.

[20] Stuart, J. M., Segal, E., Koller, D., and Kim, S. K., A gene-coexpression network for global discovery of conserved genetic modules, *Science*, 302, 249–255, 2003.

[21] Watson, M., CoXpress: differential co-expression in gene expression data, *BMC Bioinformatics*, 7:509, doi:10.1186/1471-2105-7-509, 2006.

[22] Wilcox, R.R., *Introduction to Robust Estimation and Hypothesis Testing*, Academic Press, 1997.

[23] Wu, Z-K., Zhang, Z-Y., Zhang, L-W., and Horimoto, K., Revealing disease related interactions by correlation analysis, *The Second International Symposium on Optimization and Systems Biology* (OSB 08), 341–349, 2008.

[24] Zhang, J., Ji, Y., and Zhang, L., Extracting three-way gene interactions from microarray data, *Bioinformatics*, 23(21): 2903–2909, 2007.

# GENE REGULATORY NETWORK CLUSTERING FOR GRAPH LAYOUT BASED ON MICROARRAY GENE EXPRESSION DATA

KANAME KOJIMA
kaname@ims.u-tokyo.ac.jp

SEIYA IMOTO
imoto@ims.u-tokyo.ac.jp

MASAO NAGASAKI
masao@ims.u-tokyo.ac.jp

SATORU MIYANO
miyano@ims.u-tokyo.ac.jp

*Human Genome Center, Institute of Medical Science, University of Tokyo, 4-6-1 Shirokanedai, Minato-ku, Tokyo 108-8639, Japan*

We propose a statistical model realizing simultaneous estimation of gene regulatory network and gene module identification from time series gene expression data from microarray experiments. Under the assumption that genes in the same module are densely connected, the proposed method detects gene modules based on the variational Bayesian technique. The model can also incorporate existing biological prior knowledge such as protein subcellular localization. We apply the proposed model to the time series data from a synthetically generated network and verified the effectiveness of the proposed model. The proposed model is also applied the time series microarray data from HeLa cell. Detected gene module information gives the great help on drawing the estimated gene network.

*Keywords*: gene regulatory network estimation; network module extraction; graph layout; variational Bayesian method.

## 1. Introduction

Estimation of gene regulatory networks plays an important role for understanding biological systems, and so far, Bayesian networks [7], vector autoregressive models [3, 8, 11], and state space models [5, 15, 16] have been used for the topic. In the gene regulatory network estimation, some models such as state space model [5, 15, 16], module Bayesian network [10], group Granger model [4] assume the modularity of the networks and estimate regulatory network among gene modules. In the state space model and module Bayesian network, strong similarity of expression profiles is assumed for genes in the same modules. On the other hand, group Granger model assumes that genes in the same module are more densely connected, while genes in different modules are sparsely connected.

In this study, we propose a statistical model termed module ARD for estimation of gene regulatory networks from time series gene expression data with the assumption of the modularity of the network and the dense connection of genes in the same module. The statistical model of our method is comprised of two components: vector auto regressive model representing gene regulation and binomial distribution model for module detection proposed in Homan and Wiggins [6]. Since the length of available time series microarray gene expression data (less than 50) is usually more than the number of genes (more than 100), i.e., data sample is much less than the number of parameters, sparse learning techniques such as L1 regularization [3, 8, 11–13, 17] and automatic relevance determination (ARD) [1] are necessary for the stable estimation of the parameters. As a sparse learning technique, we propose an extended ARD method, with which vector auto regressive model and binomial distribution model can be combined and estimated simultaneously.

The performance of our proposed model is verified by comparing to the method based on L1 regularization and dynamic Bayesian network using data from a simulation network. We also analyze the time series microarray gene expression data from HeLa cell cycle model [14].

## 2. Methods

### 2.1. *Model setting*

The module ARD is comprised of linear regression model with automatic relevance determination and graph module detection model. Suppose that we have gene expression profiles of $p$ genes during $T$ time points $\{y_1, \ldots, y_T\}$, The former model is represented by a first order vector autoregressive model with priors:

$$y_t = Ay_{t-1} + \varepsilon_t,$$

where $A$ is a $p \times p$ autoregressive (AR) coefficient matrix and $\varepsilon_t$ is a $p$-dimensional noise vector normally distributed with mean $\mathbf{0}$ and variance $\Sigma = \text{diag}[\sigma_i^2]$. In the automatic relevance determination, the following prior distributions are given for $a_{ij}$, the elements of $A$ and the variance $\sigma_i^2$, respectively,

$$p(a_{ij}; \sigma_i^2, \alpha_{ij}) = p_\mathcal{N}(a_{ij}; 0, \sigma_i^2/\alpha_{ij}), \tag{1}$$
$$p(\sigma_i^2; u_0, \zeta_0) = p_{\mathcal{IG}}(\sigma_i^2; u_0, \zeta_0),$$

where $\alpha_{ij}$, $u_0$, and $\zeta_0$ are hyperparameter, $p_\mathcal{N}$ represents the density of normal distribution, and $p_{\mathcal{IG}}$ represents the density of inverse gamma distribution given by

$$p_{\mathcal{IG}}(x; u, \zeta) = \frac{\zeta^u}{\Gamma(u)} (1/x)^{u+1} \exp(-\frac{\zeta}{x}).$$

These hyperparameters are repeatedly updated to increase marginal probability, or Bayesian score. Since some $\alpha_{ij}$'s converge to infinity in update step, corresponding $a_{ij}$'s converge to zero and then AR coefficients are estimated in a sparse manner. Instead of prior distribution in Equation (1), the module ARD uses the following prior distribution for $a_{ij}$

$$p(a_{ij}; \sigma_i^2, \alpha, M, e_{ij}) = \left\{ p_\mathcal{N}(a_{ij}; 0, \sigma_i^2/\alpha) \right\}^{e_{ij}} \left\{ p_\mathcal{N}(a_{ij}; 0, \sigma_i^2/(\alpha M)) \right\}^{1-e_{ij}},$$

where $\alpha$ is a hyperparameter, $M$ is a constant large value and $e_{ij}$ is a hidden binary variable taking 0 or 1. Since $M$ is large, if $e_{ij} = 0$, $a_{ij}$ is shrunk to 0. Thus, $e_{ij}$ indicates the existence of edges and sparsity of $A$ is controlled by $e_{ij}$.

Next, we explain the latter model, graph module detection model. Graph module detection model is originally proposed by Hofman and Wiggins [6]. The probability of the existence of the edge between $g_i$ and $g_j$ is given by $\theta_c$ if $g_i$ and $g_j$ are in the same module or $\theta_d$ if $g_i$ and $g_j$ are in different modules. Also the probability of module assignment of gene $g_i$ to module $\mu$ is given by $\pi_\mu$. The probability of the existence of edges $E$, which is a set of edge existence $e_{ij}$ is given by

$$p(E; \theta_c, \theta_d, \boldsymbol{\pi}) = \theta_c^{c+}(1 - \theta_c)^{c-}\theta_d^{d+}(1 - \theta_d)^{d-} \prod_{\mu=1}^{K} \pi_\mu^{n_\mu},$$

where $c_+ = \sum_{i,j} e_{ij}\delta_{i,j}$, $c_- = \sum_{i,j}(1 - e_{ij})\delta_{i,j}$, $d_+ = \sum_{i,j} e_{ij}(1 - \delta_{i,j})$, $d_- = \sum_{i,j}(1 - e_{ij})(1 - \delta_{i,j})$, and $K$ is the number of modules. Here, $\delta_{i,\mu}$ is a binary variable and take 1 if gene $g_i$ is in module $\mu$ and 0 otherwise. Also, $\delta_{i,j}$ is a binary variable and take 1 if gene $g_i$ and $g_j$ are in the same module and 0 otherwise, and thus $\delta_{i,j} = \sum_\mu \delta_{i,\mu}\delta_{j,\mu}$. If $\theta_c > \theta_d$ holds genes in the same module are more densely connected than genes in difference modules. We can estimate a gene regulatory network and its modules simultaneously by estimating edge existence $e_{ij}$ and module assignment $\delta_{i,\mu}$.

We estimate $e_{ij}$ and $\delta_{i,\mu}$ by marginalizing likelihood with parameters $a_{ij}$, $\sigma_i^2$, $\theta_c$, $\theta_d$, and $\pi_\mu$. For parameters $\theta_c$, $\theta_d$, and $\boldsymbol{\pi} = (\pi_1, \ldots, \pi_K)$, following prior distributions are used:

$$p(\theta_c) = p_B(\theta_c; \tilde{c}_{+0}, \tilde{c}_{-0}),$$
$$p(\theta_d) = p_B(\theta_d; \tilde{d}_{+0}, \tilde{d}_{-0}),$$
$$p(\boldsymbol{\pi}) = p_D(\boldsymbol{\pi}; \boldsymbol{n}_0),$$

where $p_B$ and $p_D$ indicate the densities of the beta distribution and Dirichlet distribution, respectively. The complete likelihood of the module ARD is given by

$$p(\boldsymbol{y}_t, \Theta) = p(\boldsymbol{y}|A, \Sigma) \left\{ \prod_i p(A_i|\sigma_i^2, E_i)p(\sigma_i^2) \right\} p(E; \theta_c, \theta_d, \boldsymbol{\pi})p(\theta_c)p(\theta_d)p(\boldsymbol{\pi}),$$

where $\Theta$ is all the parameters including hidden variables in the module ARD model, $A_i = (a_{i1}, \ldots, a_{ip})'$, and $E_i = (e_{i1}, \ldots, e_{ip})'$. Since calculating marginal probability of the above equation is difficult, we obtain the upper bound of the log negative marginal probability by the Gibbs inequality:

$$-\log \int_\Theta p(\boldsymbol{y}_t, \Theta)d\Theta = -\log \int_\Theta q(\Theta)\frac{p(\boldsymbol{y}_t, \Theta)}{q(\Theta)}d\Theta$$
$$\leq -\int_\Theta q(\Theta) \log \frac{p(\boldsymbol{y}_t, \Theta)}{q(\Theta)}d\Theta. \tag{2}$$

Then, we calculate the upper bound of the log negative marginal probability by using variational EM method [2]. Under the assumption that $q(\Theta)$ can be factorized as

$$q(\Theta) = \left( \prod_i q(A_i|\sigma_i^2)q(\sigma_i^2) \right) \left( \prod_{i,j} q(e_{ij}) \right) q(\theta_c)q(\theta_d)q(\pi_\mu) \left( \prod_{i,\mu} q(\delta_{i,\mu}) \right),$$

variational EM method searches a local minimum of Equation (2) by repeating two steps: variational E-step and variation M-step. We show the calculation of these two steps in the following two sections.

## 2.2. *Variational M-step*

In variational M-step, variational distributions of $A_i$, $\sigma_i^2$, $\theta_c$, $\theta_d$, and $\pi_\mu$ under expectation of $q(e_{ij} = 1)$, $q(e_{ij} = 0)$, and $q(\delta_{i,\mu})$ are calculated as follows.

- Variational distributions of $A_i$ is given by $\sigma_i^2$ is given by

$$q(A_i | \sigma_i^2) = p_{\mathcal{N}}(A_i; \boldsymbol{\mu}_{A_i}, \sigma_i^2 T_{A_i}^{-1}),$$

  where

$$T_{A_i} = \sum_{t=2}^{T} \boldsymbol{y}_{t-1} \boldsymbol{y}'_{t-1} + \mathrm{diag}[\alpha(e_{ij} + M(1 - e_{ij}))],$$

$$\boldsymbol{\mu}_{A_i} = T_{A_i}^{-1} E_i \sum_{t=2}^{T} \boldsymbol{y}_{t,i} \boldsymbol{y}_{t-1}.$$

- Variational distribution of $\sigma_i^2$ is given by

$$q(\sigma_i^2) = p_{\mathcal{IG}}(\sigma_i^2; u_i, \zeta_i),$$

  where

$$u_i = u_0 + (T - 1)/2$$

$$\zeta_i = \zeta_0 + \frac{1}{2}\left(\sum_{t=2}^{T} y_{t,i}^2 - \mu'_{A_i} T_{A_i} \mu_{A_i}\right).$$

- Variational distribution of $\theta_c$ is given by

$$q(\theta_c) = p_{\mathcal{B}}(\theta_c; \tilde{c}_+, \tilde{c}_-),$$

  where

$$\tilde{c}_+ = \tilde{c}_{+0} + \sum_{ij} q(e_{ij} = 1) q(\delta_{i,j})$$

$$\tilde{c}_- = \tilde{c}_{-0} + \sum_{ij} \{1 - q(e_{ij} = 1)\} q(\delta_{\omega_i, \omega_j}).$$

- Variational distribution of $\theta_d$ is given by

$$q(\theta_d) = p_{\mathcal{B}}(\theta_d; \tilde{d}_+, \tilde{d}_-),$$

  where

$$\tilde{d}_+ = \tilde{d}_{+0} + \sum_{ij} q(e_{ij} = 1)\{1 - q(\delta_{i,j})\}$$

$$\tilde{d}_- = \tilde{d}_{-0} + \sum_{ij} \{1 - q(e_{ij} = 1)\}\{1 - q(\delta_{i,j})\}.$$

- Variational distribution of $\boldsymbol{\pi}$ is given by

$$q(\boldsymbol{\pi}) = p_{\mathcal{D}}(\boldsymbol{\pi}; \tilde{n}_1, \ldots, \tilde{n}_K),$$

where

$$\tilde{n}_\mu = \tilde{n}_0 + \sum_i q(\delta_{i,\mu}).$$

## 2.3. *Variational E-step*

In variational E-step, we calculate the distributions of hidden variables $q(e_{ij})$ and $q(\delta_{i,\mu})$. Since $e_{ij}$ follows binomial distribution, we have

$$q(e_{ij} = 1) = \frac{\tilde{q}(e_{ij} = 1)}{\tilde{q}(e_{ij} = 1) + \tilde{q}(e_{ij} = 0)},$$

where $\tilde{q}(e_{ij} = 1)$ and $\tilde{q}(e_{ij} = 0)$ are given by

$$\tilde{q}(e_{ij} = 1) = \exp\left( \mathrm{E}_{q(a_{ij}, \sigma_i^2)}[\log p_{\mathcal{N}}(a_{ij}; 0, \sigma_i^2/\alpha)] \right.$$

$$\left. + 2q(\delta_{i,j})\mathrm{E}_{q(\theta_c)}[\log \theta_c] + 2(1 - q(\delta_{i,j}))\mathrm{E}_{q(\theta_d)}[\log \theta_d] \right) \quad (3)$$

$$\tilde{q}(e_{ij} = 0) = \exp\left( \mathrm{E}_{q(a_{ij}, \sigma_i^2)}[\log p_{\mathcal{N}}(a_{ij}; 0, \sigma_i^2/(\alpha M))] \right.$$

$$\left. + 2q(\delta_{i,j})\mathrm{E}_{q(\theta_c)}[\log(1 - \theta_c)] + 2(1 - q(\delta_{i,j}))\mathrm{E}_{q(\theta_d)}[\log(1 - \theta_d)] \right). \quad (4)$$

Here, $q(\delta_{i,j}) = \sum_\mu q(\delta_{i,\mu})q(\delta_{j,\mu})$ holds. In addition, since $\delta_{i,\mu}$ follows multinomial distribution, we also have

$$q(\delta_{i,\mu}) = \frac{\tilde{q}(\delta_{i,\mu})}{\sum_{\mu'} \tilde{q}(\delta_{i,\mu'})},$$

where $\tilde{q}(\delta_{i,\mu})$ is given by

$$\tilde{q}(\delta_{i,\mu}) = \exp\left( \sum_{j \neq i} q(e_{ij} = 1)q(\delta_{j,\mu})(\mathrm{E}_{q(\theta_c)}[\log \theta_c] - \mathrm{E}_{q(\theta_d)}[\log \theta_d]) \right.$$

$$\left. + \sum_{j \neq i} q(e_{ij} = 1)q(\delta_{j,\mu})(\mathrm{E}_{q(\theta_c)}[\log(1 - \theta_c)] - \mathrm{E}_{q(\theta_d)}[\log(1 - \theta_d)] + \mathrm{E}_{q(\pi_\mu)}[\log \pi_\mu] \right).$$

$$(5)$$

For the calculation of Equations (3), (4), and (5), the calculation of following expectations are required

$$E_{q(\theta_c)}[\log \theta_c] = \psi(\tilde{c}_+) - \psi(\tilde{c}_+ + \tilde{c}_-)$$

$$E_{q(\theta_c)}[\log \theta_d] = \psi(\tilde{d}_+) - \psi(\tilde{d}_+ + \tilde{d}_-)$$

$$E_{q(\theta_d)}[\log(1 - \theta_c)] = \psi(\tilde{c}_-) - \psi(\tilde{c}_+ + \tilde{c}_-)$$

$$E_{q(\theta_d)}[\log(1 - \theta_d)] = \psi(\tilde{d}_-) - \psi(\tilde{d}_+ + \tilde{d}_-)$$

$$E_{q(\pi_\mu)}[\log(q(\pi_\mu))] = \psi(\tilde{n}_\mu) - \psi(\sum_\nu \tilde{n}_\nu)$$

$$E_{q(\sigma_i^2)}[1/\sigma_i^2] = u_i/\zeta_i$$

$$E_{q(A_i)}[A_i] = = \mu_{A_i}$$

$$E_{q(A_i,\sigma_i^2)}\left[\frac{1}{\sigma_i^2}A_i\right] = \frac{u_i}{\zeta_i}\mu_{A_i}$$

$$E_{q(A_i,\sigma_i^2)}\left[\frac{1}{\sigma_i^2}A_i A_i'\right] = \frac{u_i}{\zeta_i}\mu_{A_i}\mu_{A_i}' + T_{A_i}^{-1},$$

where $\psi()$ is digamma function.

## 2.4. *Updating hyperparameters*

In variational EM method, the hyperparameters are updated to increase the lower bound of marginal probability. In our model, we update only $\alpha$, $u_0$, and $\zeta_0$: $\alpha$ is updated by maximizing the following equation

$$\hat{\alpha} = \arg\max_\alpha \sum_{i,j} \int \int q(A_i, \sigma_i^2)\{q(e_{ij} = 1) \log p_\mathcal{N}(A_i; 0, \sigma_i^2/\alpha)$$

$$+ q(e_{ij} = 0) \log p_\mathcal{N}(a_{ij}; 0, \sigma_i^2/(\alpha M))\}dA_i d\sigma_i^2$$

$$= \arg\max_\alpha \frac{1}{2}\log \alpha - \sum_{i,j} \frac{q(e_{ij} = 1)\alpha + q(e_{ij} = 0)\alpha M}{2}E_{q(A_i,\sigma_i^2)}[a_{ij}^2/\sigma_i^2].$$

Thus, by taking the derivative and setting it equal to zero, we have

$$\hat{\alpha} = p^2/\left\{\sum_{i,j}(q(e_{ij} = 1) + q(e_{ij} = 0)M)E_{q(A_i,\sigma_i^2)}[a_{ij}^2/\sigma_i^2]\right\}.$$

Also, $u_0$ and $\zeta_0$ are updated by maximizing the following equation:

$$(\hat{u}_0, \hat{\zeta}_0) = \arg\max_{(u_0,\zeta_0)} \sum_i \int q(\sigma_i^2) \log p_{\mathcal{IG}}(\sigma_i^2; u_0, \zeta_0)d\sigma_i^2$$

$$= \arg\max_{(u_0,\zeta_0)} \sum_i \int q^{(s)}(r_i)\left((k-1)\log q_i + k\log \zeta_0 - r_i\zeta_0 - \log\Gamma(k)\right)dr_i$$

$$= \arg\max_{(u_0,\zeta_0)} (u_0 - 1)\sum_i E_{q(\sigma_i^2)}[\log \sigma_i^2] + mk\log \zeta_0 - \zeta_0\sum_i \bar{r}_i - p\log\Gamma(u_0),$$

where $\mathrm{E}_{q(\sigma_i^2)}[1/\sigma_i^2][\log \sigma_i^2] = \psi(u_i) - \log(\zeta_i)$ By taking the derivative and setting it equal to zero, we have

$$\sum_i \mathrm{E}_{q(\sigma_i^2)}[\log \sigma_i^2] - p \log \zeta_0 + p\psi(u_0) = 0$$

$$p\frac{u_0}{\zeta_0} - \sum_i \mathrm{E}_{q(\sigma_i^2)}[1/\sigma_i^2] = 0$$

Since there is no closed form solution of $u_0$, $u_0$ is obtained by Newton's method as in the following iteration

$$\hat{u}_0^{(t+1)} = \frac{\log u_0^{(t)} - \psi(u_0) - s}{1/u_0^{(t)} - \psi'(u_0^{(t)})},$$

where $\psi'$ is trigamma function and $s$ is given by

$$s = \log\left(\frac{1}{p}\sum_i \mathrm{E}_{q(\sigma_i^2)}[1/\sigma_i^2]\right) + \frac{1}{p}\sum_i \mathrm{E}_{q(\sigma_i^2)}[\log \sigma_i^2].$$

The above procedure is repeated until convergence. After convergence, $\hat{\zeta}_0$ is given by $p\hat{u}_0/\left(\sum_i \mathrm{E}_{q(\sigma_i^2)}[1/\sigma_i^2]\right)$.

### 2.5. Calculating free energy (Bayesian score)

Upper bound of the Bayesian score in Equation (2) is given by

$$
\begin{aligned}
B(\boldsymbol{y}; K) &= -\int_\Theta q(\Theta) \log \frac{p(\boldsymbol{y}, \Theta)}{q(\Theta)} d\Theta \\
&= \sum_{i,j} q^{(s)}(e_{ij} = 1) \log q^{(s)}(e_{ij} = 1) + \sum_{i,j} q^{(s)}(e_{ij} = 0) \log q^{(s)}(e_{ij} = 0) \\
&+ \sum_{i,\mu} q^{(s)}(\delta_{i,\mu}) \log q^{(s)}(\delta_{i,\mu}) - D_{KL}(q^{(s)}(\boldsymbol{\pi}), q^{(s+1)}(\boldsymbol{\pi})) \\
&- D_{KL}(q^{(s)}(\theta_c), q^{(s+1)}(\theta_c)) - D_{KL}(q^{(s)}(\theta_d), q^{(s+1)}(\theta_d)) \\
&- \sum_i \mathrm{E}_{q^{(s)}(\sigma_i^2)}[D_{KL}(q^{(s)}(A_i), q^{(s+1)}(A_i))] + \sum_i D_{KL}(q^{(s)}(\sigma_i^2), q^{(s+1)}(\sigma_i^2)),
\end{aligned}
$$

where $D_{KL}$ is Kullback-Leibler divergence between two distributions and indices $(s)$ and $(s + 1)$ represent the updated step number of variational functions. $D_{KL}(q^{(s)}(\theta_c), q^{(s+1)}(\theta_c))$, $D_{KL}(q^{(s)}(\theta_d), q^{(s+1)}(\theta_d))$, and $D_{KL}(q^{(s)}(\boldsymbol{\pi}), q^{(s+1)}(\boldsymbol{\pi}))$ can be calculated by Kullback-Leibler divergence between Dirichlet distributions

$$D_{KL}(p_D(\boldsymbol{x}; \boldsymbol{\alpha}), p_D(\boldsymbol{x}; \boldsymbol{\alpha}')) = \log \frac{B(\boldsymbol{\alpha}')}{B(\boldsymbol{\alpha})} + \sum_i (\alpha_i' - \alpha_i)\cdot\psi(\sum_i \alpha_i) - \sum_i (\alpha_i' - \alpha_i)\psi(\alpha_i).$$

By using a formula for Kullback-Leibler divergence between multivariate normal distributions, we have

$$D_{KL}(q^{(s)}(A_i), q^{(s+1)}(A_i)) = \frac{1}{2} \log \frac{|T_{A_i}^{(s)}|}{|T_{A_i}^{(s+1)}|} + \frac{1}{2} \text{tr}(T_{A_i}^{(s)}(T_{A_i}^{(s+1)})^{-1})$$
$$+ \frac{1}{2\sigma_i^2} \left( \mu_{A_i}^{(s+1)} - \mu_{A_i}^{(s)} \right)' (T_{A_i}^{(s+1)})^{-1} \left( \mu_{A_i}^{(s+1)} - \mu_{A_i}^{(s)} \right) - \frac{p}{2}.$$

Thus, $E_{q^{(s)}(\sigma_i^2)}[D_{KL}(q^{(s)}(A_i), q^{(s+1)}(A_i))]$ is obtained by plugging the expectation of $1/\sigma_i^2$ to the above equation. Finally, from a formula for Kullback-Leibler divergence between inverse gamma distributions, we have

$$D_{KL}(q^{(s)}(\sigma_i^2), q^{(s+1)}(\sigma_i^2)) = \log \frac{\Gamma(u_i^{(s+1)})}{\Gamma(u_i^{(s)})} - \log \frac{(\zeta_i^{(s+1)})^{u_i^{(s+1)}}}{(\zeta_i^{(s)})^{u_i^{(s)}}}$$
$$+ (u_i^{(s)} - u_i^{(s+1)})(\psi(u_i^{(s)}) - \log \zeta_i^{(s)}) + u_i^{(s)} \frac{\zeta_i^{(s+1)} - \zeta_i^{(s)}}{\zeta_i^{(s)}}.$$

## 3. Experiments

### 3.1. *Experiments using simulated data*

For the evaluation of the module ARD, we compare its performance to those of two existing approaches; G1DBN [9], an approach based on dynamic Bayesian network and an first order vector autoregressive model with the Elastic Net regularization [17] by using data generated from a synthetic network. The synthetic network of 100 nodes and 137 edges is generated as follows: (i) 100 nodes are randomly divided to three modules. (ii) A tree of 100 nodes is generated. (iii) edges between node $i$ and $j$, $i \neq j$ are added in probability 0.01 if $i$ and $j$ are in the same module and 0.001 otherwise. (iv) Self regulation edges are added to root nodes of the network. Fig. 1 shows the generated synthetic network, where three gene modules are indicated by coloring genes with black, white, and gray. For generating data from linear autoregressive regulation, AR coefficients are uniformly assigned to the edges from $\{-0.9, -0.8, -0.7, -0.6, -0.5, 0.5, 0.6, 0.7, 0.8, 0.9\}$. Innovation noise is set to follow a normal distribution of mean $\mathbf{0}$ and variance $I$. From the model, time series data of 20, 30, 50, 100 time points are generated.

The numbers of true positives, false positives for estimated networks by module ARD, G1DBN, VAR (Elastic Net) are summarized in Table 1. For the threshold value of the edge detection in G1DBN, the default value $\alpha = 0.5$ is employed. The regularization parameters in Elastic Net are selected by AICc (the corrected Akaike information criterion). From the comparison, results of module ARD contain much less false positive edges than those of other two methods although module ARD has slightly worse results on the number of true positives comparing to others. In the network used in the simulation, the modularity is assumed, i.e., genes in the same module are more densely connected than genes in different modules. Since the mod-

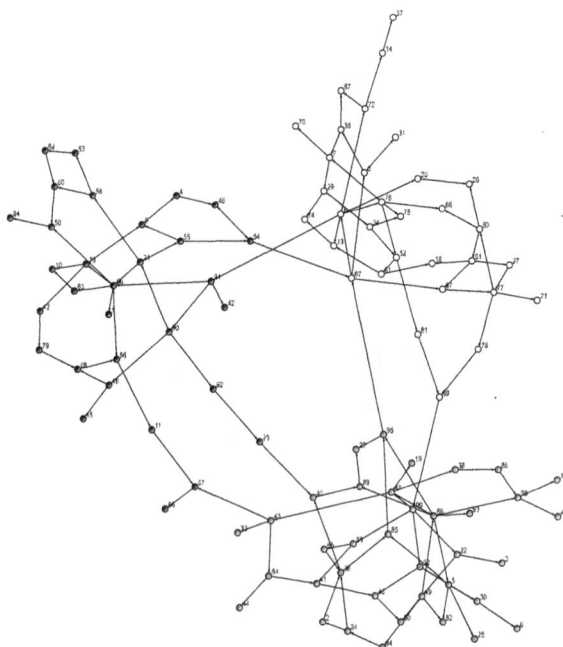

Fig. 1.   A synthetic network of 100 nodes and 137 edges. The nodes in the same modules are in the same color.

ule ARD considers modularity of the estimated network, the more accurate results can be obtained, comparing to other methods that do not assume the modularity.

Table 1.   The numbers of true positives (# TP) and false positives (# FP) of the module ARD, G1DBN, and VAR (Elastic Net).

| Algorithm | Module ARD | | G1DBN | | VAR (Elastic Net) | |
|---|---|---|---|---|---|---|
| # of time points | # TP | # FP | # TP | # FP | # TP | # FP |
| 100 | 136 | 22 | 137 | 1374 | 137 | 611 |
| 50 | 131 | 55 | 137 | 1264 | 137 | 535 |
| 30 | 112 | 18 | 131 | 1031 | 132 | 425 |
| 20 | 32 | 4 | 87 | 25 | 116 | 249 |

Fig. 2 shows the estimated network and gene modules by the module ARD. From the figure, gene module patterns on the estimated network is similar to those of true network.

## 3.2. Application of expression data of human HeLa cell

We apply the proposed method to the time series microarray gene expression data from HeLa cell [14]. The expression data were measured at 48 time points. We used 94 genes selected by Fujita et al. [3] and constructed a gene regulatory network of these genes.

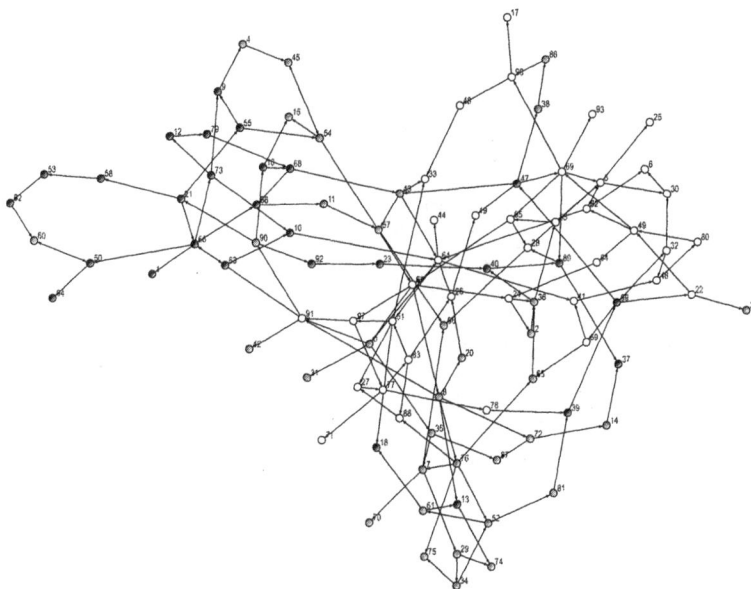

Fig. 2. The estimated gene regulatory network and gene modules by the module ARD from the simulation time series data of 100 time points.

We apply the module ARD to the data with several types of module numbers $K$. Since genes are assigned to only two modules in the modules ARD with $K > 2$, we determined that the number of module is two.

Fig. 3 shows the estimated gene regulatory network and two gene modules by the module ADR. NF$\kappa$B (in the blue module), A20 (in the black module), PKR (in the black module), STAT3 (in the black module), and EKI1 (in the black module) are isolated from the main part of the corresponding module. They should be classified to their opposite module. In the estimated network, we focus on genes with more than or equal to five degrees; FRG1-FGF, Killer/DR5, FGF7, FGF1, FGF5, PDGFRA, IL-1RA, PIDD, PAI, MASPIN, Faz(CD95), IRF-2, FGF12B, BRAC2, GDF1, TRF4-2, and FGF20. Except for FGF5, in the estimated network, these genes are connected to genes in the other module as well as gene in the same module. We should note that most of the edges between modules are placed between these genes. This observation implies that high degree genes tend to work as the connector between modules. If gene regulatory networks are comprised of several gene modules and each module has some specific functions, signal transduction between modules is important and critical for maintaining the biological systems. Thus, the above result suggests that high degree genes play a key role of signal transduction between gene modules for the robust biological systems.

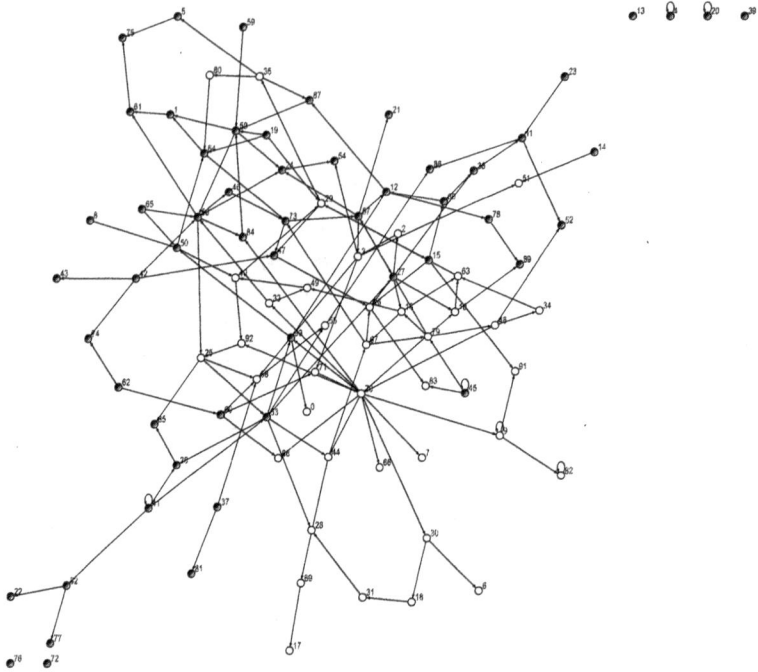

Fig. 3.   The resulting gene regulatory network from the module ARD using time series gene expression data of HeLa cell. Two modules are represented by the color of nodes (black or white).

## 4. Discussion

We proposed a new statistical method, module ARD, for estimating gene regulatory networks and gene modules from time series gene expression data. Since the length of time series gene expression data is short, a newly derived automatic relevance determination performs a sparse learning for the stable parameter estimation. From the experiments using simulation data, we verified the performance of the module ARD by comparing with those of other existing approaches.

The module ARD was applied to time series gene expression data from HeLa cell and a gene regulatory network and gene modules with automatically determined module number were estimated. In this work, the modularity of gene regulatory networks is only assumed for the estimation of networks and gene module detection, By considering the functional or locational similarity among proteins whose corresponding genes to the prior distribution of module assignment, more accurate estimation is expected. In addition, the hyperparameters for beta and Dirichlet distributions are not updated in the variational EM process. We would like to develop full search method to optimize all hyperparameters from data.

# References

[1] Archambeau, C. and Bach. F., Sparse Probabilistic Projections, *the 21th Annual Conference on Neural Information Processing Systems*, 8-11, 2008.

[2] Beal, M.J. and Ghahramani, Z., The variational Bayesian EM algorithm for incomplete data: with application to scoring graphical model structures, *Bayesian Statistics*, 7:453-464, 2003.

[3] Fujita, A., Sato, J.R., Garay-Malpartida, H.M., Morettin, P.A., Yamaguchi, R., Miyano, S., Sogayar, M.C., and Ferreira, C.E., Modeling gene expression regulatory networks with the sparse vector autoregressive model, *BMC Systems Biology*, 1:39, 2007.

[4] Fujita, A., Sato, J.R., Kojima, K., Gomes, L.R., Nagasaki, M., Sogayar, M.C., and Miyano, S. Identification of Granger causality between gene sets. *Journal of Bioinformatics and Computational Biology*, (accepted).

[5] Hirose, O., Yoshida, R., Imoto, S., Yamaguchi, R., Higuchi, T., Stephen, D., Chamock-Jones, C., and Miyano, S., Statistical inference of transcriptional module-based gene networks fro time course gene expression profiles by using state space models, *Bioinformatics*, 24(7):932–942, 2008.

[6] Hofman, J.M. and Wiggins, C.H. Bayesian approach to network modularity. *Physical Review Letters*, 100:258701–258705, 2008.

[7] Kim, S., Imoto, S., and Miyano, S., Dynamic Bayesian network and nonparametric regression for nonlinear modeling of gene networks from time series gene expression data, *Biosystems*, 75(1–3): 57–65, 2004.

[8] Kojima, K., Fujita, A., Shimamura, T., Imoto, S., and Miyano, S., Estimation of nonlinear gene regulatory networks via L1 regularized NVAR from time series gene expression data, *Genome Informatics*, 20:37-51, 2008.

[9] Lebre, S., Inferring dynamic Bayesian networks with low order dependencies, 2007.

[10] Segal, E., Pe'er, D., Regev, A., Koller, D., and Friedman, N., Learning module networks, *Journal of Machine Learning Research* 6:557-588, 2005.

[11] Shimamura, T., Imoto, S., Yamaguchi, R., Fujita, A., Nagasaki, M., and Miyano, S., Recursive elastic net for inferring large-scale gene networks from time course microarray data, *BMC Systems Biology*, 3:41, 2009.

[12] Shimamura, T., Imoto, S., Yamaguchi, R., and Miyano, S., Weighted lasso in graphical Gaussian modeling for large gene network estimation based on microarray data, *Genome Informatics*, 19:142-153, 2007.

[13] Tibshirani, R., Regression shrinkage and selection via the lasso, *Journal of Royal Statistical Society, Series B*, 5(1):267–288, 1996.

[14] Whitfield, M.L., Sherlock, G., Saldanha, A.J., Murray, J.I., Ball C.A., Alexander, K.E., Matese, J.C., Perou, C.M., Hurt, M.M., Brown, P.O., and Botstein, D., Identification of genes periodically expressed in the human cell cycle and their expression in tumors, *Molecular Biology of the Cell*, 13: 1977–2000, 2002.

[15] Yamaguchi, R. Imoto, S., Yamauchi, M., Nagasaki, M, Yoshida, R., Shimamura, T., Hatanaka, Y., Ueno, K., Higuchi, T., Gotoh, N., and Miyano, S., Predicting differences in gene regulatory systems by state space models, *Genome Informatics*, 21:101-113, 2008.

[16] Yamaguchi, R. Yoshida, R., Imoto, S. Higuchi, T., and Miyano, S., Finding module-based gene networks with state-space models-Mining high-dimensional and short time-course gene expression data, *IEEE Signal Processing Magazine*, 24(1):37-46, 2007.

[17] Zou, H. and Hastie, T., Regularization and variable selection via the Elastic Net. *Journal of the Royal Statistical Society, Series B*, 67(2), 301–320, 2005.

# FLUXVIZ — CYTOSCAPE PLUG-IN FOR VISUALIZATION OF FLUX DISTRIBUTIONS IN NETWORKS

MATTHIAS KÖNIG
matthias.koenig@charite.de

HERMANN-GEORG HOLZHÜTTER
hergo@charite.de

*Institute of Biochemistry, Medical Faculty of the Humboldt University, Charité,
Monbijoustr.2, 10117 Berlin, Germany*

**Motivation**: Methods like FBA and kinetic modeling are widely used to calculate fluxes in metabolic networks. For the analysis and understanding of simulation results and experimentally measured fluxes visualization software within the network context is indispensable.

**Results**: We present FluxViz, an open-source Cytoscape plug-in for the visualization of flux distributions in molecular interaction networks. FluxViz supports (i) import of networks in a variety of formats (SBML, GML, XGMML, SIF, BioPAX, PSI-MI) (ii) import of flux distributions as CSV, Cytoscape attributes or VAL files (iii) limitation of views to flux carrying reactions (flux subnetwork) or network attributes like localization (iv) export of generated views (SVG, EPS, PDF, BMP, PNG). Though FluxViz was primarily developed as tool for the visualization of fluxes in metabolic networks and the analysis of simulation results from FASIMU, a flexible software for batch flux-balance computation in large metabolic networks, it is not limited to biochemical reaction networks and FBA but can be applied to the visualization of arbitrary fluxes in arbitrary graphs.

**Availability**: The platform-independent program is an open-source project, freely available at http://sourceforge.net/projects/fluxvizplugin/ under GNU public license, including manual, tutorial and examples.

*Keywords*: Cytoscape; fluxes; visualization; systems biology; metabolic network; FBA.

## 1. Introduction

Software to visually explore biological networks plays a key role in the development of integrative biology, systems biology and bioinformatics. Many tools for visualization of biological networks are available including widely-used examples such as Cytoscape, VisANT, Pathway Studio and Patika [1, 18]. These tools serve, besides their main task to support visual exploration of network structure, also as platform for integration and visualization of data from experiments, simulations and bioinformatic analysis. One such data type is flux information in biological networks like metabolic fluxes in metabolic networks or information fluxes in signal transduction networks. The flux distributions can be based on experimental methods like pulse labeling or flux sensors [14] or result from simulations like FBA [8, 12] or kinetic modeling [13]. Visualization tools for flux information in the network context are essential and should implement

(i) *import* of networks and flux distributions in a variety of formats.

(ii) *batch* analysis of multiple flux distributions in a consistent network layout with simple switch between the different flux distributions.

(iii) generation of *subnetwork views* based on varying network attributes like flux values or localization.

(iv) a *flexible mapping system* between flux values and visual network properties like edge weight or color.

(v) support for the *integration and visualization of additional information* like localization or gene expression data.

(vi) *export* of generated network views in a variety of formats.

Based on the stated requirements the available tools all have mayor limitations (Table 1). CellNetAnalyzer [10] has only minor visualization capabilities and no support for batch analysis, subnetwork views, flexible mappings, visualization of additional data and export of network views. Furthermore it is based on the commercial software package Matlab. FBA-SimVis [3], a VANTED [7] plug-in for FBA simulations with integrated visualization, lacks batch analysis and export, subnetwork views and flexible mappings. FaBina [11] implements only basal subnetwork generation features like flux within single user-defined pathways or compartments. Consistent layouts between different flux distributions, flexible mapping functions and advanced subnetworks are missing. The VisANT flux visualization tool (FVT) [17] and YANAsquare [15] lack among other things crucial import and export features. Specialized tools for the visualization of biological networks like Cytoscape [9] or VisANT [6] have advanced mapping and import systems but do not implement visualization of flux distributions.

## 2. Results

We present FluxViz, an open-source Cytoscape plug-in for the visualization of flux distributions in networks. Cytoscape is an open-source bioinformatics software platform for visualizing molecular interaction networks and integrating these interactions with gene expression profiles and other state data. FluxViz extends the basic Cytoscape capabilities with flux visualization features (Table 1). FluxViz was primarily developed as a tool for the visualization of fluxes in metabolic networks and as frontend for FASIMU [5], a flexible software for flux-balance computation series in large metabolic networks, and uses the generated output files as input for visualization. FluxViz is not limited to biochemical reaction networks and FBA but can be applied for the visualization of fluxes in arbitrary graphs. The general workflow using FluxViz is depicted in Fig. 1. FluxViz visualization is solely based on network structure and flux data and therefor independent of the underlying simulation platform or experimental setup which generated the flux data. After import a standard NetworkView is generated which can be adapted and modified by application of layouts (automatic layout algorithms or manual), by changing the mapping properties between network attributes (data associated with network nodes and edges)

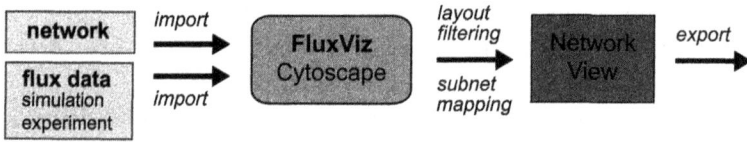

Fig. 1. FluxViz workflow. Network information and flux data can be imported in a variety of formats. A standard NetworkView is generated for the flux distributions in the network context. The NetworkView can be modified by application of layout algorithms, filtering network data, generation of subnetworks based on arbitrary node attributes or flux information and definition of additional visual mappings. The used (flux edge attribute) mapping function can be adapted and additional information like gene expression data can be mapped on visual network attributes. The resulting NetworkViews can be exported in a variety of formats.

and visual network attributes (visual representation of nodes and edges like color or size) and by selecting subnetworks based on filters. The resulting NetworkViews can be interactively analyzed in Cytoscape or be exported as images.

## 2.1. *FluxViz features*

### Import

Flux distributions can be imported from CSV, Cytoscape attribute or FASIMU val files. Networks can be imported as SBML, GML, XGMML, SIF, BioPAX, PSI-MI or can be manually generated in Cytoscape. Existing modeling tools like Matlab or FASIMU can be easily adapted to generate the CSV flux formats.

### Layout

Multiple automatic layout algorithms are available in Cytoscape and can be applied in FluxViz. Manual layouts are also supported. The layout information is a global property for all flux distributions and the network, which enables a simple comparison between flux distributions (see Fig. 2 for an example).

### Batch

FluxViz supports the work with multiple flux distributions in one session. Furthermore batch import of files and batch export of NetworkViews has been implemented. In response to selection of a flux distribution the corresponding NetworkView is generated on the fly.

### Subnetworks and filtering

FluxViz supports the filtering of NetworkViews based on flux values or node attributes and the generation of subnetworks based on the filter selections. Hereby the NetworkView can for example be constraint to the flux containing subgraph

Table 1. FluxViz features in comparison with alternative visualization tools. [+] feature supported, [−] feature not supported, [±] feature partially supported or only basal implementation.

| Feature | CellNet-Analyzer | FBA-SimVis | fa-BINA | VisANT FVT | YANA square | Cyto-scape | Flux-Viz | FluxViz Details |
|---|---|---|---|---|---|---|---|---|
| *Network import* | − | + | + | − | + | + | + | many formats (SBML, GML, XGMML, SIF, BioPAX, PSI−MI) |
| *Flux data import* | + | + | + | − | − | − | + | CSV format, FASIMU val files, Cytoscape attributes |
| *Export flux distribution views* | − | + | + | + | − | + | + | many formats (PDF, SVG, EPS, JPEG, PNG, BMP) |
| *Batch export* | − | − | + | − | − | − | + | batch export of selected flux distributions in many formats |
| *Filtering and subnetwork views* | − | ± | ± | − | ± | − | + | flux containing network attribute based subnetworks (like compartment or pathway) flux containing attribute networks |
| *Flux mapping on visual attributes* | − | + | + | + | + | − | + | all node and edge attributes can be utilized edge size, direction and tooltip used for default visualization |
| *Flexible mapping functionality* | − | − | ± | + | − | + | + | node and edge attributes for visualization of additional data like localization as node color or gene expression as node size. |
| *Adaptable mapping functions* | − | + | − | + | − | + | + | global (all distributions) vs. local (single distribution) settings linear and stepwise linear mappings based on setpoints |
| *Batch support for flux distributions* | − | − | ± | − | − | − | + | batch import and export cycling through flux distributions on the fly generation of views with consistent layout |
| *Functional enrichment* | − | ± | ± | + | ± | + | + | many Cytoscape plug−ins available, large community simple enhancement through plug−in architecture |

or can be limited based localization information to generate compartment subnetworks. The flux distributions can be analyzed in small subgraphs separately. This feature is especially important for the analysis of large-scale networks (Fig. 3) which are difficult to analyze as complete graph.

*Flexible mapping*

FluxViz supports flexible mappings of flux information to the visual node and edge attributes of the NetworkView. In the standard mapping the edge size, direction and the tooltip are used to represent the flux through the corresponding edge. All node and edge attributes can be used to represent additional network information like localization of nodes or gene expression for proteins. In this process the mapping functions between network attributes and visual network attributes can be adapted. After selection of flux distributions the visual edge and node attributes of

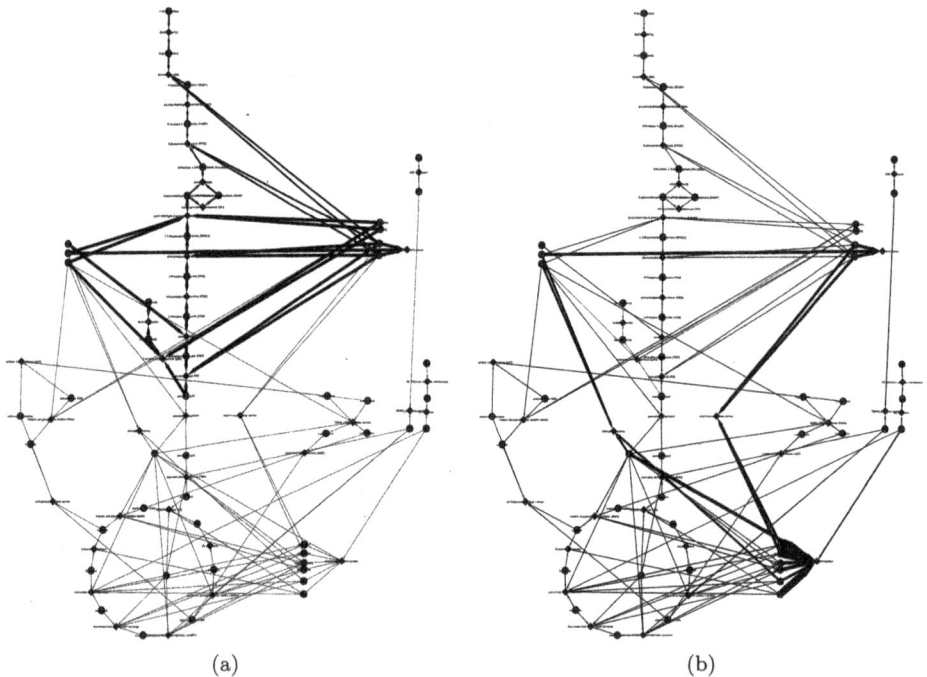

(a)                              (b)

Fig. 2.   Example of FluxViz visualization of flux distributions. (a) Glucose utilization under aerob conditions. ATP is only generated in glycolysis, the resulting pyruvate is converted to lactate and exported. (b) Glucose utilization under anaerob conditions. ATP is mainly generated by oxidative phosphorylation, no lactate is exported, much less glucose is needed for the same ATP production. Flux distributions are FASIMU FBA results in human hepatocyte network consisting of glycolysis, gluconeogenesis, pentose phosphate pathway and citrate cycle (supplementary information). Flux minimization [4] was used as objective function with ATP production as target flux under varying oxygen availability. Manually generated consistent layout for comparison, linear (linear flux edge weight) global mapping function.

the NetworkView are changed according to the selected distribution (Fig. 2, Fig. 3).

*Export*

The generated NetworkViews can be exported in a variety of formats (PDF, SVG, EPS, JPEG, PNG, BMP). Batch export is supported.

*Integration into well established platform*

FluxViz is implemented as an extension of the well established visualization and analysis platform Cytoscape [9]. Therefore many existing features and plug-ins, like network loaders, automatic layout algorithms, advanced filtering mechanisms or network analysis tools, are available and can be utilized in combination with FluxViz.

## 2.2. *Application*

FluxViz has been applied for kinetic models simulated in Matlab and FBA simulations in FASIMU (Fig. 2, Fig. 3). Furthermore, FluxViz was used for the reconstruction and analysis of a human hepatocyte core network (Fig. 3(a)) and for the visualization of FBA simulations in the reconstruction of the complete human hepatocyte network [2].

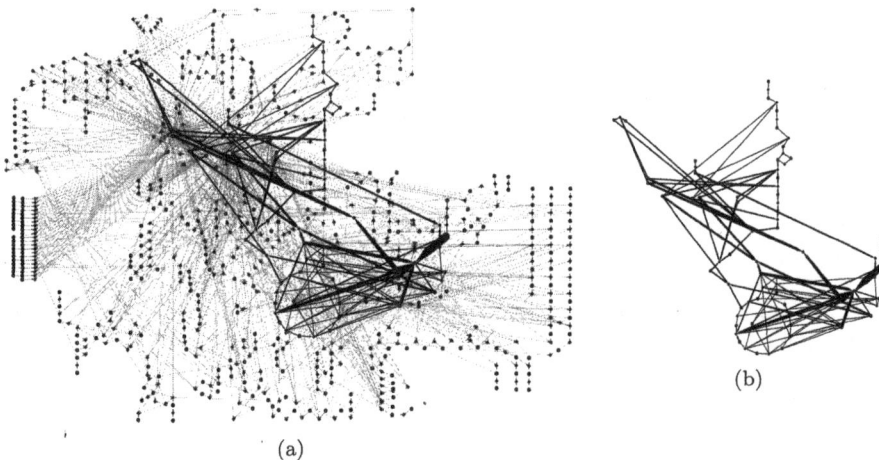

(a)

(b)

Fig. 3.   Filtering and subnetwork features. For the analysis of flux distributions in medium to large-scale networks limitation of network views based on selected attributes is crucial. (a) Flux distribution in the full network (b) flux subnetwork with much lower complexity. Flux distribution is FASIMU FBA result in a reconstructed human hepatocyte network (supplementary information). Flux Minimization [4] was used as objective function in the analysis of glucose formation from oxalacetate. Manually generated Layout, linear (linear flux  edge weight) mapping function. Subnetwork is flux subnetwork and consists of edges and nodes with non-zero flux. FluxViz subnetworks can be generated based on arbitrary network attributes like flux values or node localization.

## 3. Summary

FluxViz provides the necessary tools for the visualization of flux distributions in networks. FluxViz is a visualization tool solely based on network structure and flux data and is therefore independent of used simulation platforms or experimental setups which generate the flux distributions. As a result FluxViz can easily be integrated in existing simulation workflows. Due to features like attribute- and flux-subnetworks and consistent layouts for the loaded networks even genome-scale networks can be analyzed and simulations under varying conditions be compared. Especially for large networks a visual inspection of flux simulation results is an important step in the validation of the network capabilities and the iterative improvement of network reconstructions and models. Due to comfortable mapping capabilities of additional data, flux simulations can be easily visualized in the context of additional information like gene expression values in combination with resulting fluxes in a network [16].

## References

[1] Bell, G.W. and Lewitter, F., Visualizing networks, *Methods Enzymol.*, 411:408–421, 2006.

[2] Gille, C., Boelling, C., Hoppe, A., Bulik, S., Hoffmann, S., Hübner, K., Karlstädt, A., Ganeshan, R., König, M., Rother, K., Weidlich, M., Behre, J., and Holzhütter, H.G., A metabolic network of the human hepatocyte to simulate liver functions, *Molecular Systems Biology*, submitted 2010.

[3] Grafahrend-Belau, E., Klukas, C., Junker, B.H., and Schreiber, F., FBA-SimVis: interactive visualization of constraint-based metabolic models, *Bioinformatics*, 25:2755–2757, 2009.

[4] Holzhütter, H.G., The principle of flux minimization and its application to estimate stationary fluxes in metabolic networks, *Eur. J. Biochem.*, 271:2905–2922, 2004.

[5] Hoppe, A., Hoffmann, S., König, M., Gerasch, A., Gille, C., and Holzhütter, H.G., FASIMU: flexible software for batch flux-balance computation in large metabolic networks, *Bioinformatics*, submitted 2010.

[6] Hu, Z., Hung, J., Wang, Y., Chang, Y., Huang, C., Huyck, M., and DeLisi, C., Tools for visually exploring biological networks, *Nucleic Acids Res.*, 37:W115–W121, 2009.

[7] Junker, B.H., Klukas, C., and Schreiber, F., VANTED: a system for advanced data analysis and visualization in the context of biological networks, *Bioinformatics*, 7:109, 2006.

[8] Kauffman, K.J., Prakash, P., and Edwards, J.S., Advances in flux balance analysis, *Curr. Opin. Biotechnol.*, 14:491–496, 2003.

[9] Killcoyne, S., Carter, G.W., Smith, J., and Boyle, J., Cytoscape: a community-based framework for network modeling, *Methods Mol. Biol.*, 563:219–239, 2009.

[10] Klamt, S., Saez-Rodriguez, J., and Gilles, E.D., Structural and functional analysis of cellular networks with CellNetAnalyzer, *BMC Syst. Biol.*, 1:2, 2007.

[11] Küntzer, J., Blum, T., Gerasch, A., Backes, C., Hildebrandt, A., Kaufmann, M., Kohlbacher, O., and Lenhof, H., BN$^{++}$ - a biological information system, *J. Integr. Bioinformatics*, 3(2):34, 2006.

[12] Lee, J.M., Gianchandani, E.P., and Papin, J.A., Flux balance analysis in the era of metabolomics, *Brief Bioinform.*, 7:140–150, 2006.

[13] Morgan, J.A. and Rhodes, D., Mathematical modeling of plant metabolic pathways, *Metab. Eng.*, 4:80–89, 2002.

[14] Niittylae, T., Chaudhuri, B., Sauer, U., and Frommer, W.B., Comparison of quantitative metabolite imaging tools and carbon-13 techniques for fluxomics, *Methods Mol. Biol.*, 553:355–372, 2009.

[15] Schwarz, R., Liang, C., Kaleta, C., Kühnel, M., Hoffmann E., Kuznetsov, S., Hecker, M., Griffiths, G., Schuster, S., and Dandekar, T., Integrated network reconstruction, visualization and analysis using YANAsquare, *BMC Bioinformatics*, 8:313, 2007.

[16] Shlomi, T., Cabili, M.N., Herrgård, M.J., Palsson, B.Ø., and Ruppin, E., Network-based prediction of human tissue-specific metabolism, *Nat. Biotechnol.*, 26:1003–1010, 2008

[17] Snitkin, E.S., Dudley, A.M., Janse, D.M., Wong, K., Church, G.M., and Segrè D., Model-driven analysis of experimentally determined growth phenotypes for 465 yeast gene deletion mutants under 16 different conditions, *Genome Biol.*, 9:R140, 2008.

[18] Suderman, M. and Hallett, M., Tools for visually exploring biological networks, *Bioinformatics*, 23:2651–2659, 2007.

# COMPREHENSIVE GENOMIC ANALYSIS OF SULFUR-RELAY PATHWAY GENES

MASAAKI KOTERA[1,§]
kot@kuicr.kyoto-u.ac.jp

TAKESHI KOBAYASHI[1,†,§]
koba@kuicr.kyoto-u.ac.jp

MASAHIRO HATTORI[1,¶]
hattori@kuicr.kyoto-u.ac.jp

TOSHIAKI TOKIMATSU[1]
tokimatu@kuicr.kyoto-u.ac.jp

SUSUMU GOTO[1]
goto@kuicr.kyoto-u.ac.jp

HISAAKI MIHARA[2,‡]
mihara@fc.ritsumei.ac.jp

MINORU KANEHISA[1,3,*]
kanehisa@kuicr.kyoto-u.ac.jp

[1] Bioinformatics Center, Institute for Chemical Research, Kyoto University, Uji, Kyoto, 611-0011, Japan.
[2] Laboratory of Molecular Microbial Science, Institute for Chemical Research, Kyoto University, Uji, Kyoto, 611-0011, Japan.
[3] Human Genome Center, Institute of Medical Science, University of Tokyo, Minato-ku, Tokyo, 108-8639, Japan.
†Present address: Matsumoto Yushi-Seiyaku Co., Ltd., Yao, Osaka, 581-0075, Japan.
¶Present address: School of Bioscience and Biotechnology, Tokyo University of Technology, Hachioji City, Tokyo, 192-0982, Japan.
‡Present address: College of Life Sciences, Ritsumeikan University, Kusatsu, Shiga, 525-8577, Japan.
§These authors contributed equally.
*To whom correspondence should be addressed.

Many cofactors and nucleotides containing sulfur atoms are known to have important functions in a variety of organisms. Recently, the biosynthetic pathways of these sulfur-containing compounds have been revealed, where many enzymes relay sulfur atoms. Increasing evidence also suggests that the prokaryotic sulfur-relay enzymes might be the evolutionary origin of ubiquitination and the related systems that control a wide range of physiological processes in eukaryotic cells. However, these sulfur-relay enzymes have been studied in only a small number of organisms. Here we carried out comparative genomic analysis and examined the presence and absence of sulfurtransferases utilized in the biosynthetic pathways of molybdenum cofactor (Moco), 2-thiouridine ($S^2U$), and 4-thiouridine ($S^4U$), and IscS, a cysteine desulfurase. We found that all eukaryotes and many other organisms lack the intermediate enzymes in $S^2U$ biosynthesis. It is also found that most genes lack rhodanese homology domain (RHD), a catalytic domain of sulfurtransferase. Some organisms have a conserved sequence composed of about 100 residues in the C terminus of TusA, different from RHD. Host-associated organisms have a tendency to lose Moco biosynthetic enzymes, and some organisms have MoaD-MoaE fusion protein. Our findings suggest that sulfur-relay pathways have been so diversified that some putative sulfurtransferases possibly function in other unknown pathways.

Keywords: sulfur relay; sulfurtransferase; cysteine desulfurase; Moco; rhodanese homology domain (RHD); ubiquitin.

# 1. Introduction

Sulfur is one of the essential elements, incorporated in proteins as thiol groups of cysteine residues as the result of anabolic reduction of sulfate. Sulfur also plays important roles in some cofactors such as molybdenum cofactor (Moco) and in modifications of tRNAs such as 2-thiouridine ($S^2U$) and 4-thiouridine ($S^4U$) [24]. The biosynthetic pathways of Moco, $S^2U$, and $S^4U$ have been unveiled and are found to have some features in common (Fig. 1). At the beginning of all the pathways, IscS, a cysteine desulfurase, receives a sulfur atom by the conversion of cysteine to alanine. Consequently, this sulfur atom is relayed to various enzymes, and is finally incorporated into the final product.

Moco is a complex of a molybdenum atom and molybdopterin, and activates molybdenum enzymes such as sulfite oxidase (SO) and xanthine dehydrogenase (XDH). Most molybdenum enzymes are related to metabolism of carbon, nitrogen and sulfur, indicating their importance in organisms. For example, deficiency of SO causes neurologic abnormalities and dimorphic features of the brain and head, and the increased XDH activity causes hyperuricemia [22]. $S^2U$ and $S^4U$ are thiolated uridines found in tRNAs. $S^2U$ is found at position 34 in the anticodon of tRNAs specific for lysine, glutamine and glutamate and has a role in correct codon recognition. Most organisms except *Mycoplasma* are known to have $S^2U$. $S^4U$ is found at position 8 in tRNAs. $S^4U$ undergoes a photoinduced cross-linking reaction when the bacterium is exposed to UV radiation that stops protein synthesis and allows bacteria to repair DNA damage.

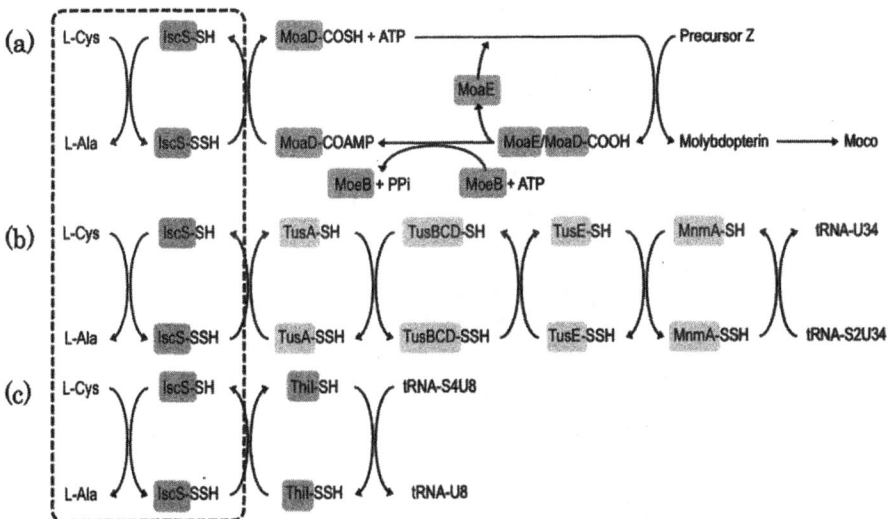

Figure 1. The sulfur-relay pathways for the synthesis of (a) Moco, (b) $S^2U$ and (c) $S^4U$. Three major pathways comprising the sulfur relay systems are shown here. (a) In Moco biosynthesis, MoaD, MoaE and MoeB play a key role in activating a Precursor Z into Molybdopterin. (b) In $S^2U$ biosynthesis the sulfur relay is carried out by Tus family and MnmA. Here, TusB, TusC and TusD form one large complex. (c) The biosynthesis of $S^4U$ needs only ThiI protein. All three pathways have a common initiating step, the sulfur transition from L-Cys to IscS, indicated by the dashed box. After this step, each pathway relays the sulfur atom to the end product following a different pathway.

In the past few years, some of these sulfur-relay enzymes in prokaryotes were found to have unexpected similarities with enzymes of ubiquitin systems in eukaryotes. MoeB, an *Escherichia coli* protein required for biosynthesis of Moco, was found to have weak but significant sequence similarity to E1, a ubiquitin activating enzyme [21]. MoeB forms complex proteins with MoaD, where MoeB transfers AMP onto the C-terminal of glycine carboxylate of MoaD. An adenylated MoaD receives a sulfur atom from IscS to form a thiocarboxylated MoaD. This uses the same mechanism as the ubiquitin-E1 complex, where E1 transfers AMP onto the C-terminal of glycine carboxylate of ubiquitin, then the adenylated ubiquitin forms a thioester bond with E1. These pathways are also preserved in other ubiquitin-like protein systems, suggesting the possibility that the eukaryotic ubiquitin and ubiquitin-like systems have evolved from prokaryotic sulfur-relay systems [11,12,17].

Despite the vigorous investigations mentioned above, studies on the sulfur-relay pathways are limited to a small number of organisms. In this study, we conducted comprehensive genomic analysis to create an overview of these pathways throughout all organisms with complete genome sequences.

## 2.    Data and Methods

We utilized complete genomes of 737 organisms including 60 eukaryotes, 625 prokaryotes, and 52 archaea, and 3,048,661 genes in the GENES section of the KEGG (Kyoto Encyclopedia of Genes and Genomes) database (Release 46.0), which compiles genetic information from organisms with sequenced genomes well annotated based on

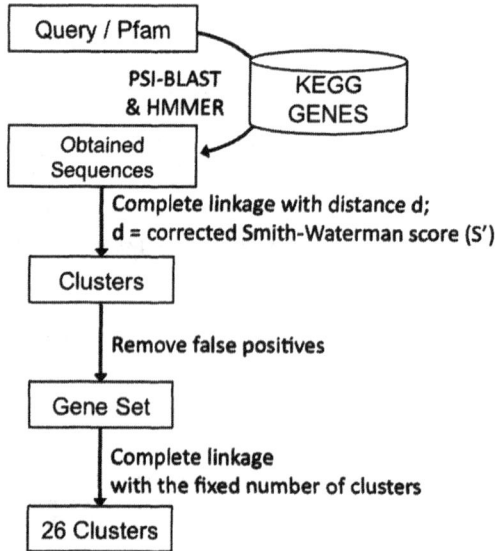

Figure 2. Schematic diagram for dataset construction and clustering.

the KO (KEGG Orthology) system [18]. Based on the phylogenetic relationships, the 737 organisms are further divided into 34 taxa (4, 14 and 4 taxa from eukaryotes, prokaryotes and archaea, respectively; see Supplementary material), of which we used the 26 taxa (4, 20 and 2 taxa, respectively; see Table 1) that contain at least 5 organisms each. The biosynthetic pathways and related enzymes examined in this study include: *(a)* MoaD, MoaE, and MoeB in Moco biosynthesis, *(b)* TusA, TusB, TusC, TusD, TusE and MnmA in $S^2U$ biosynthesis, and *(c)* ThiI in $S^4U$ biosynthesis (Fig. 1). We also examined the presence of cysteine desulfurases IscS and MJ1025, as well as Ncs6, which is thought to relate to $S^2U$ biosynthesis in eukaryotes and archaea.

E. *coli* genes were used as queries, and each gene family was processed independently in the following steps. 1,039 IscS orthologs were obtained simply by using BLAST against KEGG GENES with an e-value of $\leq 10^{-4}$. Other gene families needed the following two different methods to collect putative genes related to the sulfur relay pathways (Fig. 2). The first one is to use PSI-BLAST [1] against KEGG GENES with an e-value of $\leq 10^{-4}$, and the second is HMMER [6] with an e-value of $\leq 10^{-2}$. The Pfam entries used in HMMER include *(a)* ThiS and MoaE, *(b)* tRNA_Me_trans, ATP_bind_3, SirA, DsrH, DsrE, DsrE and DsrC, and *(c)* ThiI. Some remote homologs could not be detected with these methods for several organisms.

Table 1. The numbers of genes and organisms in the final dataset. Each number represents the number of genes of each gene family found within each taxon. Hyphens indicate that no relevant gene was found. 723 out of 737 organisms included in KEGG are reorganized into 26 taxa in which each taxon is required to have more than 4 organisms. Under this condition, 14 organisms are isolated from any taxa. These are shown in the last row under "Others" and were excluded from further analyses.

| Organism taxon | # of organisms | IscS | MoaD | MoaE | MoeB | TusA | TusB | TusC | TusD | TusE | MnmA | ThiI |
|---|---|---|---|---|---|---|---|---|---|---|---|---|
| Animals | 19 | 18 | 7 | 16 | 17 | 1 | - | - | - | - | 13 | - |
| Plants | 6 | 6 | 6 | 6 | 6 | - | - | - | - | - | 6 | - |
| Fungi | 21 | 13 | 6 | 8 | 12 | - | - | - | - | - | 5 | - |
| Protists | 14 | 13 | 1 | 2 | 14 | - | - | - | - | - | 8 | 2 |
| Gamma/enterobacteria | 51 | 51 | 42 | 43 | 43 | 47 | 47 | 48 | 48 | 49 | 51 | 45 |
| Gamma/others | 107 | 93 | 76 | 80 | 68 | 79 | 56 | 67 | 77 | 83 | 105 | 66 |
| Beta | 53 | 52 | 43 | 44 | 3 | 50 | 1 | 2 | 5 | 1 | 52 | - |
| Epsilon | 19 | 18 | 17 | 14 | 6 | 14 | - | - | - | - | 19 | 2 |
| Delta | 18 | 18 | 15 | 12 | 9 | 15 | - | 2 | 4 | 8 | 18 | - |
| Alpha/rickettsias | 23 | 21 | - | 1 | 9 | - | - | - | - | - | 23 | - |
| Alpha/rhizobacteria | 37 | 37 | 33 | 33 | 37 | 32 | - | 1 | 1 | - | 37 | - |
| Alpha/others | 24 | 23 | 23 | 23 | 24 | 17 | 1 | 1 | 1 | 1 | 23 | - |
| Bacillales | 45 | 45 | 43 | 43 | 44 | 38 | - | 2 | 2 | - | 45 | 44 |
| Lactobacillales | 49 | 46 | 2 | 2 | 2 | 5 | - | - | - | - | 46 | 44 |
| Clostridia | 30 | 30 | 14 | - | 16 | 25 | 2 | 3 | 5 | 6 | 30 | 16 |
| Mollicutes | 19 | - | - | - | - | - | - | - | - | - | 19 | 16 |
| Actinobacteria | 52 | 51 | 7 | 46 | 48 | 14 | - | 4 | 6 | 3 | 50 | 5 |
| Chlamydia | 13 | 13 | - | - | - | - | - | - | - | - | 13 | - |
| Spirochete | 10 | 5 | - | 1 | 1 | 1 | - | - | - | - | 9 | 2 |
| Cyanobacteria | 32 | 32 | 17 | 20 | 32 | 17 | - | 1 | 3 | - | 32 | - |
| Bacteroides | 12 | 6 | 2 | 3 | 10 | 1 | - | - | 1 | - | 12 | - |
| Green sulfur bacteria | 5 | 5 | - | - | 5 | 5 | 5 | 5 | 5 | 5 | 5 | - |
| Green nonsulfur bacteria | 7 | 6 | 4 | 4 | 4 | 3 | - | - | - | - | 7 | - |
| Hyperthermophilic bacteria | 7 | 5 | 2 | 1 | 3 | 4 | 3 | 1 | 4 | - | 7 | 4 |
| Euryarchaeota | 34 | 14 | 23 | 32 | 30 | 21 | 5 | 15 | 16 | 1 | - | 33 |
| Crenarchaeota | 16 | 1 | 13 | 13 | 15 | 10 | - | 7 | 7 | 6 | 1 | 14 |
| Others | 14 | 18 | 10 | 10 | 13 | 4 | 1 | 2 | 2 | 2 | 20 | 5 |
| Total number of genes | - | 1039 | 483 | 505 | 591 | 675 | 122 | 178 | 289 | 186 | 710 | 308 |

The obtained gene set inevitably included many false positive sequences, which were removed by hierarchical clustering as follows. First, all vs. all sequence alignments were performed for each gene family in the gene set to calculate the Smith-Waterman scores ($S$) [30]. We used BLOSUM50 [10] as the amino acid substitution matrix, 12 as the gap-opening cost, and 2 as the gap-extension cost. To remove the high dependency on the sequence length, this score was normalized ($S'$) with the method described in [2]. The reciprocal of the normalized score was defined as the distance ($d$) between two genes, which was used to calculate complete linkage clustering. We set the threshold so that genes were regarded as false positives and removed when their annotation were not related to the sulfur transferases.

Existence of Rhodanese homology domain (RHD) was also investigated. RHD is involved in the various transferase catalytic domains and are so diverse that some RHD sequences are not detected using a simple motif search. Therefore, we determined the existence of RHD using a combination of multiple sequence alignments with CLUSTALW [32] and a motif search with HMMER.

## 3.   Results

As a result of homology search using BLAST and HMMER, we obtained 7,015 homologous genes in total. After the removal of the genes having the annotations different from sulfertransferases, 4,201 genes were obtained in total (Table 1). In order to clarify the relationship between taxa and the types óf gene families, each gene family in the final gene set was divided using complete linkage clustering. The threshold was adjusted so that each gene family was divided into 26 clusters, corresponding to the number of taxa we defined. The presence/absence of the gene families is described in Fig. 3, and the numbers of organisms with each gene in each taxon is shown in Table 1 (see the supplementary material http://web.kuicr.kyoto-u.ac.jp/koba/ for the detail). The family analysis revealed that many clusters fit well into a taxon or into several related taxa. For example, many genes in the alpha-proteobacteria formed distinct clusters. Another example is the ThiI gene family, which formed large two clusters: one including the Gamma-proteobacteria and the other including *Firmicute*. Similarly, most eukaryotes lack ThiI. On the other hand, in *Bacillales*, MoaD and TusA are each divided among two clusters, while other gene families belong to discrete clusters.

Rhodanese homology domain (RHD) functions to transfer sulfur atoms by way of an intermediate persulfide on the active-site cysteine residue [24]. RHD of ThiI is thought to be essential in biosynthesis of $S^4U$, and ThiI without RHD is suggested to act with other enzymes having RHD [4,16,20,35]. Its presence/absence is shown in Fig. 4(a). Some organisms in evolutionarily distant lineages have RHD in MoeB, while only some Gamma-proteobacteria have RHD in ThiI, and only Bacillus have RHD in TusA. Interestingly, although RHD generally exists in the C-terminus, MoeB of *Xantomonas* and *Xylella*, which are members of Gamma/others and lack other gene families, have RHD in their N-terminus.

We found that most organisms have IscS, the cysteine desulfurase shared in the three pathways. Some prokaryotes such as *Mollicute*, *Xylella*, and *Xantomonus* and many

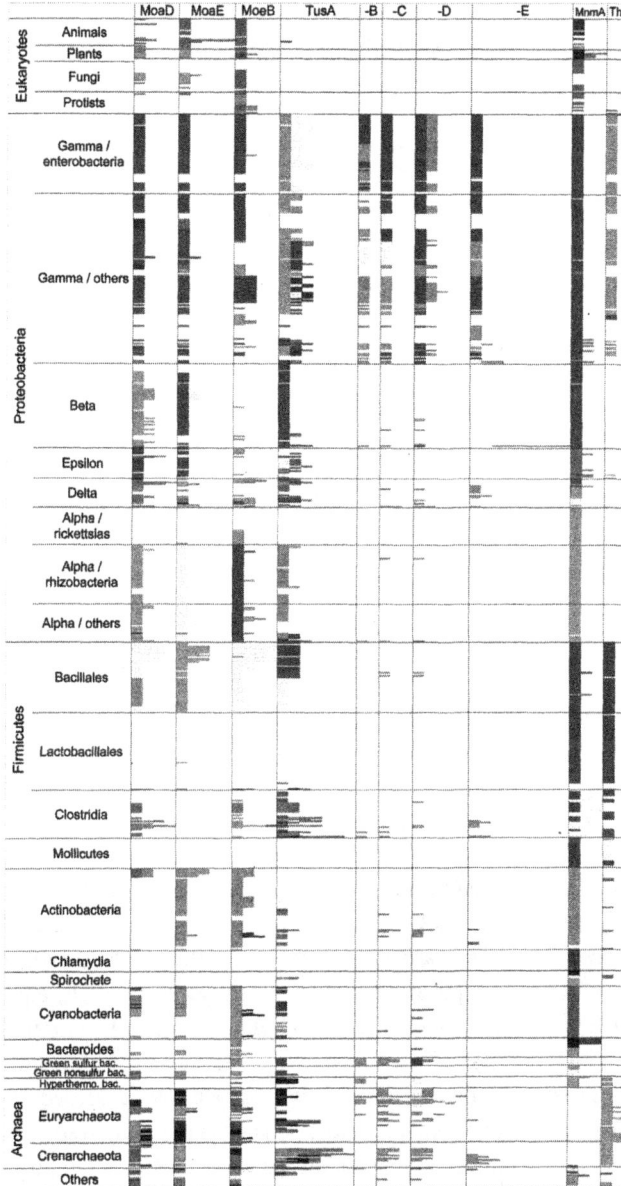

Figure 3. The resulting 26 clusters of each gene family. The columns and rows represent the gene families and organisms, respectively. TusB, TusC, TusD and TusE are abbreviated as -B, -C, -D and -E, respectively, due to space limitations. In each organism multiple columns in each gene family represents multiple putative genes. The top nine largest clusters in each gene family are colored independently to emphasis which taxa have been classified into larger clusters. If the number of members of a cluster is less than ten or if the cluster is not one of the top nine largest, then relevant genes are colored in gray. High resolution color graphics can be obtained at http://web.kuicr.kyoto-u.ac.jp/supp/koba/figure3.pdf

archaea lack IscS. We investigated the possibility of alternative enzymes. Cysteine desulfurases are classified into group I (containing IscS) and group II (containing SufS) [23], and it is reported that SufS may play the same role as IscS in sulfur transfer in Moco biosynthesis [19]. Also, another group of cysteine desulfurase MJ1025 has been recently isolated from *Methanococcus jannaschii* [31], suggesting that MJ1025 takes the role of IscS in the three pathways. Fig. 4(b) shows the presence/absence of the cysteine desulfurases group I, group II and MJ1025 homologs. It is revealed that *Mollicute* and *Xantomonas* have group II instead of group I, while they also have Moco and $S^4U$ biosynthetic enzymes, respectively (as shown in Fig. 3), suggesting that the group II enzymes work in these organisms. On the other hand, in archaea, there are several organisms without group I, group II, or MJ1025. In other words, the sulfur-relay pathways in archaea may not function in these organisms, or different types of enzymes may take the place of IscS.

We also found that most organisms lack enzymes in the intermediate stages (TusB, TusC, TusD and TusE) of $S^2U$ biosynthesis, where TusB, TusC and TusD function as a protein complex. Ikeuchi *et al.* [15] conducted a series of knockout experiments of $S^2U$ biosynthesis enzymes in *E. coli*, and reported that $S^2U$ was synthesized *in vivo* even without the intermediate step enzymes although it was inefficient. They also reported that a knockout mutant of TusA, TusB, TusC and TusD, as well as MnmA- mutant could not synthesize $S^2U$. Our study indicates that more important gene families are well conserved in the process of evolution.

## 4.   Discussion

Recently, it has been shown that *Saccharomyces cerevisiae* Ncs6 protein is related to the last step in the $S^2U$ biosynthetic pathway, and its involvement in eukaryotes and archaea $S^2U$ biosynthesis is suggested [26]. We conducted the comparisons of the presence/absence of MnmA and Ncs6. As shown in Fig. 4(c), MnmA is found in eukaryotes and prokaryotes, while Ncs6 is found in eukaryotes and archaea. Therefore, it seems that in the $S^2U$ biosynthetic pathway MnmA is utilized in prokaryotes, and Ncs6 is utilized in eukaryotes and archaea. Mtu1, a MnmA-like protein found in *Homo sapiens* and *Saccharomyces cerevisiae*, are known to be related to $S^2U$ biosynthesis only in mitochondria, and not in cytoplasm [34]. Some archaea have Ncs6, and more than one Tus protein (Fig. 3). Although *Mycoplasma* do not have $S^2U$ [3], we found that they have MnmA-like proteins (Fig. 3), which are in the same cluster as MnmA in *E. coli* with experimentally confirmed function. In addition, they have conserved cysteine residues involved in transfer of sulfur atom. Therefore, it is possible that these proteins are related to sulfur transfer in other biosynthetic pathways. The cluster including TusA genes in Epsilon-proteobacteria, Delta-proteobacteria, and *Clostridia* has a conserved sequence motif with about 100 amino acids in C terminus, different from RHD. The function of these proteins remains unclear, but we assume this is worth investigating because the motif is highly conserved (45 out of 46 genes have this motif).

Most organisms having Moco biosynthetic enzymes also were found to have molybdenum enzymes. The preceding study [36] has already reported that *Rickettiases,*

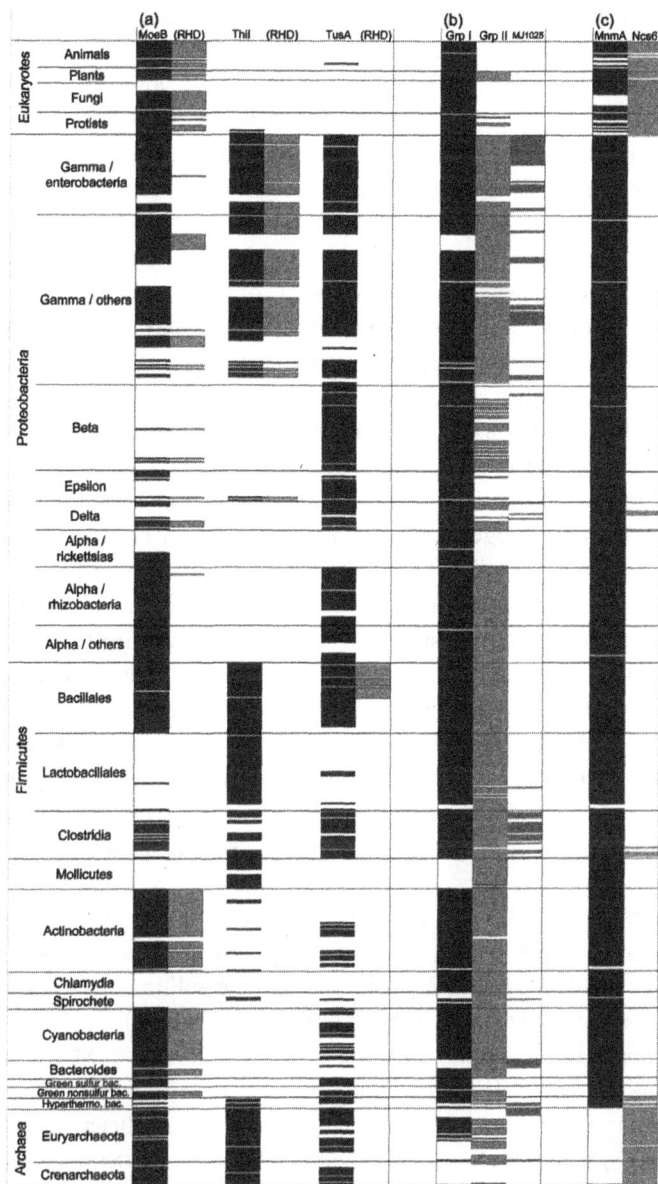

Figure 4. The presence/absence of (a) three gene families: MoeB, ThiI, and TusA, (b) the three types of cysteine desulfurases: group I, group II, and MJ1025, (c) two gene families: MnmA and Ncs6. (a) Presence of MoeB, ThiI and TusA is represented in red in the left column. These three gene families are thought to have the RHD domain. Thus, the presence of the RHD domain within each gene family is also indicated in green. (b) The presence of Grp1, Grp2, and MJ1025 homologs are depicted. (c) The results for MnmA and Ncs6. High resolution color graphics are available at http://web.kuicr.kyoto-u.ac.jp/supp/koba/figure4.pdf

*Mollicute, Chlamydia, Lactobacillus* and *Saccharomycetes* do not have sulfurtransferases related to Moco biosynthesis, of which *Saccharomyces cerevisiae* is experimentally confirmed not to have these enzymes [33]. In this study, we confirmed the results and also found that neither do *Buchnera, Francisella* and *Neisseria*. It should be noted that most of these organisms except *Saccharomycetes* are host-associated and evolutionarily distant lineages, suggesting that these organisms have lost the Moco biosynthesis pathway independently in the process of relying increasingly on their hosts. We also found a cluster including a MoaD-MoaE fusion gene in some *Actinobacteria* and archaea, which are distantly related organisms. The fusion genes are found only in this cluster.

One of the difficulties in analyzing the sulfur-relay pathways is that there are many cases 1) where alternative enzymes compensate for the pathway, 2) where ortholog enzymes have different targets, and 3) where some of them are bifunctional. As discussed above, there are three types of cysteine desulfurases that may take the same function of IscS, and there are some possible alternative enzymes for MnmA. Although MoeB functions in Moco biosynthesis in *E. coli*, it is reported that TtuC, a MoeB-like protein of *Thermus thermophilus*, is involved in various reactions producing sulfur-containing molecules such as Moco, thiamine and 2-thioribothymidine ($S^2T$), a thiolated ribothymidine at position 54 in tRNA [29]. We found that TtuC was classified in the same cluster with MoeB, suggesting that some genes in those clusters may be involved in thiamine and $S^2T$ biosynthesis. This kind of diversification may have occurred more frequently than previously expected in sulfur-relay pathways, since there are many other cases, such as thiamine biosynthesis in *Saccharomyces serevisiae* [7,9], thiocarboxylate derivatization of CysO in *Mycobacterium tuberculosis* [5], and biosynthesis of sulfur-containing thioquinolobactin siderophore in *Pseudomonas fluorescens* [8]. We found that many organisms lack the intermediate step enzymes in the $S^2U$ biosynthesis pathway. However, this alone does not prove that these organisms lack the pathway, partly because a small amount of $S^2U$ synthesis has been observed *in vitro* without the intermediary step enzymes in *E. coli*, but also because the existence of alternative enzymes are still possible.

Recent research indicates that the diversification of sulfurtransferases might be one of the evolutionary origins of ubiquitin and related systems in eukaryotes [11]. A good example of the potential evolutionary link between the sulfurtransferases and ubiquitin-related systems is a system composed of ubiquitin-related modifier 1 (Urm1) [7,9], which is similar to MoaD and ubiquitin, and Uba4, which is similar to MoeB and E1. Urm1 thiolates uridine to form $S^2U$ in the anticodon of certain tRNAs [14,25]. Urm1-Uba4 system was first found in *Saccharomyces cerevisiae*, and functions as part of an ubiquitin-related system. Uba4 has an E1-like domain and a C-terminal RHD, which is similar to many MoeB familiy proteins, although E1s for other ubiquitin-like proteins do not contain RHD. Uba4 is reported to be a bifunctional enzyme with sulfurtransferase activity [28]. The activity does not require the cysteine residue in the E1 domain *in vitro*, but the residue is necessary for conjugation to Urm1 *in vivo* [7]. In contrast, the cysteine residue in the RHD is required for the Urm1 to function [13]. The E1-like superfamily enzymes catalyze adenylation for the activation of proteins such as ubiquitin, the mechanism of which is analogous to the activation of amino acids by aminoacyl-tRNA synthases [27]. Hochstrasser [11] suggests the possibility that the sulfurtransferase activity by the RHD has become the side-reaction or has been eliminated from the E1 superfamily during the evolution of ubiquitin-related systems.

We also investigated the presence/absense of RHD in various sulfurtransferases, finding that most organisms lack RHD in MoeB, ThiI and TusA. It is suggested that ThiI without RHD may act with other enzymes having RHD [4,16,20,35]. In the case of MoeB, by analogy to the Urm1-Uba4 system, MoeB without RHD might act with other enzymes to modify particular proteins. Similary, MoaD-MoeE fusion protein may have a different function than sulfurtransferase activity. In general, MoaD and MoaE are close in genomic position, and these enzymes work as a complex. It may suggest the possibility that the fusion protein actually functions as a sulfurtransferase, although the fusion protein lacks the C-terminal in the MoaD subunit, which must be thiocarboxylated for MoaD to function. Our results also provide new starting points for further investigation of the evolutionary link between the sulfurtransferases and ubiquitin-related systems with regard to evolutionary relationships of interacting partners and ubiquitin deconjugating proteins.

## Acknowledgments

This work was supported by the Ministry of Education, Culture, Sports, Science and Technology of Japan, and the Japan Science and Technology Agency. Computational resources were provided by the Bioinformatics Centre and the Supercomputer Laboratory, Institute for Chemical Research, Kyoto University.

## References

[1]  Altschul, S.F., Madden, T.L., Schaffer, A.A., Zhang, J., Zhang, Z., Miller, W., and Lipman, D.J., Gapped BLAST and PSI-BLAST: A new generation of protein database search programs. *Nucleic Acids Res.*, 25(17):3389-3402, 1997.

[2]  Altschul, S.F. and Gish, W., Local alignment statistics. *Methods Enzymol.*, 266:460-80, 1996.

[3]  Björk, G.R., Huang, B., Persson, O.P., and Byström, A.S., A conserved modified wobble nucleoside (mcm5s2U) in lysyl-tRNA is required for viability in yeast. *RNA*, 13(8):1245-1255, 2007.

[4]  Bordo, D. and Bork, P., The rhodanese/Cdc25 phosphatase superfamily. Sequence-structure-function relations. *EMBO Rep.*, 3(8):741-746, 2002.

[5]  Burns, K.E., Baumgart, S., Dorrestein, P.C., Zhai, H., McLafferty, F.W., and Begley, T.P., Reconstitution of a new cysteine biosynthetic pathway in *Mycobacterium tuberculosis. J. Am. Chem. Soc.*, 127(33):11602-11603, 2005.

[6]  Eddy S.R., Profile hidden Marcov models. *Bioinformatics*, 14(9): 755-763, 1998.

[7]  Furukawa, K., Mizushima, N., Noda, T., and Ohsumi, Y., A protein conjugation system in yeast with homology to biosynthetic enzyme reaction of prokaryotes. *J. Biol. Chem.*, 275(11):7462-7465, 2000.

[8]  Godert, A.M., Jin, M., McLafferty, F.W., and Begley, T.P., Biosynthesis of the thioquinolobactin siderophore: an interesting variation on sulfur transfer. *J. Bacteriol.*, 189(7):2941-2944, 2007.

[9]  Goehring, A.S., Rivers, D.M., and Sprague, G.F. Jr., Attachment of the ubiquitin-related protein Urm1p to the antioxidant protein Ahp1p. *Eukaryot. Cell*, 2(5):930-936, 2003.

[10] Henikoff, S. and Henikoff, J.G., Amino acid substitution matrices from protein blocks. *Proc. Natl. Acad. Sci. USA*, 89(22):10915-10919, 1992.

[11] Hochstrasser, M., Origin and function of ubiquitin-like proteins. *Nature*, 458(7237):422-429, 2009.

[12] Hochstrasser, M., Evolution and function of ubiquitin-like protein-conjugation systems. *Nature Cell Biol.*, 2(8):E153-157, 2000.

[13] Hochstrasser, M., in *Protein Degradation: The Ubiquitin-Proteasome System* (eds Mayer, R. J., Ciechanover, A. & Rechsteiner, M.) 249-278 (Wiley, 2006).

[14] Huang, B., Lu, J., and Bystrom, A.S., A genome-wide screen identifies genes required for formation of the wobble nucleotide 5-methoxycarbonylmethyl-2-thiouridine in *Saccharomyces cerevisiae*. *RNA*, 14(10):2183-2194, 2008.

[15] Ikeuchi, Y., Shigi, N., Kato, J., Nishimura, A., and Suzuki, T., Mechanistic insights into sulfur relay by multiple sulfur mediators involved in thiouridine biosynthesis at tRNA wobble positions. *Mol. Cell*, 21(1):97-108, 2006.

[16] Iwata-Reuyl, D., An embarrassment of riches: the enzymology of RNA modification. *Curr. Opin. Chem. Biol.*, 12(2):126-133, 2008.

[17] Iyer, L.M., Burroughs, A.M., and Aravind, L., The prokaryotic antecedents of the ubiquitin-signaling system and the early evolution of ubiquitin-like beta-grasp domains. *Genome Biol.*, 7(7):R60, 2006.

[18] Kanehisa, M., Araki, M., Goto, S., Hattori, M., Hirakawa, M., Itoh, M., Katayama, T., Kawashima, S., Okuda, S., Tokimatsu, T., and Yamanishi, Y., KEGG for linking genomes to life and the environment. *Nucleic Acids Res.*, 36(Suppl. 1):D480-D484, 2008.

[19] Leimkühler, S. and Rajagopalan, K.V., A sulfurtransferase is required in the transfer of cysteine sulfur in the *in vitro* synthesis of molybdopterin from precursor Z in *Escherichia coli. J. Biol. Chem.*, 276(25):22024-22031, 2001.

[20] Matthies, A., Rajagopalan, K.V., Mendel, R.R., and Leimkühler, S., Evidence for the physiological role of a rhodanese-like protein for the biosynthesis of the molybdenum cofactor in humans. *Proc. Natl. Acad. Sci. USA*, 101(16) 5946-5951, 2004.

[21] McGrath, J.P., Jentsch, S., and Varshavsky, A., UBA1: an essential yeast gene encoding ubiquitin-activating enzyme. *EMBO J.*, 10 (1):227-236, 1991.

[22] Mendel, R.R. and Bittner, F., Cell biology of molybdenum. *Biochim. Biophys. Acta*, 763(7):621-635, 2006.

[23] Mihara, H., Kurihara, T., Yoshimura, T., Soda, K., and Esaki, N., Cysteine sulfinate desulfinase, a NIFS-like protein of *Escherichia coli* with selenocysteine lyase and cysteine desulfurase activities. Gene cloning, purification, and

characterization of a novel pyridoxal enzyme. *J. Biol. Chem.*, 272(36):22417-22424, 1997.

[24] Mueller, E.G., Trafficking in persulfides: delivering sulfur in biosynthetic pathways. *Nature Chem. Biol.*, 2(4):185-194, 2006.

[25] Nakai, Y., Nakai, M., and Hayashi, H., Thio-modification of yeast cytosolic tRNA requires a ubiquitin-related system that resembles bacterial sulfur transfer systems. *J. Biol. Chem.*, 283(41):27469-27476, 2008.

[26] Noma, A., Sakaguchi, Y., and Suzuki, T., Mechanistic characterization of the sulfur-relay system for eukaryotic 2-thiouridine biogenesis at tRNA wobble positions. *Nucleic Acids Res.*, 37(4):1335-1352, 2009.

[27] Pickart, C. M., Mechanisms underlying ubiquitination. *Ann. Rev. Biochem.*, 70:503-533, 2001.

[28] Schmitz, J., Chowdhury, M.M., Hänzelmann, P., Nimtz, M., Lee, E., Schindelin, H., and Leimkühler, S., The sulfurtransferase activity of Uba4 presents a link between ubiquitin-like protein conjugation activation of sulfur carrier proteins. *Biochemistry* 47(24):6479-6489, 2008.

[29] Shigi, N., Sakaguchi, Y., Asai, S., Suzuki, T., and Watanabe, K., Common thiolation mechanism in the biosynthesis of tRNA thiouridine and sulphur-containing cofactors. *EMBO J.* 27(24):3267-3278, 2008.

[30] Smith, T.F. and Waterman, M.S., Comparison of biosequences. *Adv. Appl. Math.*, 2(4):482-489, 1981.

[31] Tchong, S.I., Xu, H., and White, R.H., L-cysteine desulfidase: an [4Fe-4S] enzyme isolated from *Methanocaldococcus jannaschii* that catalyzes the breakdown of L-cysteine into pyruvate, ammonia, and sulfide. *Biochemistry*, 44(5):1659-1670, 2005.

[32] Thompson, J.D., Higgins, D.G., and Gibson T.J., CLUSTALW: Improving the sensitivity of progressive multiple sequence alignment through sequence weighting, position-specific gap penalties and weight matrix choice. *Nucleic Acids Res.*, 22(22):4673-4680, 1994.

[33] Truong, H.N., Meyer, C., and Daniel-Vedele, F., Characteristics of *Nicotiana tabacum* nitrate reductase protein produced in *Saccharomyces cerevisiae*. *Biochem. J.*, 278(2):393-397, 1991.

[34] Umeda, N., Suzuki, T., Yukawa, M., Ohya, Y., Shindo, H., Watanabe, K., and Suzuki, T., Mitochondria-specific RNA-modifying enzymes responsible for the biosynthesis of the wobble base in mitochondrial tRNAs. Implications for the molecular pathogenesis of human mitochondrial diseases. *J. Biol. Chem.*, 280(2):1613-1624, 2005.

[35] Waterman, D.G., Ortiz-Lombardia, M., Fogg, M.J., Koonin, E.V., and Antson, A.A., Crystal Structure of *Bacillus anthracis* ThiI, a tRNA-modifying Enzyme Containing the Predicted RNA-binding THUMP Domain. *J. Mol. Biol.*, 356(1):97-110, 2006.

[36] Zhang, Y. and Gladyshev, V.N., Molybdoproteomes and evolution of molybdenum utilization. *J. Mol. Biol.*, 379(4):881-899, 2008.

# PHYLOGENETIC ANALYSIS OF LIPID MEDIATOR GPCRs

SAYAKA MIZUTANI[1]

smizutan@kuicr.kyoto-u.ac.jp

MICHIHIRO TANAKA[1]

mtanaka@kuicr.kyoto-u.ac.jp

CRAIG E. WHEELOCK[2]

craig.wheelock@ki.se

MINORU KANEHISA[1,3]

kanehisa@kuicr.kyoto-u.ac.jp

SUSUMU GOTO[1]

goto@kuicr.kyoto-u.ac.jp

[1] *Bioinformatics Center, Institute for Chemical Research, Kyoto University, Gokasho, Uji, Kyoto 611-0021, Japan*
[2] *Department of Medical Biochemistry and Biophysics, Division of Physical Chemistry II, Karolinska Institutet, Stockholm, 171-77, Sweden*
[3] *Human Genome Center, Institute of Medical Science, University of Tokyo, 4-6-1 Shirokanedai, Minato-ku, Tokyo 108-8639, Japan*

Lipid mediator is the collective term for prostanoids, leukotrienes, lysophospholipids, platelet-activating factor, endocannabinoids and other bioactive lipids, that are involved in various physiological functions including inflammation, immune regulation and cellular development. They act by binding to their ligand-specific G-protein coupled receptors (GPCRs). Since 1990's a number of lipid GPCRs have been cloned in humans, with a few more identified in other vertebrates. However, the conservation of these receptors has been poorly investigated in other eukaryotes. Herein we performed a phylogenetic analysis by collecting their orthologs in 13 eukaryotes with complete genomes. The analysis shows that orthologs for prostanoid receptors are likely to be conserved in the 13 eukaryotes. In contrast, those for lysophospholipid and cannabinoid receptors appear to be conserved only in vertebrates and chordates. Receptors for leukotrienes and other bioactive lipids are limited to vertebrates. These results indicate that the lipid mediators and their receptors have coevolved with the development of highly modulated physiological functions such as immune regulation and the formation of the central nervous system. Accordingly, examining the presence and role of lipid mediator GPCR orthologs in invertebrate species can provide insight into the development of fundamental biological processes across diverse taxa.

*Keywords*: lipid mediators; eicosanoids; GPCRs; phylogenetic analysis; eukaryotes; invertebrates.

## 1. Introduction

Lipid mediator is a collective term for prostanoids, leukotrienes (LTs), lysophospholipids, platelet-activating factor (PAF), endocannabinoids (CB), and other bioactive lipids. These mediators have a range of biological activities and are important in multiple fundamental biological processes including reproduction, signaling processes, immune function and disease etiology and pathogenesis to name a few. Lipid mediators are synthesized from precursor membrane lipids (*e.g.*, arachidonic acid) and have relatively short half-lives, which limits their signaling functions to be autocrinic or paracrinic. These mediators act on Class I G protein-coupled receptors (GPCRs) [18]. Eicosanoids are a major class of lipid mediators that are the oxygenated products of arachidonic acid [17, 5]. Following its release from membrane phospholipids, arachidonic acid can be modified by 3 distinct enzymatic pathways including cyclooxygenase (COX), lipoxygenase (LOX) and cytochrome P450 (P450) pathways. Eicosanoids are further divided into subgroups based upon these pathways, with prostaglandins (PGs) being synthesized via COX-dependent pathways in combination with terminal synthases and LTs being synthesized via LOX-dependent pathways. The different eicosanoids bind to distinct GPCRs that demonstrate receptor-specific binding affinities. There are currently 5 known prostanoid receptors: EP FP, DP, IP and TP, which preferentially respond to PGE2, PGF2α, PGD2 and PGI2 and thromboxane A2 (TXA2), respectively. The EP group is further divided into 4 subtypes termed EP1-4. The physiological effects of each PG have been well-studied in mammals [5]. PGE2 and PGI2 induce several inflammatory responses including fever, pain, and vasodilatation. PGF2α and PGE2 have essential roles in reproduction including oocyte maturation and uterine contraction in mammals [22]. TXA2 acts in platelet aggregation.

To date, four different LT receptors have been identified: cysteinyl leukotriene receptor 1 and 2 (CysLT1/2) and leukotriene B4 (LTB4) receptor 1 and 2 (BLT1/2). The primary ligands for CysLT1/2 are the glutathione peptide conjugated LTs (LTC4, LTD4 and LTE4) and the main ligand for the BLT1/2 is LTB4. However, recently, a COX-catalyzed product 12-HHT (12S-hydroxy-5Z,8E,10E-heptadecatrienoic acid) was found to be a ligand of BLT2 [13]. Leukotriene receptors are highly expressed in neutrophils and eosinophils, and are involved in the chemotaxis of these cells to the inflamed site. Recently it has been elucidated that they are also expressed in immune cells, regulating immune responses [24]. Lysophospholipids are structurally categorized into lysophosphatidic acids (LPAs) and sphingosine 1-phosphates (S1Ps), both of which generate a large number of molecular species depending on the *sn-1* and *sn-2* positions of the lipid architecture and the number and the positions of double bonds in the acyl chains. Recently 6 distinct receptors for LPAs, LPA1-6, and five for S1Ps, S1P1-5, have been identified. Lysophospholipids are involved in the regulation of cellular responses including cell proliferation, differentiation, migration, adhesion and morphogenesis. Several pathological studies have revealed their essential roles in the formation of the central nervous system, angiogenesis and the regulation of vascular contraction [2, 11, 19]. S1Ps are also involved in the regulation of lymphocyte migration [14]. Finally, two CB receptors, CB1 and CB2, have been identified; CB1 is expressed mainly in the central nervous system and CB2 in immune cells [18].

Although the functional roles of these lipid mediators have been intensively revealed in humans and rodents, much less is known in other eukaryotes, especially in invertebrates. In order to understand the roles of bioactive lipids in an evolutionary aspect, the invertebrate conservation of these signaling molecules should be elucidated. It has been suggested that the production of some lipid mediators is widely conserved in invertebrates. Genes in the biosynthesis of some lipid mediators have been identified in a few invertebrate species [15]. However, most of them have not been elucidated yet. In addition, those of the receptors are poorly understood. To address this point, we performed a phylogenetic analysis to investigate the conservation of lipid mediator GPCRs in 13 eukaryotic organisms with complete genomes covering a wide range of invertebrate phyla. Our results show an asymmetric conservation pattern between prostanoid receptors and other lipid mediator GPCRs in invertebrates. This can be the first step to correlate the emergence of lipid signaling functions and the development of physiological activities organisms have gained in the evolutionary course.

## 2.   Materials and Methods

### 2.1.   *Obtaining sequences of experimentally characterized lipid mediator GPCRs*

Forty-six amino acid sequences (26 in *H. sapiens*, 10 in *M. musculus*, seven in *R. norvegicus*, two in *X. laevis* and one in *D. rerio*) of experimentally annotated lipid mediator GPCRs were obtained from the UniProt database [25]. The UniProt IDs and annotations of the query sequences are listed in Table 1.

### 2.2.   *BLAST search for lipid mediator GPCR candidates*

Using the query sequences we performed a BLAST search [1] for 16 eukaryotic organisms with complete genomes, *H. sapiens* (hsa), *M. musculus* (mmu), *R. norvegicus* (rno), *X. laevis* (xla), *D. rerio* (dre), *B. floridae* (bfo), *C. intestinalis* (cin), *S. purpuratus* (spu), *D. melanogaster* (dme), *C. elegans* (cel), *B. malayi* (bmy), *N. vectensis* (nve), *T. adhaerens* (tad), *A. thaliana* (ath), *S. cerevisiae* (sce) and *M. brevicollis* (mbr). Amino acid sequences of the 16 organisms were downloaded from the KEGG GENES database [9]. Since a large part of the search results consisted of olfactory receptor sequences, each of the 46 search results was given a specific E-value threshold (listed in Table 1) so that BLAST hits whose E-values are higher than those of annotated olfactory receptors can be removed. This is based on an assumption that candidate lipid GPCRs orthologs should have higher degrees of similarity with the query lipid mediator GPCRs than with olfactory receptors.

### 2.3.   *Identifying ortholog candidates using phylogenetic trees*

For each of the organisms the refined search results were merged with the 46 lipid query sequences. The merged sequences were used to perform a multiple sequence alignment with the program E-INS-I in MAFFT version 5.8 [10] and to construct a rooted phylogenetic tree using the neighbor-joining (NJ) method [16] using the program

QuickTree [7] with bootstrap values. Nine human olfactory receptor sequences were used as an outgroup in order to identify the root of the tree. In order to identify ortholog candidates, query sequences were grouped into seven classes, prostanoid receptors (PG), leukotriene receptors (LT), fatty acid receptors (FA), lysophospholipid receptors (LPA & S1P), cannabinoid receptors (CB), platelet-activating factor receptors (PAF) and bile acid receptors (BA). In this process subsets that belong to the same class, such as receptors for PGE2 and PGF2α in the prostanoid receptor class (PG), were not distinguished from each other. Since the BLAST search results contained many non-lipid GPCRs whose similarities are higher than olfactory receptors, we also considered 16 known human non-lipid receptors, which were also in the BLAST hits, so that the annotated part of the search results were increased. On the constructed trees ortholog candidates of the query sequences were manually extracted by the procedures (i) and (ii).

(i)      For each class of queries, define ortholog cluster(s) via the following two steps;

        Step 1.  For each query, if the nearest query node(s) belongs to the same class, all the internal candidates are considered as orthologs.

        Step 2.  Repeat Step 1 until no such nearest queries are found.

(ii)     To expand each ortholog cluster defined in (i), candidates in the sibling cluster are tested for validity by BLAST search against the human genome; if queries in the defined ortholog cluster are listed as the top hits, the candidates are considered as orthologs.

Fig. 1 shows the overall flow chart of the methods described in the sections 2.1-2.3.

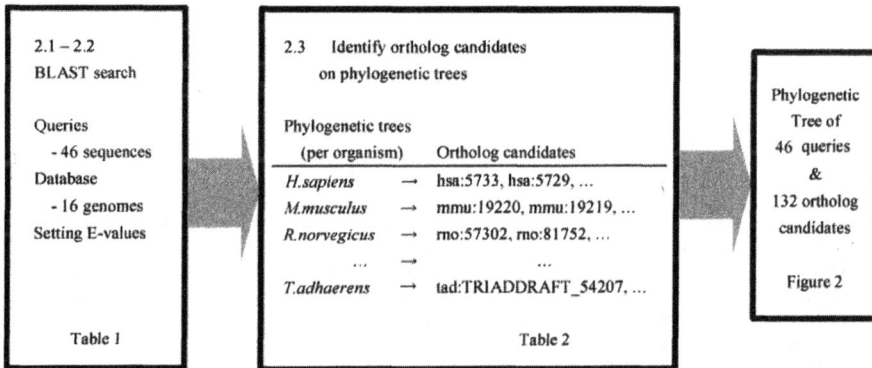

Fig 1. Flow chart of the overall method.

## 3.   Results

### 3.1. *BLAST search results*

The total of 3134 hits was obtained from the BLAST search results with selective E-values. The E-values and the number of BLAST hits for each query are listed in Table 1 .

Table 1. The number of BLAST hits for 46 query sequences with selective E-values

| UniProt ID | KEGG ID[a] | UniProt annotation | E-val[b] | # of BLAST hits | Pubmed ID |
|---|---|---|---|---|---|
| Prostanoid receptors | | | | | |
| P34995 | hsa:5731 | Prostaglandin E2 receptor EP1 subtype | - | 38 | 8253813 |
| P35375 | mmu:19216 | Prostaglandin E2 receptor EP1 subtype | - | 41 | 7690750 |
| P43116 | hsa:5732 | Prostaglandin E2 receptor EP2 subtype | - | 65 | 8078484 |
| Q62053 | mmu:19217 | Prostaglandin E2 receptor EP2 subtype | - | 42 | 7556658 |
| P35408 | hsa:5734 | Prostaglandin E2 receptor EP4 subtype | $10^{-1}$ | 829 | 8163486 |
| P43114 | rno:84023 | Prostaglandin E2 receptor EP4 subtype | - | 942 | 8185583 |
| Q9R261 | rno:498475 | Prostaglandin D2 receptor | $10^{-1}$ | 81 | 10448933 |
| P43118 | rno:25652 | Prostaglandin F2-alpha receptor | - | 124 | 7972878 |
| P43119 | hsa:5739 | Prostacyclin receptor | $10^{-2}$ | 73 | 7512962 |
| P43252 | mmu:19222 | Prostacyclin receptor | $10^{-2}$ | 91 | 7511597 |
| P43253 | rno:292661 | Prostacyclin receptor | $10^{-2}$ | 96 | 7803522 |
| P21731 | hsa:6915 | Thromboxane A2 receptor | - | 55 | 1825698 |
| P30987 | mmu:21390 | Thromboxane A2 receptor | $10^{-1}$ | 45 | 1375456 |
| P34978 | rno:24816 | Thromboxane A2 receptor | - | 56 | 7635958 |
| Leukotriene receptors | | | | | |
| Q15722 | hsa:1241 | Leukotriene B4 receptor 1 | $10^{-1}$ | 1116 | 8702478 |
| Q9NPC1 | hsa:56413 | Leukotriene B4 receptor 2 | - | 237 | 10913346 |
| Q9Y271 | hsa:10800 | Cysteinyl leukotriene receptor 1 | $10^{-3}$ | 1225 | 10462554 |
| Q99JA4 | mmu:58861 | Cysteinyl leukotriene receptor 1 | $10^{-2}$ | 1956 | 11705452 |
| Q9NS75 | hsa:57105 | Cysteinyl leukotriene receptor 2 | $10^{-6}$ | 923 | 11093801 |
| Fatty acid receptors | | | | | |
| O14842 | hsa:2864 | Free fatty acid receptor 1 | - | 237 | 12496284 |
| Q76JU9 | mmu:233081 | Free fatty acid receptor 1 | $10^{-2}$ | 67 | 12629551 |
| O15552 | hsa:2867 | Free fatty acid receptor 2 | $10^{-3}$ | 418 | 12684041 |
| Q8VCK6 | mmu:233079 | Free fatty acid receptor 2 | $10^{-2}$ | 474 | 12684041 |
| O14843 | hsa:2865 | Free fatty acid receptor 3 | $10^{-2}$ | 868 | 12711604 |
| Q8TDS5 | hsa:165140 | Oxoeicosanoid receptor 1 | $10^{-3}$ | 589 | 12065583 |
| Lysophospholipid receptors | | | | | |
| Q92633 | hsa:1902 | Lysophosphatidic acid receptor 1 | $10^{-7}$ | 351 | 9070858 |
| Q9PU17 | xla:373591 | Lysophosphatidic acid receptor 1 | $10^{-6}$ | 295 | 11278944 |
| Q9PU16 | xla:379403 | Lysophosphatidic acid receptor 1 | $10^{-6}$ | 286 | 11278944 |
| Q9HBW0 | hsa:9170 | Lysophosphatidic acid receptor 2 | $10^{-1}$ | 611 | 15143197 |
| Q9UBY5 | hsa:23566 | Lysophosphatidic acid receptor 3 | $10^{-3}$ | 704 | 10488122 |
| Q99677 | hsa:2846 | Lysophosphatidic acid receptor 4 | $10^{-6}$ | 1331 | 12724320 |
| P43657 | hsa:10161 | Lysophosphatidic acid receptor 6 | $10^{-4}$ | 1375 | 18297070 |
| P21453 | hsa:1901 | Sphingosine 1-phosphate receptor 1 | $10^{-6}$ | 486 | 9488656 |
| Q9DDK4 | dre:64617 | Sphingosine 1-phosphate receptor 1 | $10^{-3}$ | 655 | 11112429 |
| O95136 | hsa:9294 | Sphingosine 1-phosphate receptor 2 | $10^{-7}$ | 279 | 10617617 |
| Q99500 | hsa:1903 | Sphingosine 1-phosphate receptor 3 | $10^{-10}$ | 166 | 10617617 |
| O95977 | hsa:8698 | Sphingosine 1-phosphate receptor 4 | $10^{-1}$ | 244 | 10753843 |
| Q9JKM5 | rno:60399 | Sphingosine 1-phosphate receptor 5 | - | 252 | 10799507 |
| Cannabinoid receptors | | | | | |
| P21554 | hsa:1268 | Cannabinoid receptor 1 | $10^{-7}$ | 382 | 15620723 |
| P47746 | mmu:12801 | Cannabinoid receptor 1 | $10^{-7}$ | 372 | 8777318 |
| P34972 | hsa:1269 | Cannabinoid receptor 2 | $10^{-6}$ | 405 | 10051546 |
| P47936 | mmu:12802 | Cannabinoid receptor 2 | $10^{-7}$ | 264 | 8679694 |
| PAF receptors | | | | | |
| P25105 | hsa:5724 | Platelet-activating factor receptor | $10^{-1}$ | 1109 | 1656963 |
| Q62035 | mmu:19204 | Platelet-activating factor receptor | $10^{-1}$ | 997 | 8670084 |
| P46002 | rno:58949 | Platelet-activating factor receptor | $10^{-2}$ | 1249 | 8168510 |

Table 1 (contd.)

| Bile acid receptor | | | | | |
|---|---|---|---|---|---|
| Q8TDU6 | hsa:151306 | G-protein coupled bile acid receptor 1 | - | 8 | 12419312 |

[a] hsa: *H. sapiens*, mmu: *M musculus*, rno: *R. norvegicus*, xla: *X laevis*, dre: *D. rerio*, bfo: *B floridae*, cin: *C. intestinalis*, spu: *S. purpuratus*, dme: *D. melanogaster*, cel: *C elegans*, bmy: *B. malayi*, nve: *N vectensis*, tad: *T. adhaerens*. [b] Hyphens represents no E-values set.

## 3.2. *Frequency of candidate lipid mediator GPCRs in 13 eukaryotic genomes*

132 ortholog candidates were identified from the 13 constructed phylogenetic trees. No BLAST search hits were obtained for *A. thaliana*, *S. cerevisiae* and *M. brevicollis*. Table 2 shows the number of orthologs for each class of lipid mediator GPCR. Ortholog candidates for PG GPCRs were detected in all organisms except for nematodes. Those for LPA receptors and CB receptors were extracted in vertebrates and chordates. Ortholog candidates for other receptors including LT receptors and FA receptors were obtained only in vertebrates.

Table 2. Conservation of lipid mediator GPCRs.

| Organism categories | Species | Prostanoid receptors | Leukotriene receptors | Fatty acid receptors | Lysophospho-lipid receptors | Cannabinoid receptors | PAF receptor | Bile acid receptor |
|---|---|---|---|---|---|---|---|---|
| Vertebrata | *H. sapiens* (hsa) | 8 (5) | 4 (4) | 4 (4) | 10 (9) | 2 (2) | 1 (1) | 1 (1) |
| | *M. musculus* (mmu) | 8 (4) | 4 (1) | 3 (2) | 10 | 2 (2) | 1 (1) | 0 |
| | *R. norvegicus* (rno) | 9 (5) | 4 | 3 | 11 (1) | 2 | 1 (1) | 0 |
| | *X. laevis* (xla) | 4 | 0 | 0 | 4 (2) | 2 | 1 | 0 |
| | *D. rerio* (dre) | 13 | 12 | 10 | 14 (1) | 3 | 2 | 1 |
| Chordata | *B. floridae* (bfo) | 11 | 0 | 0 | 0 | 1 | 0 | 0 |
| | *C. intestinalis* (cin) | 2 | 0 | 0 | 2 | 1 | 0 | 0 |
| Echinodermata | *S. purpuratus* (spu) | 3 | 0 | 0 | 0 | 0 | 0 | 0 |
| Arthropoda | *D. melanogaster* (dme) | 1 | 0 | 0 | 0 | 0 | 0 | 0 |
| Nematoda | *C. elegans* (cel) | 0 | 0 | 0 | 0 | 0 | 0 | 0 |
| | *B. malayi* (bmy) | 0 | 0 | 0 | 0 | 0 | 0 | 0 |
| Cnidaria | *N. vectensis* (nve) | 1 | 0 | 0 | 0 | 0 | 0 | 0 |
| Placozoa | *T. adhaerens* (tad) | 2 | 0 | 0 | 0 | 0 | 0 | 0 |

Numbers include both the newly detected ortholog candidates and query sequences. Queries, if any, are indicated in parentheses.

### 3.3. *Proposed phylogenetic tree of query sequences and detected candidates in 13 eukaryotic organisms.*

We constructed a phylogenetic tree using the 46 queries and 132 newly detected ortholog candidates (Fig. 2). It mostly agreed with the classification of lipid GPCR orthologs; members of the same class had higher sequence similarities with each other than to members of other groups. Lysophospholipid receptors were separated into two groups, the fact which agrees with the observation in a previous study with a phylogenetic tree of experimentally annotated human lipid GPCRs [18].

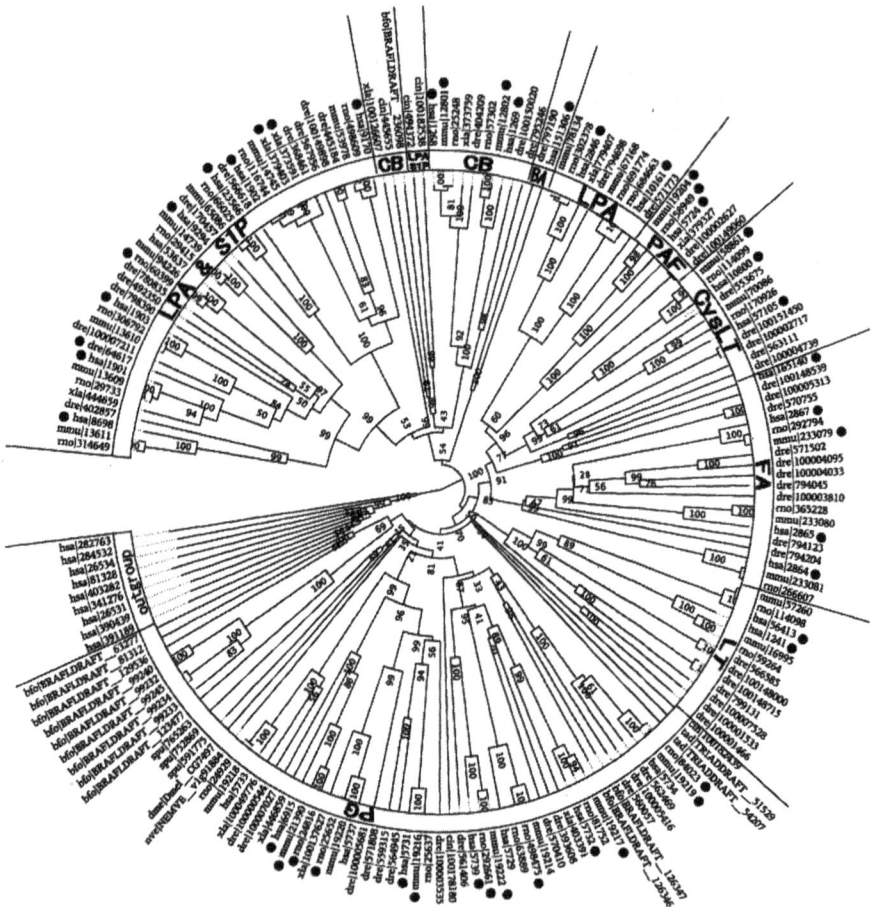

Fig. 2. Proposed phylogenetic tree of lipid mediator GPCRs in 13 eukaryotes. Queries are indicated with solid circles. Sequences are represented by KEGG IDs using three letter organism codes used in the KEGG database (see Table 2). PG: prostanoid receptors, LT: leukotriene receptors, FA: fatty acid receptors, LPA: lysophosphatidic acid receptors, S1P: sphingosine 1-phosphate receptors, CB: cannabinoid receptors, PAF: platelet-activating factor receptors, BA: bile acid receptors.

## 4. Discussions

Table 2 shows an asymmetric conservation pattern of lipid mediator GPCRs in 13 eukaryotic organisms. Prostanoid receptors were conserved in all eukaryotes except in nematodes. In contrast a limited conservation in other lipid mediator receptors was observed. Lysophospholipid and CB receptors appear to be conserved in vertebrates and chordates. Other receptors including LT, FA, PAF and BA receptors were only found in vertebrates. Herein we discuss this tendency in a functional aspect focusing on the asymmetric conservation of these mediator molecules in invertebrates.

### 4.1. *Conservation of prostanoid receptors in invertebrates*

Prostanoid receptors were conserved widely in the eukaryotic genomes. Synthesis and biological effects of prostanoids have been reported in *C. intestinalis* [8]. The detected candidates would be potential receptors for the *Ciona* prostanoid molecules. Generally the roles of lipid mediators are largely unknown in non-chordate invertebrates. However, there is a fair amount of research that supports the signaling functions of PGs and other eicosanoids in invertebrates. In insects detailed analyses have revealed that PGs and other eicosanoids play essential roles in egg laying and in cellular immunity against microbial invasions [20]. These compounds have also been shown to have a role in reproduction in non-insectan invertebrates [21]. Most of the ortholog candidates detected in *C. intestinalis*, *B. floridae*, *S. purpuratus*, *D. melanogaster*, *N. vectensis* and *T. adhaerens* form independent clusters on the phylogenetic tree (Fig. 2). Invertebrates have been reported to have their own forms of prostaglandins as well as the classic vertebrate forms (PGE2, PGD2, etc.) [15]. Thus the identified ortholog candidates could be the potential receptors for these molecules. It is remarkable that PG receptors were not found in *C. elegans*, a free living worm, and *B. malayi*, a parasitic worm. *B. malayi* has been reported to release PGs, which causes the inhibition of platelet aggregation in the host blood cells [12]. The absence of PG receptors in these organisms can be understood by their limited usage in the host cells, but not in their innate physiology. Collectively, these results suggest a hypothesis regarding the functional roles of PG receptors in invertebrates. However, in order to fully understand the conservation and evolution of PGs and other eicosanoid functions, the variation in their biosynthetic pathways, in particular the genomic conservation of COX and LOX, should also be investigated.

### 4.2. *Leukotriene, lysophospholipid and cannabinoid receptors in invertebrates*

Leukotriene receptors appear to be conserved only in vertebrates. In mammals the well-known function of LT receptors is the chemotactic effects of leukocytes to the inflamed sites. Interestingly, although the phagocytotic hemocytes are found in many invertebrates, no LT receptors were detected in the invertebrate genomes tested. Lysophospholipid receptor orthologs were detected in vertebrates and in *C. intestinalis*, but not in other invertebrates. One of the two candidates detected in *C. intestinalis* (hsa:494372 in Fig. 2) has been previously annotated as EDG-3-like protein [6, 23]. Cannabinoid receptor orthologs were also detected in the chordates. One ortholog (cin:445655) has previously been reported [4] with equal similarities to the mammalian CB1 and CB2. Interestingly, the one we detected in *B. floridae* (bfo:BRAFLDRAFT_236098) was different from what has been previously identified by another group [3]. In mammals, lysophospholipids are known to be involved in the formation of the central nervous system, angiogenesis and the regulation of the vascular and the immune systems. This information may collectively explain the coevolution of lysophospholipid and cannabinoid signaling functions with the development of highly modulated nervous, immune and vascular systems in vertebrates as the origin of these systems are thought to be traced back into Chordates.

### 4.3. *Possible improvements on our method*

Our method searches for sequences with higher similarities to the queries than to olfactory receptor sequences in each genome. This is based on the assumption that if orthologs of queries had been emerged in the common ancestor, their sequence similarities could be conserved with a higher degree to the query sequences than to other GPCRs in a genome. Although this process successfully reduces the number of false positives in the BLAST search hits, which makes the following phylogenetic analyses much easier, it also limits the possibility of detecting candidate genes that may have been evolved independently of the queries. In addition, in the obtained BLAST search hits there were many sequences that remained unannotated especially in the invertebrates. Annotation of these sequences by known non-lipid GPCRs such as amine receptors and peptide receptors should help in characterizing these elements.

## 5.    Conclusions

We performed a phylogenetic analysis to search for mammalian homologs of lipid mediator GPCRs. Our results show that orthologs for the PG receptors are conserved widely in all tested eukaryotic organisms except in nematodes. Lysophospholipid receptors and CB receptors are observed in vertebrates and chordates. Leukotriene receptors, PAF receptors and BA receptors appear to be vertebrate-specific. The eukaryotic conservation of PG receptors can be explained by their reproductive function, which has been demonstrated by previous investigations in insects and other invertebrates. The absence of other receptors in non-chordate invertebrates indicates that lipid signaling via GPCRs may have evolved with the development of highly modulated vascular, immune and nervous systems in vertebrates. These results can give new insight into our understanding of lipid signaling in invertebrates.

## Acknowledgements

This research was partly funded by a joint Japan Society for the Promotion of Science (JSPS) and VINNOVA scientific co-operation grant. CEW was supported by a Centre for Allergy Research (CfA) research fellowship.

## References

[1] Altschul, S.F., Madden, T.L., Schäffer, A.A., Zhang, J., Zhang, Z., Miller, W., and Lipman, D.J., Gapped BLAST and PSI-BLAST: a new generation of protein database search programs, *Nucleic Acids Res.*, 25:3389-3402, 1997.

[2] Contos, J.J., Fukushima, N., Weiner, J.A., Kaushal, D., and Chun, J., Requirement for the lpA1 lysophosphatidic acid receptor gene in normal suckling behavior, *Proc. Natl. Acad. Sci.*, 97:13384-13389, 2000.

[3] Elphick, M.R., BfCBR: a cannabinoid receptor ortholog in the cephalochordate *Branchiostoma floridae* (Amphioxus), *Gene,* 399:65-71, 2007.

[4] Elphick, M.R., Satou, Y., and Satoh, N., The invertebrate ancestry of endocannabinoid signaling: an orthologue of vertebrate cannabinoid receptors in the urochordate *Ciona intestinalis*, *Gene,* 302:95-101, 2003.

[5] Funk, C.D., Prostaglandins and leukotrienes: advances in eicosanoid biology, *Science*, 294:1871-1875, 2001.

[6] Hla, T., Genomic insights into mediator lipidomics, *Prostaglandins & Other Lipid Mediators*, 77:197-209, 2005.

[7] Howe, K., Bateman, A., and Durbin, R., QuickTree: building huge Neighbour-Joining trees of protein sequences, *Bioinformatics,* 18:1546-1547, 2002.

[8] Jarving, R., On the evolutionary origin of cyclooxygenase (COX) isozymes: characterization of marine invertebrate *COX* genes points to independent duplication events in vertebrate and invertebrate lineages, *J. Biol. Chem.,* 279:13624-13633, 2004.

[9] Kanehisa, M., Araki, M., Goto, S., *et al.*, KEGG for linking genomes to life and the environment. *Nucleic Acids Res.,* 36:D480-D484, 2008.

[10] Katoh, K. and Toh, H., Recent developments in the MAFFT multiple sequence alignment program. *Briefings in Bioinformatics,* 9:286-298, 2008.

[11] Kingsbury, M.A., Rehen, S.K., Contos, J.J., Higgins, C.M., and Chun, J., Non-proliferative effects of lysophosphatidic acid enhance cortical growth and folding, *Nature Neurosci.,* 6:1292-1299, 2003.

[12] Liu, L.X. and Weller, P.F., Intravascular Filarial parasites inhibit platelet aggregation. Role of parasite-derived prostanoids, *J. Clin. Invest.*, 89:1113-1120, 1992.

[13] Okuno, T., Iizuka, Y., Okazaki, H., Yokomizo, T., Taguchi, R., and Shimizu, T., 12(S)-hydroxyheptadeca-5Z, 8E, 10E–trienoic acid is a natural ligand for leukotriene B4 receptor 2, *J. Exp. Med.*, 205:759-766, 2008.

[14] Rosen, H. and Goetzl, E., Sphingosine 1-phosphate and its receptors: an autocrine and paracrine network, *Nature Rev. Immunol.*, 5:560-570, 2005.

[15] Rowley, A.F., Vogan, C.L., Taylor, G.W., and Clare, A.S., Prostaglandin in non-insectan invertebrates: recent insights and unsolved problems, *J. Exp. Biol.*, 208:3-14, 2005.

[16] Saitou, N. and Nei, M., The neighbor-joining method: a new method for reconstructing phylogenetic trees, *Mol. Biol. Evol.*, 4:406-425, 1987.

[17] Samuelsson, B., Dahlén, S.E., Lindgren, J.A., Rouzer, C.A., and Serhan, C.N., Leukotrienes and lipoxins: structures, biosynthesis, and biological effects, *Science*, 237:1171-1176, 1987.

[18] Shimizu, T., Lipid mediators in health and disease: enzymes and receptors as therapeutic targets for the regulation of immunity and inflammation, *Annu. Rev. Pharmacol. Toxicol.*, 49:123–150, 2009.

[19] Spiegel, S. and Milstien, S., Sphingosine-1-phosphate: an enigmatic signaling lipid, *Nature Rev. Mol. Cell. Biol.*, 4:397-407, 2003.

[20] Stanley, D., Prostaglandins and other eicosanoids in insects: biological significance, *Annu. Rev. Entomol.*, 51:25-44, 2006.

[21] Stanley, D.W. and Miller, J.S., Eicosanoids in animal reproduction: what can we learn from invertebrates? In *Eicosanoids and Related Compounds in Plants and Animals* (ed. Rowley, A.F., Kuhn, H and Schewe, T), pp. 183-196, 1998.

[22] Tanaka, C. and Kato R., *New Pharmacology*, 4th edition, Nankodo, Japan, 2002.

[23] Terajima, D., Identification and sequence of seventy-nine new transcripts expressed in hemocytes of *Ciona intestinalis*, three of which may be involved in characteristic cell-cell communication, *DNA research*, 10:203-212, 2003.

[24] Terawaki, K., Yokomizo, T., Nagase, T., Toda, A., Taniguchi, M., Hashizume, K., Yagi, T., and Shimizu, T., Absence of leukotriene B4 receptor 1 confers resistance to airway hyperresponsiveness and Th2-type immune responses, *J. Immunol.*, 175:4217-4225, 2005.

[25] The UniProt Consortium, The Universal Protein Resource (UniProt) in 2010, *Nucleic Acids Research*, 38:D142-D148, 2010.

# GENOME-WIDE ANALYSIS OF PLANT UGT FAMILY BASED ON SEQUENCE AND SUBSTRATE INFORMATION

YOSUKE NISHIMURA[1]        TOSHIAKI TOKIMATSU[1]        MASAAKI KOTERA[1]
yosuke@kuicr.kyoto-u.ac.jp    tokimatu@kuicr.kyoto-u.ac.jp    kot@kuicr.kyoto-u.ac.jp

SUSUMU GOTO[1]        MINORU KANEHISA[1,2]
goto@kuicr.kyoto-u.ac.jp    kanehisa@kuicr.kyoto-u.ac.jp

[1] *Bioinformatics Center, Institute for Chemical Research, Kyoto University, Uji, Kyoto 611-0011, Japan*
[2] *Human Genome Center, Institute of Medical Science, University of Tokyo, 108-8639, Japan*

UGTs (UDP glycosyltransferase) are the largest glycosyltransferase gene family in higher plants, modifying secondary metabolites, hormones, and xenobiotics. This gene family plays an important role in the vast diversity of plant secondary metabolites specific to species. Experimental data of biochemical activities and physiological roles of plant UGTs are increasing but most UGTs are not still functionally characterized. To understand their catalytic specificity and function from sequence data, phylogenetic analyses have been achieved mainly in *Arabidopsis*, but massive and comprehensive approach covering various species has not been applied yet. In this study, we collected 733 UGT sequences derived from 96 plant species and 252 substrate specificity data. We constructed a phylogenetic tree and divided most part of these genes into nine sequence groups, which are characterized by biochemical specificity. Furthermore, we performed genome-wide analysis of seven plant species UGTs by mapping them into these groups. We propose this is the first step to understand whole glycosylated secondary metabolites of each plant species from its genome information.

*Keywords*: UGT; plant secondary metabolites; substrate specificity; genome-wide analysis.

## 1. Introduction

Plant UGTs (UDP glycosyltranferases) catalyze the glycosylation of various chemical substances, such as secondary metabolites, hormones, and xenobiotics. This glycosylation alters the hydrophilicity of the acceptors, their stability and chemical properties, their subcellular localization and their bioactivity, so glycosylation plays a major role both in cellular homeostasis and in buffering the impact of xenobiotic challenges on the plant. UGTs also play an important role in the vast diversity of plant secondary metabolites, mostly ensured by multiple modifications of a common skeleton by hydroxylation, methylation, acylation or conjugation to small molecules. For instance, the diversity of the more than 4000 known flavonoids originates from such combinatorial modifications of their common aromatic structure [8].

Plant UGTs catalyze various secondary metabolites, including terpenoids, alkaloids, cyanogenic glucosides and glucosinolates as well as flavonoids, isoflavonoids and other phenylpropanoids. Plant UGTs transfer nucleotide diphosphate-activated sugars to low molecular-weight substrates. The activated sugar form is usually UDP-glucose (UDP-Glc) but UDP-galactose (UDP-Gal), UDP-rhamnose (UDP-Rha), UDP-xylose (UDP-Xyl), UDP-glucuronic acid (UDP-GlcA), and UDP-arabinose (UDP-Ara) are also found [2, 12, 47]. In general, plant UGTs show very high specificity for the sugar donor. Substrate specificity of UGTs is defined by the combination of a sugar acceptor, a UDP-sugar donor, and a reaction regioselectivity. This combination contributes to a wide diversity of plant secondary metabolites.

Higher plants have notably many UGTs in their genomes, in contrast with the other eukaryotic genomes [13]. Complete genome sequences are now available for several plants, such as *Arabidopsis thaliana*, *Oryza sativa* (rice), *Vitis vinifera* (grapevine), *Medicago truncatula* (barrel medic). These genomes have been estimated to possess more than 100 UGT genes, respectively [46]. On the other hand, human genome has been computationally estimated to possess 21 UGT genes, most of which are included in major two families, use UDP-glucuronic acid to glucuronidate bilirubin, steroids, bile acids, drugs, and many other endogenous chemicals and xenobiotics [13, 27].

Phylogenetic comparisons of UGTs from plants, animals, fungi, bacteria, and viruses reveal that plant UGTs represent three distinct clades, which consist of one major clade and two minor clades. The major clade has diverged after the bifurcation of the kingdoms and the two minor clades (UGT80 and UGT81) contain the sterol and lipid glycosyltransferases and show more homology to bacteria or fungi sequences [37]. UGT belongs to GT1 family in the CAZy database, which is a knowledge-based resource specialized in the enzymes that build and breakdown complex carbohydrates and glycoconjugates [4]. The glycosyltransferases (GTs) are currently divided into 92 families based on sequence identity scores and GT1 family includes 1971 eukaryote sequences [51].

Although, the physiological roles and biochemical activities of most plant UGTs are still elusive because both of a high copy number of UGT genes and the sugar acceptors diversity, experimental confirmation of their specificity from various plant species are progressively increasing. Phylogenetic studies are efficient for sequence based prediction of substrate specificity and have ever been performed. For example, in *Arabidopsis*, phylogenetic studies of UGT including screening catalytic specificity *in vitro* have been achieved [3, 5, 21, 22, 23]. Furthermore, several researches revealed that sequence similarity of flavonoid UGTs of plants generally correspond to the reaction regioselectivity [34, 35, 36, 39, 44, 48]. These analyses surely provided interesting functional and evolutionary knowledge of UGTs but it is still difficult to predict an accurate substrate specificity because a small number of amino acid differences are often responsible for the functional differentiation of plant enzymes [35].

In our study, we performed a large phylogenetic analysis over plant species to understand plant UGTs more comprehensively and accurately, dividing most of these genes into several sequence groups, which are characterized by biochemical specificity. This can be a supportive evidence to suggest that the biochemical specificity of plant UGTs is partly classified by their sequences. Furthermore, we performed genome-wide analysis of seven plant species UGTs by mapping them into these groups, aimed at revealing differnce of glycosylated secondary metabolites among these species. We propose this is the first step to provide species-specific catalogues of glycosylated secondary metabolites from genome information.

## 2.    Methods and Results

### 2.1   *Dataset collection from the CAZy database, Uniprot, and papers*

To obtain the plant UGT sequences comprehensively, we collected the UGT sequences from GT1 family of the CAZy database and extracted 733 plant complete sequences whose length is from 350 to 600. These sequences were collected from 96 plant species including only one gymnosperm. We also obtained 252 functional annotations (*e.g.* anthocyanidin 3-*O*-glucosyltransferase) and family ID (*e.g.* UGT72A4) from CAZy, Uniprot [51], and several published papers [1, 7, 9, 10, 11, 14, 15, 18, 20, 23, 24, 25, 26, 28, 29, 30, 31, 32, 33, 33, 34, 38, 39, 41, 42, 43, 45, 46]. Top ten species possessing many UGT sequences in the dataset are listed in Table1. *O. sativa* and *A. thaliana* have long been investigated by many researchers and therefore their genome sequences have been read much earlier than other species. Major substrates and the number of genes that catalyze substrates of these clusters are shown in Table 2. Flavonoids and hormones are two major substrate groups in the dataset.

Table 1. Top 10 species in the dataset

| Species | #genes |
|---|---|
| *Oryza sativa* | 159 |
| *Arabidopsis thaliana* | 116 |
| *Zea mays* | 86 |
| *Vitis vinifera* | 68 |
| *Lycium barbarum* | 24 |
| *Dianthus caryophyllus* | 21 |
| *Medicago truncatula* | 17 |
| *Sorghum bicolor* | 13 |
| *Stevia rebaudiana* | 12 |
| *Nicotiana tabacum* | 10 |

Table 2.  Major substrate group

| substrate group | #genes |
|---|---|
| Flavonoids | 130 |
| Hormones | 63 |
| Terpenoids and Steroids | 25 |
| Phenylpropanoids | 24 |
| Others | 17 |

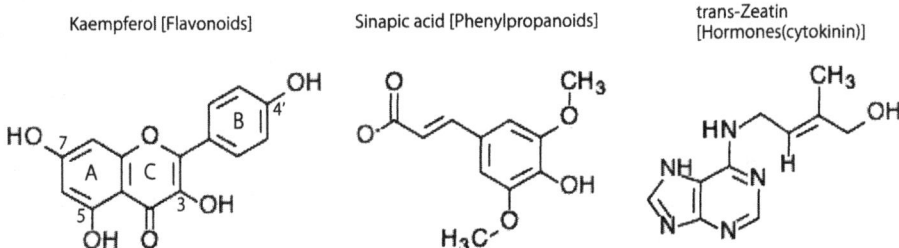

Fig. 1. Three chemical structure examples of substrates, with their glycosylation sites highlighted

Three chemical structure examples of substrates are shown in Fig. 1, with their glycosylation sites highlighted. Kaempferol have four known glycosylation sites: 3-OH on the C-ring, 5- and 7-OH on the A-ring, and 4'-OH on the B-ring. Sinaptic acid has an ester-forming site of the carboxyl cluster and *trans*-zeatin has a *N*-glycosylation site, in addition to an *O*-glycosylation site.

## 2.2   Phylogenetic tree construction and clustering, leading to defining sequence groups of characteristic catalytic specificity

Multiple alignments of amino acid sequences were constructed using MAFFT (FFT-NS-2) [17]. Phylogenetic tree was constructed using the Neighbor-joining (NJ) method provided by the ClustalX program [19]. Bootstrapping was performed 100 times to obtain support values for each branch. Interactive Tree Of Life (iTOL) was used to visualize and divide into 56 clusters by its automatically collapse option with average distance to leaves less than 0.3 [49]. Two clusters were fused to another cluster respectively because these clusters had low bootstrap values. We finally obtained 54 clusters of sequence similarity, including six singletons (Fig. 2, Table 3). Fig. 2 represents 54 clusters of the phylogenetic tree and sequence groups we defined, described with boxes (Details of these groups and abbreviations are shown in table 4). Any group is supported by at least 90/100 bootstrap value and F5G (flavonoid 5-*O*-glycosyltransfearse) group, covering cluster 43 and 44 is nested by ester-forming group, means that cluster 43 and 44 share features of groups. Table 3 shows the number of genes in the cluster, bootstrap value, UGT family nomenclature class, and the number of genes catalyzing major substrate groups.

The number of genes in each cluster with corresponding groups is shown in Table 4 where each column represents catalytic specificity. These groups are characterized in terms of catalytic activities such as regioselectivity of flavonoids and cytokinins, selective recognition on the sugar residues of glycoside for additional glycosylation, and ester-forming specificity on various metabolites possessing carboxyl cluster, including acid hormones, terpenoids, steroids, phenylpropanoids. Three rightmost columns show specificity on ester-forming, sugar residue, and cytokinin *N*-glycosylation, respectively, and each column approximately corresponds to a group. Two cytokinin *O*-GT groups for the clusters 9, 10 and cluster 54 and F3G (flavonoid 3-*O*-glycolyltransferase) group clearly show substrate specificities. The remaining two groups, various I and II show extremely wide variety of substrate specificity. The main difference between the two groups is regiospecificity of flavonoids; it is shown in the table that the genes of various I group are limited to 7-*O* or B-ring sites, while various II group has wider regioselectivity.

In Fig. 2, clusters for which any groups were not defined (1, 2, 11-21, 38, 39, 47-49, 52, and 53) contained relatively small number of genes and had little experimental data. In fact, 633 out of 733 genes were included in some group. This means that the current knowledge on UGT substrates can be classified into several groups and it is applicable to the genomic-scale annotation of those groups of secondary metabolites. However, many genome projects are in progress for diverse plant species and much amount of sequences will be available. Therefore we also need to analyze more UGTs in the future.

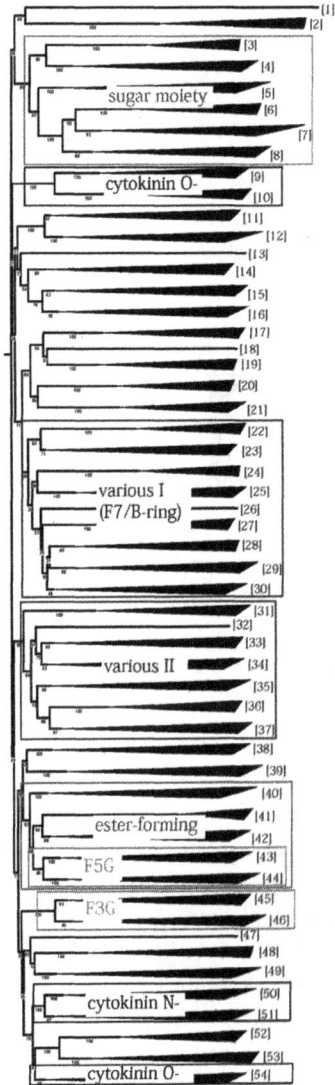

Fig. 2. Phylogenetic tree representing gene clusters and catalytic activity groups

(Tree labels: sugar moiety; cytokinin O-; various I (F7/B-ring); various II; ester-forming; F5G; F3G; cytokinin N-; cytokinin O-; branches [1]–[54])

Table 3. Phylogenetic clusters and the number of genes catalyzing major substrate groups (Singletons are omitted except cluster 26.)

| Cluster | #genes | Bootstrap | UGT family | Flavonoids | Hormones | Terpenoids & Steroids | Phenyl-propanoids |
|---|---|---|---|---|---|---|---|
| 2 | 3 | 100 | UGT80 | | | 1 | |
| 3 | 2 | 100 | UGT94A | 1 | | | |
| 4 | 5 | 100 | UGT94B,D,E | 2 | | | 1 |
| 5 | 25 | 100 | UGT91 | 2 | | | |
| 6 | 4 | 100 | UGT79A | 3 | | | |
| 7 | 4 | 92 | UGT79C | | | | |
| 8 | 21 | 99 | UGT79B | 3 | | | |
| 9 | 8 | 100 | UGT93A | | 4 | | |
| 10 | 16 | 100 | UGT93B | | 3 | | |
| 11 | 5 | 87 | UGT92A | | | | |
| 12 | 8 | 100 | UGT92B,C,D,E,F | | 1 | | |
| 14 | 3 | 69 | - | | | | |
| 15 | 7 | 83 | UGT89 | 1 | | 2 | |
| 16 | 9 | 98 | UGT96 | | | | |
| 17 | 2 | 100 | UGT90B | | | | |
| 19 | 3 | 100 | UGT90A,C | | | | |
| 20 | 2 | 100 | - | | | | |
| 21 | 4 | 100 | UGT97A | | | | |
| 22 | 7 | 100 | UGT701 | | | 1 | |
| 23 | 39 | 74 | UGT73C,D,E,H/98,99 | | 4 | | |
| 24 | 2 | 100 | UGT704 | 1 | | | |
| 25 | 9 | 100 | UGT705 | | | 2 | |
| 26 | 1 | - | UGT73J | 1 | | | |
| 27 | 4 | 100 | UGT703 | | | | |
| 28 | 4 | 99 | UGT73F,K | 2 | | 1 | |
| 29 | 5 | 98 | UGT73L | | | | |
| 30 | 35 | 46 | UGT73A,B,G,M,P,Q | 17 | 3 | | 2 |
| 31 | 8 | 100 | UGT708 | 3 | | | |
| 33 | 8 | 99 | UGT72A,C,D,E | 1 | | | |
| 34 | 32 | 63 | UGT72B,F,H,J | 2 | | | |
| 35 | 56 | 98 | UGT88,706 | 15 | | | 1 |
| 36 | 11 | 100 | UGT707 | 5 | | | |
| 37 | 51 | 97 | UGT71 | 7 | 4 | 1 | |
| 38 | 7 | 100 | UGT86 | 1 | | | |
| 39 | 7 | 100 | UGT87 | | 2 | | |
| 40 | 43 | 100 | UGT74 | 1 | 11 | 3 | |
| 41 | 3 | 100 | UGT84B | | 1 | | |
| 42 | 35 | 88 | UGT84A,C,D,F | 2 | 1 | 10 | 11 |
| 43 | 12 | 100 | UGT75E,F,G,H,J,K | 1 | 2 | | |
| 44 | 36 | 100 | UGT75A,B,C,D,L | 13 | 3 | 2 | |
| 45 | 8 | 93 | UGT77 | 5 | | | |
| 46 | 36 | 96 | UGT78 | 31 | | | |
| 48 | 3 | 100 | UGT82 | | | | |
| 49 | 20 | 100 | UGT83 | | 1 | | |
| 50 | 15 | 100 | UGT710 | | 1 | | |
| 51 | 31 | 87 | UGT76 | 4 | 2 | 1 | |
| 52 | 3 | 100 | - | | | | |
| 53 | 10 | 100 | UGT709 | 1 | 1 | | |
| 54 | 56 | 96 | UGT85 | 1 | 7 | 1 | |

Table 4. Catalytic specificity of UGTs in each group
(Clusters having genes in any column are omitted.)

| Cluster | Group | Flavonoids 3-O | 5-O | 7-O | B-ring | sugar residue | Hormones CK | AH | T&S | PP | other specificity ester-forming | sugar residue | cytokinin N- |
|---|---|---|---|---|---|---|---|---|---|---|---|---|---|
| 3 | sugar residue | 1 | | | | | | | | | | | |
| 4 | | | | | | | 2 | | | | 1 | 3 | |
| 5 | | 2 | | | | | | | | | | | |
| 6 | | | | | | | 2 | | | | | 2 | |
| 8 | | | | | | | 2 | | | | | 2 | |
| 9 | cytokinin O-GT | | | | | | 4 | | | | | | |
| 10 | | | | | | | 3 | | | | | | |
| 22 | | | | | | | 1 | | | | | | |
| 23 | various I (F7/B-ring) | | | 1 | | | 6 | | | | | | |
| 24 | | | | 1 | | | | | | | | | |
| 25 | | | | | | | 2 | | | | | | |
| 28 | | | | 1 | | | | | | 1 | | | |
| 29 | | | | | | | | | | | | | |
| 30 | | | | 4 | 5 | | 3 | | | | 2 | 2 | |
| 31 | various II | 2 | | | | | | | | | | | |
| 33 | | 1 | | | | | | | | | | | |
| 34 | | | | | | | | | | 1 | | | |
| 35 | | 9 | 6 | 2 | 2 | | | | | 1 | | | |
| 36 | | 5 | 2 | | | | | | | | | | |
| 37 | | 1 | | 1 | | | | | 3 | 1 | 2 | | |
| 40 | ester-forming | | | | | | | 10 | 4 | | 12 | | |
| 41 | | | | | | | | 1 | | | 1 | | |
| 42 | | | | | | | | 1 | 10 | 11 | 12 | | |
| 43 | F5G | | 1 | | | | | | 2 | | | | |
| 44 | | | 9 | 2 | | | | | 2 | 2 | | 3 | |
| 45 | F3G | 5 | | | | | | | | | | | |
| 46 | | 30 | | 1 | 1 | | | | | | | | |
| 50 | cytokinin N-GT | | | | | | 1 | | | | | | 1 |
| 51 | | | 2 | | | | 2 | | 1 | | | 1 | 2 |
| 54 | cytokinin O-GT | | | | | | 7 | | | 1 | | | |

3-O: flavonoid 3-O-GT
5-O: flavonoid 5-O-GT
7-O: flavonoid 7-O-GT
B-ring: flavonoid B-ring GT
CK: cytokinins GT
AH: acid hormones GT (e.g. auxin, abcsisic acid, benzoate)
T&S: terpenoids and steroids GT
PP: phenylpropanoids GT
F7/B-ring: flavonoid 7-O-GT/flavonoid B-ring GT
F5G: flavonoid 5-O-GT
F3G: flavonoid 3-O-GT

## 2.3 Genome-wide analysis of UGTs using seven plant species

To understand genome-wide distribution of UGTs in several species, we further collected UGTs of complete genomes from KEGG GENES. There are seven entries of plant species in the database (shown in Table 5) but *A. thaliana* and *O. sativa* UGTs were covered in the first datasets already.

We additionally collected UGTs of five species except these two, by using PSI-BLAST search with a seed gene VvGT1 from grapevine (UniProt ID: O22304) in cluster 46 of the phylogenetic tree. These genes were added to the first set 733 genes and the total number of UGT sequences reached to 1306. *V. vinifera* has the largest number of UGTs and thus may have more glycosylated secondary metabolites than any other species. It may be related to the fact that the past research of comparative genomics revealed that a selective amplification of genes belonging to terpenes and tannins synthesis pathway occurred in the grapevine genome, in contrast with other plant genomes [16].

Table 5. The number of additional UGT sequences and whole entries of seven species in KEGG GENES

| Species | #entries | #UGT genes | code |
|---|---|---|---|
| *Arabidopsis thaliana* | 27,217 | 116 | ath |
| *Oryza sativa* | 26,937 | 156 | osa |
| *Vitis vinifera* | 22,511 | 241 | vvi |
| *Populus trichocarpa* | 42,307 | 137 | pop |
| *Sorghum bicolor* | 33,002 | 171 | sbi |
| *Zea mays* | 17,818 | 86 | zma |
| *Ricinus communis* | 31,221 | 97 | rcu |

We reconstructed the phylogenetic tree in the way described above. The cluster numbers were assigned by genes of the first dataset. All the clusters were conserved but only cluster 29, 30 were mixed and united into one cluster (Table 6). This table represents the number of genes of the seven species in each cluster. Widespread distributions of each species UGTs and difference of gene appearance pattern between dicots and monocots are shown. However, there are few drastic differences about the number of genes occupied by each group between the species comparing among dicots and monocots respectively. Hence, it may be difficult to compare the structual variation of glycosylated secondary metabolites between species by only sequence information.

Table 6. The number of genome-wide UGT genes
(Abbreviations are the same as Table 4.)

| Cluster | Group | ath | vvi | pop | rcu | osa | sbi | zma |
|---|---|---|---|---|---|---|---|---|
| 3 | | | | | | 1 | | 1 |
| 4 | | | 5 | | 1 | | | |
| 5 | sugar residue | 3 | 12 | 8 | | 8 | 7 | 4 |
| 6 | | | | | | | | |
| 8 | | 11 | 5 | | | | | |
| 9 | cytokinin *O*-GT | | 2 | 3 | 2 | | | |
| 10 | | | | | | 7 | 10 | 5 |
| 22 | | | | | | 3 | 4 | 3 |
| 23 | | 8 | 8 | 3 | 5 | 10 | 10 | 7 |
| 24 | various I | | | | | 1 | 1 | 1 |
| 25 | (F7/B-ring) | | | | | 5 | 4 | 4 |
| 28 | | | | | | | | |
| 29,30 | | 6 | 4 | 8 | 12 | | | |
| 31 | | | | | 1 | 3 | 3 | 5 |
| 33 | | 8 | 5 | 3 | | | | |
| 34 | various II | 3 | 5 | 5 | 1 | 8 | 8 | 7 |
| 35 | | 1 | 23 | 8 | 2 | 16 | 21 | 11 |
| 36 | | | | | | 7 | 13 | 4 |
| 37 | | 14 | 38 | 15 | 5 | | | |
| 40 | | 7 | 17 | 12 | 10 | 7 | 11 | 6 |
| 41 | | 3 | 1 | 2 | 3 | | | |
| 42 | ester-forming | 4 | 4 | 1 | 3 | 2 | 4 | 3 |
| 43 | F5G | | | | | 7 | 9 | 4 |
| 44 | | 4 | 21 | 2 | 1 | | | |
| 45 | F3G | | | | | 2 | 2 | 5 |
| 46 | | 4 | 10 | 8 | 3 | | | |
| 50 | cytokinin *N*-GT | | | | | 8 | 10 | 9 |
| 51 | | 21 | 6 | 4 | 4 | | | |
| 54 | cytokinin *O*-GT | 6 | 20 | 28 | 16 | 17 | 16 | 7 |

## 3.   Discussion

In this study, we achieved a comprehensive phylogenetic analysis of plant UGTs and defined nine sequence groups, characterized by each catalytic specificity. Most of the genes in the dataset are divided into groups, thus we are able to deduce catalytic specificity of plant UGTs to some extent from their sequences. The cytokinin *O*-GT, the cytokinin *N*-GT, and F3G sequence group are specific to one substrate group. On the other hand, the sugar moiety and ester-forming sequence group are not. The various I and II groups show broad substrate specificity, and will be related to various function, including pathogen resistance and detoxyfication of xenobiotics. The various I group has TOGT (cluster 30), which glycosylating phenylpropanoids and benzoic acid derivatives and its biological role is related to response against tobacco mosaic virus [6, 7]. The various II group has interesting two genes, UGT72B1 (cluster 34) and UGT71G1 (cluster 37) highlighting the catalytic plasticity of the group. UGT72B1 is involved in xenobiotic metabolism and shows bifunctional activity as *N*- and *O*-glucosyltransferase [3] and UGT71G1 glycosylates certain flavonoid with higher efficiency than triterpenes in vitro, but its biological role is terpenoid biosynthesis determined by transcript and metabolite profiling [1].

Additionally, we discovered a characteristic conserved glutamine residue in ester-forming group, whose position is Q151 of UGT84A2 (cluster 42). We analyzed this alignment position of 733 genes in the dataset and found that most of genes in the ester-forming group have the Q residue. In contrast, only five genes outside of the group have this residue. Interestingly, three Brassicaceae UGTs, do not have this residue in the ester-forming group (cluster 40) glycosylate thiohydroxymate *S*-GT [46, 52], not ester-forming. The position specific sequence analysis suggests a possibility that the glutamine residue has an important role in ester-forming catalytic specificity.

The genome-wide analysis revealed that widespread distributions of each species UGTs and there are few drastic differences about the number of genes occupied by each group between the species when comparing among dicots and monocots, respectively. We consider that an important extension for understanding glycosylated secondary metabolites is to combine our knowledge with metabolome or transcriptome. Glycosides detected in metabolome will help to determine the catalytic specificity of UGTs. The KNApSAcK database [40, 50] is a comprehensive species-metabolite relationship database. In the database, 59 glycosylated metabolites detected from *Arabidopsis* were registered and 19, 43, and 22 from rice, grapevine, and maize, respectively. Many of those glycosides are indeterminable which UGT catalyzes its glycosylation by the knowledge of substrate specificity annotation in the datasets, for example, Glc(1->2)Rha, Glc(1->2)Xyl, and Glc(1->6)Glc branches of flavonoid 3-*O* position in *Arabidopsis*, flavonoid 3-*O*-Gal in rice, flavonoid 3-*O*-Rha and flavonoid 3-*O*-GlcA in grapevine. In our results, it will be predictable from the groups that Glc(1->2)Rha, Glc(1->2)Xyl, and Glc(1->6)Glc are glycosylated by genes in the sugar moiety group, and 3-*O*-Gal, 3-*O*-Rha, and 3-*O*-GlcA are by genes in F3G group. Thus, omics studies, such as metabolome and transcriptome information will be a next step to understand whole glycosylated secondary metabolite of each plant. Furthermore, comparative genomics of not only UGTs but also whole genes related to secondary metabolic pathways will provide an alternative approach for understanding UGTs function.

## References

[1] Achnine, L., Huhman, D.V., Farag, M.A., Sumner, L.W., Blount, J.W., and Dixon, R.A., Genomics-based selection and functional characterization of triterpene glycosyltransferases from the model legume *Medicago truncatula*, *Plant. J.*, 41:875-887, 2005.

[2] Bowles, D., Isayenkova, J., Lim, E.K., and Poppenberger, B., Glycosyltransferases: managers of small molecules, *Curr. Opin. Plant. Biol.*, 8:254-263, 2005.

[3] Brazier-Hicks, M., Offen, W.A., Gershater, M.C., Revett, T.J., Lim, E.K., Bowles, D.J., Davies, G.J., and Edwards, R., Characterization and engineering of the bifunctional *N*- and *O*-glucosyltransferase involved in xenobiotic metabolism in plants, *Proc. Natl. Acad. Sci. USA.*, 104:20238-20243, 2007.

[4] Cantarel, B.L., Coutinho, P.M., Rancurel, C., Bernard, T., Lombard, V., and Henrissat, B., The Carbohydrate-Active EnZymes database (CAZy): an expert resource for Glycogenomics, *Nucleic Acids Res.*, 37:D233-238, 2008.

[5] Caputi, L., Lim, E.K., and Bowles, D.J., Discovery of new biocatalysts for the glycosylation of terpenoid scaffolds, *Chemistry*, 14:6656-6662, 2008.

[6] Chong, J., Baltz, R., Schmitt, C., Beffa, R., Fritig, B., and Saindrenan, P., Downregulation of a pathogen-responsive tobacco UDP-Glc:phenylpropanoid glucosyltransferase reduces scopoletin glucoside accumulation, enhances oxidative stress, and weakens virus resistance, *Plant Cell*, 14:1093-1107, 2002.

[7] Fraissinet-Tachet, L., Baltz, R., Chong, J., Kauffmann, S., Fritig, B., and Saindrenan, P., Two tobacco genes induced by infection, elicitor and salicylic acid encode glucosyltransferases acting on phenylpropanoids and benzoic acid derivatives, including salicylic acid, *FEBS Lett.*, 437:319-323, 1998.

[8] Gachon, C.M., Langlois-Meurinne, M., and Saindrenan, P., Plant secondary metabolism glycosyltransferases: the emerging functional analysis, *Trends Plant Sci.*, 10:542-549, 2005.

[9] Griesser, M., Hoffmann, T., Bellido, M.L., Rosati, C., Fink, B., Kurtzer, R., Aharoni, A., Muñoz-Blanco, J., and Schwab, W., Redirection of flavonoid biosynthesis through the down-regulation of an anthocyanidin glucosyltransferase in ripening strawberry fruit, *Plant Physiol.*, 146:1528-1539, 2008.

[10] Griesser, M., Vitzthum, F., Fink, B., Bellido, M.L., Raasch, C., Munoz-Blanco, J., and Schwab, W., Multi-substrate flavonol O-glucosyltransferases from strawberry (*Fragaria x ananassa*) achene and receptacle, *J. Exp. Bot.*, 59:2611, 2008.

[11] Hall, D. and De Luca, V., Mesocarp localization of a bi-functional resveratrol/hydroxycinnamic acid glucosyltransferase of Concord grape (*Vitis labrusca*), *Plant J.* 49:579-591, 2007.

[12] Hansen, E.H., Osmani, S.A., Kristensen, C., Møller, B.L., Hansen, and J., Substrate specificities of family 1 UGTs gained by domain swapping, *Phytochemistry*, 70:473-482, 2009.

[13] Hashimoto, K., Tokimatsu, T., Kawano, S., Yoshizawa, A.C., Okuda, S., Goto, S., and Kanehisa, M., Comprehensive analysis of glycosyltransferases in eukaryotic genomes for structural and functional characterization of glycans, *Carbohydr. Res.*, 344:881-887, 2009.

[14] Imanishi, S., Hashizume, K., Nakakita, M., Kojima, H., Matsubayashi, Y., Hashimoto, T., Sakagami, Y., Yamada, Y., and Nakamura, K., Differential induction by methyl jasmonate of genes encoding ornithine decarboxylase and other enzymes involved in nicotine biosynthesis in tobacco cell cultures, *Plant Mol. Biol.*, 38:1101-1111, 1998.

[15] Isayenkova, J., Wray, V., Nimtz, M., Strack, D., and Vogt, T., Cloning and functional characterisation of two regioselective flavonoid glucosyltransferases from *Beta vulgaris*, *Phytochemistry*, 67:1598-1612, 2006.

[16] Jaillon, O., *et. al.,* French-Italian Public Consortium for Grapevine Genome Characterization, The grapevine genome sequence suggests ancestral hexaploidization in major angiosperm phyla, *Nature*, 449:463-467, 2007.

[17] Katoh, K., and Toh, H., Recent developments in the MAFFT multiple sequence alignment program, *Brief Bioinform.,* 9:286-298, 2008.

[18] Kohara, A., Nakajima, C., Hashimoto, K., Ikenaga, T., Tanaka, H., Shoyama, Y., Yoshida, S., and Muranaka, T., A novel glucosyltransferase involved in steroid saponin biosynthesis in *Solanum aculeatissimum*, *Plant Mol. Biol.* 57:225-239, 2005.

[19] Larkin, M.A., Blackshields, G., Brown, N.P., Chenna, R., McGettigan, P.A., McWilliam, H., Valentin, F., Wallace, I.M., Wilm, A., Lopez, R., Thompson, J.D., Gibson, T.J., and Higgins, D.G., Clustal W and Clustal X version 2.0, *Bioinformatics*, 23:2947-2948, 2007.

[20] Lee, H.I. and Raskin, I., Purification, cloning, and expression of a pathogen inducible UDP-glucose:Salicylic acid glucosyltransferase from tobacco, *J. Biol. Chem.*, 274:36637-36642, 1999.

[21] Li, Y., Baldauf, S., Lim, E.K., and Bowles, D.J., Phylogenetic analysis of the UDP-glycosyltransferase multigene family of *Arabidopsis thaliana*, *J. Biol. Chem.*, 9:4338-4343, 2001.

[22] Lim, E.K., Baldauf, S., Li, Y., Elias, L., Worrall, D., Spencer, S.P., Jackson, R.G., Taguchi, G., Ross, J., and Bowles, D.J., Evolution of substrate recognition across a multigene family of glycosyltransferases in *Arabidopsis*, *Glycobiology*, 13:139-145, 2003.

[23] Lim, E.K., Doucet, C.J., Li, Y., Elias, L., Worrall, D., Spencer, S.P., Ross, J., and Bowles, D.J., The activity of Arabidopsis glycosyltransferases toward salicylic acid, 4-hydroxybenzoic acid, and other benzoates, *J. Biol. Chem.*, 277:586-592, 2002.

[24] Lim, E.K., Jackson, R.G., and Bowles, D.J., Identification and characterisation of *Arabidopsis* glycosyltransferases capable of glucosylating coniferyl aldehyde and sinapyl aldehyde, *FEBS Lett.*, 579:2802-2806, 2005.

[25] Lunkenbein, S., Bellido, M., Aharoni, A., Salentijn, E.M., Kaldenhoff, R., Coiner, H.A., Muñoz-Blanco, J., and Schwab, W., Cinnamate metabolism in ripening fruit. Characterization of a UDP-glucose:cinnamate glucosyltransferase from strawberry, *Plant Physiol.*, 140:1047-1058, 2006.

[26] Ma, L.Q., Liu, B.Y., Gao, D.Y., Pang, X.B., Lü, S.Y., Yu, H.S,. Wang, H., Yan, F., Li, Z.Q., Li, Y.F., and Ye, H.C., Molecular cloning and overexpression of a novel UDP-glucosyltransferase elevating salidroside levels in *Rhodiola sachalinensis*, *Plant Cell Rep.*, 989-999, 2007.

[27] Mackenzie, P.I., Rogers, A., Treloar, J., Jorgensen, B.R., Miners, J.O., and Meech, R. Identification of UDP glycosyltransferase 3A1 as a UDP N-acetylglucosaminyltransferase, *J. Biol. Chem.*, 283:36205-36210, 2008.

[28] Matsuba, Y., Okuda, Y., Abe, Y., Kitamura, Y., Terasaka, K., Mizukami, H., Kamakura, H., Kawahara, N., Goda, Y., and Sasaki, N., Enzymatic preparation of 1-*O*-hydroxycinnamoyl-beta-D-glucoses and their application to the study of 1-*O*-hydroxycinnamoyl-beta-D-glucose-dependent acyltransferase in anthocyanin-producing cultured cells of *Daucus carota* and *Glehnia littoralis*, *Plant Biotech.*, 25:369-375, 2008.

[29] McCue, K.F., Allen, P.V., Shepherd, L.V., Blake, A., Maccree, M.M., Rockhold, D.R., Novy, R.G., Stewart, D., Davies, H.V., and Belknap, W.R., Potato glycosterol rhamnosyltransferase, the terminal step in triose side-chain biosynthesis, *Phytochemistry*, 68:327-334, 2007.

[30] Meesapyodsuk, D., Balsevich, J., Reed, D.W., and Covello, P.S., Saponin biosynthesis in Saponaria vaccaria. cDNAs encoding beta-amyrin synthase and a triterpene carboxylic acid glucosyltransferase, *Plant Physiol.*, 143:959-969, 2007.

[31] Milkowski, C., Baumert, A., and Strack, D., Cloning and heterologous expression of a rape cDNA encoding UDP-glucose:sinapate glucosyltransferase, *Plnata*, 211:883-886, 2000.

[32] Mittasch, J., Strack, D., and Milkowski, C., Secondary product glucosyltransferases in seeds of *Brassica napus*, *Planta*, 225:515-522, 2007.

[33] Morita, Y., Hoshino, A., Kikuchi, Y., Okuhara, H., Ono, E., Tanaka, Y., Fukui, Y., Saito, N., Nitasaka, E., Noguchi, H., and Iida, S., Japanese morning glory *dusky* mutants displaying reddish-brown or purplish-gray flowers are deficient in a novel glycosylation enzyme for anthocyanin biosynthesis, UDP-glucose:anthocyanidin 3-*O*-glucoside-2"-*O*-glucosyltransferase, due to 4-bp insertions in the gene, *Plant J.*, 42:353-363, 2005.

[34] Noguchi, A., Fukui, Y., Iuchi-Okada, A., Kakutani, S., Satake, H., Iwashita, T., Nakao, M., Umezawa, T., and Ono, E., Sequential glucosylation of a furofuran lignan, (+)-sesaminol, by *Sesamum indicum* UGT71A9 and UGT94D1 glucosyltransferases, *Plant. J.*, 54:415-427, 2008.

[35] Noguchi, A., Horikawa, M., Fukui, Y., Fukuchi-Mizutani, M., Iuchi-Okada, A., Ishiguro, M., Kiso, Y., Nakayama, T., and Ono, E., Local differentiation of sugar donor specificity of flavonoid glycosyltransferase in Lamiales, *Plant Cell*, 21:1556-1572, 2009.

[36] Noguchi, A., Nobuhiro, S., Masahiro, N., Harukazu, F., Seiji, T., Tokuzo, N., Toru, N., J. Mol. and Catal. B, cDNA cloning of glycosyltransferases from Chinese wolfberry (*Lycium barbarum* L.) fruits and enzymatic synthesis of a catechin glucoside using a recombinant enzyme (UGT73A10), *J. Mol. Catal. B*, 55:84-92, 2008.

[37] Paquette, S., Møller, B.L., and Bak, S., *Phytochemistry*, 62:399-413.

[38] Richman, A., Swanson, A., Humphrey, T., Chapman, R., McGarvey, B., Pocs, R., and Brandle, J., Functional genomics uncovers three glucosyltransferases involved in the synthesis of the major sweet glucosides of *Stevia rebaudiana*, *Plant J.* 41:56-67, 2005.

[39] Sawada, S., Suzuki, H., Ichimaida, F., Yamaguchi, M.A., Iwashita, T., Fukui, Y., Hemmi, H., Nishino, T., and Nakayama, T., UDP-glucuronic acid:anthocyanin

glucuronosyltransferase from red daisy (*Bellis perennis*) flowers. Enzymology and phylogenetics of a novel glucuronosyltransferase involved in flower pigment biosynthesis., *J. Biol. Chem.*, 280:899-906, 2005.

[40] Shinbo, Y., Nakamura, Y., Altaf-Ul-Amin, M., Asahi, H., Kurokawa, K., Arita, M., Saito, K., Ohta, D., Shibata, D., and Kanaya, S. (2006) in Plant Metabolomics (Saito, K., Dixon, R., and Willmitzer, L., eds) Vol. 57, pp. 165–184, *Springer Verlag*, Heidelberg, Germany

[41] Suzuki, H., Hayase, H., Nakayama, A., Yamaguchi, I., Asami, T., and Nakajima, M., Identification and characterization of an *Ipomoea nil* glucosyltransferase which metabolizes some phytohormones, *Biochem. Biophys. Res. Commun.*, 361:980-986, 2007.

[42] Ueyama, Y., Katsumoto, Y., Fukui, Y., Fukuchi-Mizutani, M., Ohkawa, H., Kusumi, T., Iwashita, T., and Tanaka, Y., Molecular characterization of the flavonoid biosynthetic pathway and flower color modification of *Nierembergia* sp., *Plant Biotech.*, 23:19-24, 2006.

[43] Umemura, K., Satou, J., Iwata, M., Uozumi, N., Koga, J., Kawano, T., Koshiba, T., Anzai, H., and Mitomi, M., Contribution of salicylic acid glucosyltransferase, OsSGT1, to chemically induced disease resistance in rice plants, *Plant J.*, 57:463-472, 2009.

[44] Vogt, T. and Jones, P., Glycosyltransferases in plant natural product synthesis: characterization of a supergene family, *Trends Plant Sci.*, 5:380-386, 2000.

[45] Yamazaki, M., Gong, Z., Fukuchi-Mizutani, M., Fukui, Y., Tanaka, Y., Kusumi, T., and Saito, K., Molecular cloning and biochemical characterization of a novel anthocyanin 5-*O*-glucosyltransferase by mRNA differential display for plant forms regarding anthocyanin, *J. Biol. Chem.*, 12:7405-7411, 1999.

[46] Yonekura-Sakakibara, K., Functional genomics of family 1 glycosyltransferases in *Arabidopsis*, *Plant Biotech.*, 26:267-274, 2009.

[47] Yonekura-Sakakibara, K., Tohge, T., Matsuda, F., Nakabayashi, R., Takayama, H., Niida, R., Watanabe-Takahashi, A., Inoue, E., and Saito, K., Comprehensive flavonol profiling and transcriptome coexpression analysis leading to decoding gene-metabolite correlations in *Arabidopsis*, *Plant Cell*, 20:2160-2176, 2008.

[48] Yonekura-Sakakibara, K., Tohge, T., Niida, R., and Saito, K., Identification of a flavonol 7-*O*-rhamnosyltransferase gene determining flavonoid pattern in *Arabidopsis* by transcriptome coexpression analysis and reverse genetics, *J. Biol. Chem.*, 282:14932-14941, 2007.

[49] http://itol.embl.de/

[50] http://kanaya.naist.jp/KNApSAcK/

[51] http://www.cazy.org/

[52] http://www.uniprot.org/

# ROBUST GENE NETWORK ANALYSIS REVEALS ALTERATION OF THE STAT5a NETWORK AS A HALLMARK OF PROSTATE CANCER

ANUPAMA REDDY[1]
anupamar@gmail.com

C. CHRIS HUANG[2]
chuang4@its.jnj.com

HUIQING LIU[2]
HLiu32@its.jnj.com

CHARLES DELISI[3]
charlesdelisi@gmail.com

MARJA T. NEVALAINEN[4]
Marja.Nevalainen@jefferson.edu

SANDOR SZALMA[5]
sszalma@its.jnj.com

GYAN BHANOT[1,6,7,*]
gyanbhanot@gmail.com

[1] BioMaPS Institute, Rutgers University, Piscataway, NJ, USA
[2] Centocor R&D, Inc., 145 King of Prussia Rd, Radnor, PA, USA
[3] Center for Advanced Genomic Technology, Boston University, Boston, MA, USA
[4] Kimmel Cancer Center, Thomas Jefferson University, Philadelphia 19107 P, USA
[5] Centocor R&D, Inc., 3210 Merryfield Row, San Diego, CA, USA
[6] Department of Biology and Biochemistry, Department of Physics, Rutgers University, Piscataway, NJ, USA
[7] Simons Center for Systems Biology, Institute for Advanced Study, Princeton, NJ, USA
* Corresponding author

We develop a general method to identify gene networks from pair-wise correlations between genes in a microarray data set and apply it to a public prostate cancer gene expression data from 69 primary prostate tumors. We define the degree of a node as the number of genes significantly associated with the node and identify hub genes as those with the highest degree. The correlation network was pruned using transcription factor binding information in VisANT (http://visant.bu.edu/) as a biological filter. The reliability of hub genes was determined using a strict permutation test. Separate networks for normal prostate samples, and prostate cancer samples from African Americans (AA) and European Americans (EA) were generated and compared. We found that the same hubs control disease progression in AA and EA networks. Combining AA and EA samples, we generated networks for low ($<7$) and high ($\geq 7$) Gleason grade tumors. A comparison of their major hubs with those of the network for normal samples identified two types of changes associated with disease: (i) Some hub genes increased their degree in the tumor network compared to their degree in the normal network, suggesting that these genes are associated with gain of regulatory control in cancer (e.g. possible turning on of oncogenes). (ii) Some hubs reduced their degree in the tumor network compared to their degree in the normal network, suggesting that these genes are associated with loss of regulatory control in cancer (e.g. possible loss of tumor suppressor genes). A striking result was that for both AA and EA tumor samples, STAT5a, CEBPB and EGR1 are major hubs that gain neighbors compared to the normal prostate network. Conversely, HIF-1α is a major hub that loses connections in the prostate cancer network compared to the normal prostate network. We also find that the degree of these hubs changes progressively from normal to low grade to high grade disease, suggesting that these hubs are master regulators of prostate cancer and marks disease progression. STAT5a was identified as a central hub, with ~120 neighbors in the prostate cancer network and only 81 neighbors in the normal prostate network. Of the 120 neighbors of STAT5a, 57 are known cancer related genes, known to be involved in functional pathways associated with tumorigenesis. Our method is general and can easily be extended to identify and study networks associated with any two phenotypes.

*Keywords*: network analysis; microarray analysis; prostate cancer; STAT5a; cancer progression.

## 1.    Introduction

It is estimated that about 1.5 million new cases of cancer will be diagnosed in 2009 in the US alone of which about 766,000 will be in men and about 713,000 in women. 25% of all male cancers will be in the prostate (www.cancer.org). Improved diagnosis methods and effective surgical intervention followed by radiation and/or hormone treatment have reduced mortality to ~20%, however, prostate cancer is still second in cancer-related mortality in men, after lung/bronchial cancers. Prostate-specific antigen (PSA) is one of the few reliable early indicators for prostate cancer and almost half the diagnosed prostate cancers are identified due to biopsy following elevated PSA levels. PSA is also used, with significantly less success, to monitor progression and recurrence.

A significant concern relates to the fact that prostate cancer is often diagnosed on autopsy in young accident victims, suggesting that we might be over diagnosing/treating prostate cancer. This suggests that molecular identification of markers associated with aggressive tumors may suggest which tumors to treat and which to leave alone. Another concern is that once the tumor becomes refractory to hormone treatment, it is very difficult to treat and often becomes metastatic. Hence, a major clinical quandary at diagnosis is which tumors to treat and which to leave alone? In such a complex clinical setting, identifying the key markers which track disease progression becomes an urgent need.

Most clinical assays use single gene markers to track disease stage, grade and progression. Gene expression studies on the other hand, generate long lists of genes significantly associated with disease but have difficulty identifying biological mechanisms or key genes associated with disease progression which might reliably complement PSA and Gleason grade (Gleason grade is a measure used to assess disease progression on the basis of histology). The goal of our paper is to develop a simple approach which uses correlations between genes in expression datasets to identify the network associated with normal prostate and low and high grade disease, and to identify the key nodes associated with progression. To reduce the number of false positive hubs, we apply the transcription factor binding network from VisANT (http://visant.bu.edu) as a filter.

There are two popular approaches to building networks:

(i) *knowledge-based networks*: These are networks built from mining the literature. Some examples are protein-protein interaction (PPI) networks [19], transcription factor networks [29], pathways databases (KEGG, BIOCARTA, etc.)

(ii) *high dimensional data-driven networks*: these are networks built from molecular or genetic data These methods use correlation [3, 18, 14, 23]. Li and Horvath [15] proposed a robust measure called Topological Overlap measure (TOM) for identifying modules associated with disease in microarray data. Yeung and colleagues [30] built gene networks by reverse engineering methods from microarray data and then used these networks as filters for predicting outcome in prostate cancer.

Knowledge based networks are generally reliable, but they tend not to suggest novel targets or mechanisms because of their bias towards "popular" genes/proteins. Data-driven networks use an unbiased approach but tend to be noisy and generate many false positives. A possible improved approach to network analysis involves combining knowledge-based and data-based networks (see [4], [27] for some examples of this approach). In this paper we propose a novel mixed approach which combines knowledge-based and data-driven networks to find a small number of reliable and biologically relevant markers of progression in high throughput data.

## 2. Methods

### 2.1. Preprocessing of microarray data

We used data from previously published molecular profiling of prostate cancer by Ambs *et al.* [2, 28]. The data consisted of expression levels for 69 prostate cancer samples, 18 adjacent normal samples and 2 pooled normal prostate samples on Affymetrix U133A_2 chips. The 69 prostate cancer samples consisted of 33 African American (AA) and 36 European American (EA) patients. CEL files were downloaded from GEO (GSE6569) and RMA normalized [13]. Probe annotations were updated using the CDF file provided from MBNI version 12, Ensembl gene database [7, 22]. The MBNI algorithm re-annotates the Affymetrix probes to the current build of the human genome and retains only those probes which map to unique regions. Finally FARMS I/NI filtering [24] was performed to remove uninformative and noisy probes.

### 2.2. Co-regulation of genes

Correlation (co-regulation) between pairs of genes was measured using Spearman Rank Correlation Coefficient. We chose this measure because it is more insensitive to noise in the data than others (e.g. Pearson correlation coefficient). To assess the significance of the correlation, i.e., probability that the correlation is not by chance, we created empirical null distributions by randomly permuting the samples for a given pair of genes. These permutation tests were run 10,000 times for every pair of genes. The correlation between each pair of genes was considered significant if the p-value of finding the same or higher correlation in the empirical null distribution were < 0.1, i.e., for each gene-gene pair, less than 10% of the correlations in randomized permutation tests are larger than the actual correlation. Using such an empirical p-value to assess the significance of the correlation is better than using a threshold on the correlations, because it allows every gene pair to have its own significance level based on the noise inherent to that pair. It is also the preferred way to compare different datasets.

### 2.3. Network analysis

The correlations are input into a matrix $C$, called the "connectivity matrix" of size $N_g \times N_g$ where $N_g$ = number of genes. The entry $C_{ij}$ was set to 1 if genes $i$, $j$ interact (using a criterion defined below) and 0 otherwise. In our definition, two genes were connected (and the corresponding entry in C was 1) if they were significantly co-regulated at p-value < 0.1 relative to their data-inferred null distribution. Note that we do not distinguish between positive and negative correlations or between association and actual (direct physical) connections. To distinguish between these, it is necessary to use additional (biological or other) information as a filter. In this paper we used the transcription factor networks [10] as implemented in VisANT (http://visant.bu.edu; [11]) for this purpose. The transcription factor network was overlaid on the co-regulation network obtained from gene expression data, and only the intersection was retained. We identified hubs as genes with degree > 15 in the resulting networks.

### 2.4. Stability analysis by bootstrapping

Bootstrap analysis was used to quantify the stability of hubs under sample perturbation. The network analysis was repeated 100 times for subsets of 80% of the samples, and the degree and connections of the hubs were determined for each subset. This analysis also provided the mean and 95% confidence interval for the degree of each hub, which provided a measure of significance for changes in the hub degree when comparing between two phenotypes (e.g. normal vs. disease).

### 3.   Results

We downloaded gene-expression data from [2], filtered probes using MBNI probe annotations, normalized the data using RMA, and filtered the informative genes using FARMS I/NI. MBNI retains only those probes which map to a unique region in the human genome. This reduced the reduced the number of probes from ~ 22,000 to 11,000. FARMS I/NI filtering for informative probes further reduced this number to 5,961 probes. Separate networks were built using the genes corresponding to this reduced set of probes for African American prostate cancer (AA) samples, and European American prostate cancer (EA) samples. Independently, networks were also built from normal prostate, low and high Gleason grade tumor datasets. The major hubs in each of these phenotypes were compared to the others to understand their role in disease networks.

### 3.1. Hubs in AA & EA prostate cancer networks are similar

The detailed networks for African American and European American prostate cancer samples are shown in Supplementary Figure 1. The reliable hubs in the AA & EA networks have similar degrees (Fig. 1a) suggesting that at least to the accuracy of this analysis, race does not distinguish disease progression in AA or EA patients. This result was validated by principal component analysis on the combined dataset (Fig. 1b). We see that the AA (blue) & EA (green) samples do not form separate clusters when projected onto their first two principle components in Fig. 1b. Thus, to the accuracy of this dataset, the network structure and topology for both normal prostate and prostate cancer does not depend on ethnicity. Consequently, in the subsequent analysis, we combine the AA & EA samples within each phenotype when comparing networks associated with progression from normal prostate to low and high grade disease.

### 3.2. Prostate cancer and normal networks show similar genes and structure

The networks for the 18 normal prostate (adjacent tissue) samples, 18 low Gleason grade (<7) samples and 51 high Gleason grade (≥ 7) samples are shown in Supplementary Figures 2 a, b and c respectively. Even though these networks were built independently, they show a very similar overall structure in terms of the genes involved in the network and the relative importance of the hubs. This suggests that progression to prostate cancer does not require a complete change in the topology of the normal prostate network. Rather, prostate cancer initiation and progression is measured by which how many genes are co-regulated by the key hubs active in normal prostate tissue.

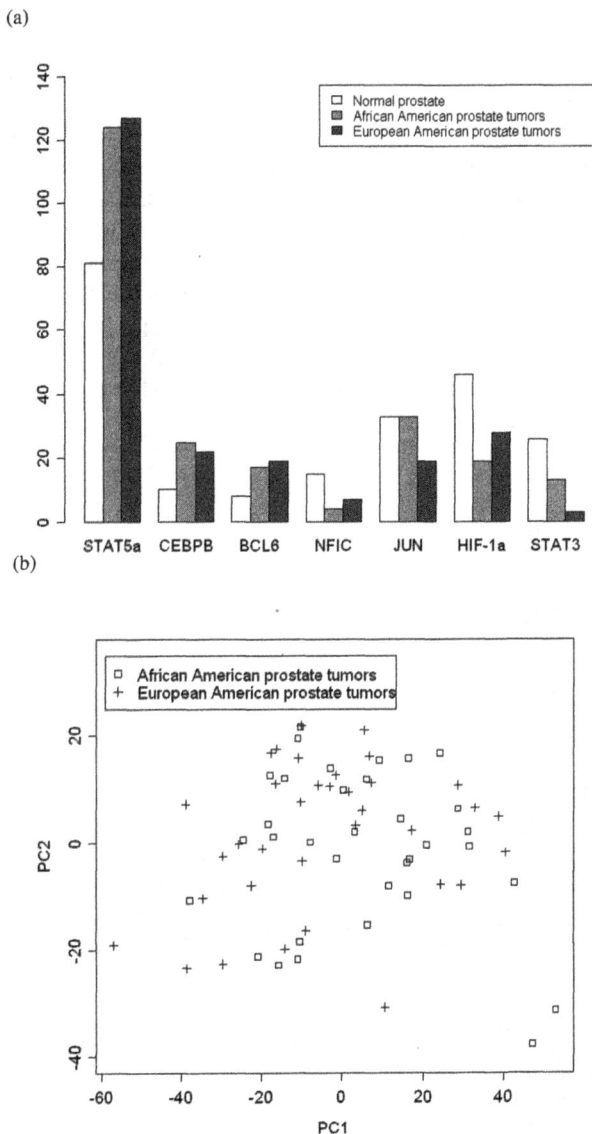

(a)

(b)

Fig. 1. The degree of key hubs in African American (AA), European American (EA) and normal prostate networks. (a) The degree of the hubs for AA and EA networks are not significantly different. This means that that race is not a confounding issue in tumor progression. (b) PCA plot of AA and EA prostate cancer samples. AA & EA samples cluster together in the projection of their first two PCs.

### *3.3. STAT5a is the largest hub in prostate cancer networks*

The major result of our analysis is the strong identification of STAT5a as the largest hub gene altered in disease and disease progression (Fig. 2). STAT5a is a transcription factor on chromosome 17 involved in the JAK/STAT pathway and known to be strongly associated with the initiation and progression of prostate cancer [6, 27, 29]. The degree of STAT5a in the normal, low grade and high grade prostate cancer networks is 81, 116 and 122 respectively. It has by far the highest degree (number of neighbors) compared to all other hubs in the prostate networks. Of the 120 neighbors of STAT5a in prostate cancer networks, 57 are known cancer related genes, involved in known functional pathways associated to tumorigenesis. Interestingly, the levels of STAT5a can distinguish normal samples from low grade tumors (p-value = 0.036) and high grade tumors (p-value = 0.016). However, although its connectivity changes significantly from low to high grade, STAT5a expression is not significantly different between high and low grade tumors. This suggests that the recruitment of additional components to the STAT5a node, once the tumor is established, does not depend on the level of STAT5a; i.e. prostate cancer requires changes in the level of STAT5a to cause the changes in its network which are necessary tumor establishment but loses addiction to STAT5a level after the tumor is established. This may mean that any therapy based on STAT5a antagonists may be most effective mostly in low grade disease.

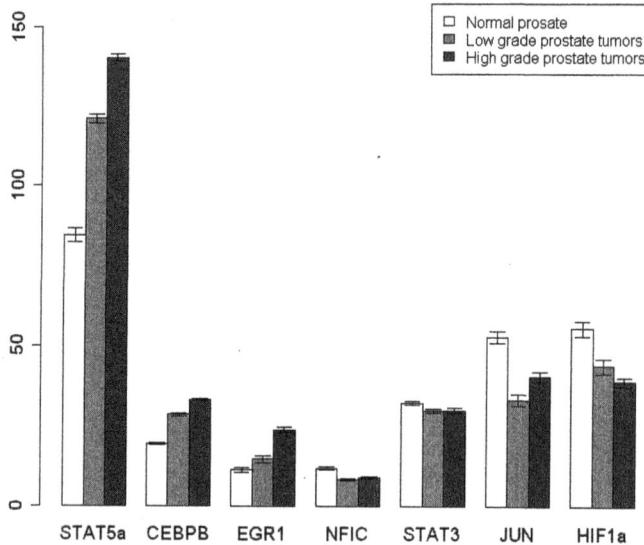

Fig. 2. Hubs associated with initiation and progression of prostate cancer. Key gene hubs were identified for normal prostate (white), low (< 7) Gleason grade tumors (light grey) and high (> 7) Gleason grade tumors (dark grey). Hubs define highly co-regulated genes which had more than 15 other genes significantly associated with them determined by a combination of expression and transcription factor data. The plot shows the mean degree (number of co-regulated genes) of highly connected hubs. The error bars are 95% confidence intervals using sample bootstrap experiments. Note that the degree of some of the hubs is monotone with respect to progression of disease, i.e. the degree in low grade tumors is between the degree in high grade tumors and normal prostate. STAT5a, CEBPB, EGR1 are "gain hubs", where the tumor progressively gains connections compared to normal prostate. HIF-1a is a "loss hub", where the tumor progressively loses connections compared to the normal prostate.

## 3.4. Hubs are associated with progression of prostate cancer

Fig. 2 compares the degrees for the significant and highly connected hubs (degree > 15) in the normal prostate and low and high Gleason grade tumor networks. The genes at the center of these hubs are STAT5a, CEBPB, EGR1, NFIC, STAT3, JUN and HIF1a. There seem to be two types of disease associated changes in hubs: (i) Some hubs (e.g. STAT5a, CEBPB and EGR1) increase their degree in tumor samples compared to normal samples, suggesting their association with gain of regulatory control in cancer (possible turning on of oncogenes or other tumorigenic processes). (ii) Some other hubs (we identify HIF-1α as one of these) decrease their degree in tumor samples compared to normal samples, suggesting their association with loss of regulatory control in cancer (possible loss of tumor suppressor genes or turning off of tumor suppressive processes). Significantly, the degree of these hubs changes progressively from normal to low grade to high grade disease, suggesting their strong association with disease initiation and progression and identifying them genes as potential targets for therapeutic intervention. Table 1 summarizes the hubs we identified. It also gives p-values for hub identification

Table 1. Differential expression of hubs identified in regulatory networks. Significance level of hubs associated with low and high grade tumors using single-gene analysis only. Student's t-test p-values for differentiating pair-wise between normal prostate, low grade tumor and high grade tumor are shown with those significant at p-value <0.05 shown in bold (these hubs can be identified by single link analysis using the t-test). However, the rest of the hubs (those with poor p-values) can only be identified by the co-regulation analysis we present here. They are not identifiable using single gene analysis. The functional pathways associated with these genes are presented in Supplementary Table 1.

| Gene Name | Type of hub | Normal vs. Low grade tumors | Normal vs. High grade tumors | Low vs. High grade tumors |
|---|---|---|---|---|
| | | t-test p-value | | |
| STAT5a signal transducer and activator of transcription 5A | Gain | **0.036** | **0.016** | 0.819 |
| CEBPB CCAAT/enhancer binding protein (C/EBP), beta | Gain | 0.570 | **0.041** | 0.237 |
| EGR1 early growth response 1 | Gain | 0.601 | 0.362 | 0.142 |
| NFIC nuclear factor I/C (CCAAT-binding transcription factor) | | 0.897 | 0.820 | 0.776 |
| STAT3 signal transducer and activator of transcription 3 | | **0.001** | **0.005** | **0.002** |
| JUN jun oncogene | | 0.173 | 0.070 | 0.707 |
| HIF-1a hypoxia inducible factor 1, alpha subunit | Loss | **0.026** | 0.050 | 0.464 |

using a t-test to identify single genes discriminating between normal vs low grade and low grade vs high grade tumors. The results in Table 1 show that whereas a simple t-test can identify some of the hubs, it does not identify them all. Nor can it identify the genes associated with the hubs.

### 3.5. The STAT5a hub gains neighbors in prostate cancer (Fig 3)

The STAT5a hub gained 38 connections from normal to low grade tumors (Table 2a, Fig. 3a,b). These include genes known to be involved in many cancer related pathways: MAPK signaling (AKT1, CACNA2D2, FGFR2, DUSP8), ERBB signaling (AKT1, CAMK2B), mTOR signaling (AKT1), Wnt signaling (CAMK2B, FZD10), focal adhesion (AKT1, IGF1R), autophagy (ATG4B), etc. STAT5a gained another 21 connections when transitioning from low grade to high grade tumors (Table 2b, Fig. 3a,c), including genes in MAPK signaling (CACNA1H, FGFR3), apoptosis and toll-like receptor signaling

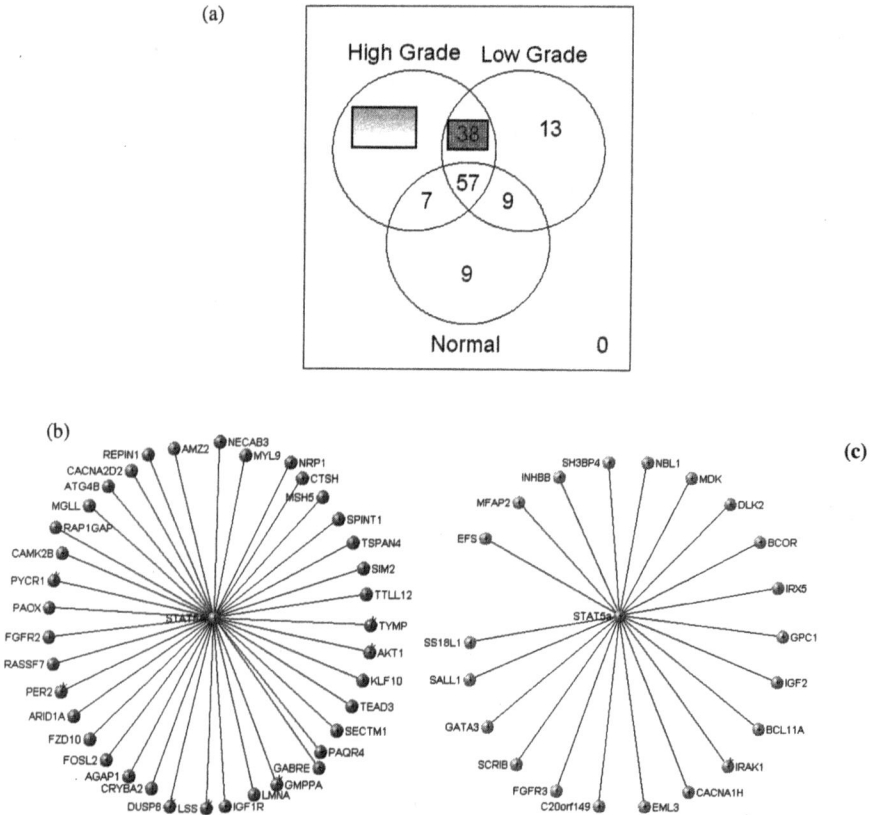

Fig. 3. STAT5a hub in prostate cancer. (a) Venn diagram showing the number of genes connected to STAT5A in normal, low grade and high grade tumors and their intersections. (b) Network connections gained by STAT5a in transformation from normal prostate and low or high grade tumors. (c) Network connections gained by STAT5a in progression from low grade to high grade prostate cancer.

(IRAK1), TGFβ signaling (INHBB), etc. Of these neighbors gained AKT1, FGFR2, FGFR3 are known oncogenes, while IGF2 is a known tumor suppressor gene. In Table 2 we also provide the p-value for differential expression pair-wise between normal, low grade and high grade tumors. Note that not all the genes are differentially expressed between normal samples and low grade tumors, suggesting that the complexity of cancer progression is not captured by expression analysis alone.

Table 2. Gene connections gained by STAT5a from (a) normal prostate to low grade tumor and (b) in low grade to high grade progression. The p-values shown are from using single gene analysis and the Student's t-test, with significant p-values (< 0.05) marked in bold. Note again that single gene analysis is unable to identify many of these associations (those with poor p-values). However, the co-regulation analysis and transcription factor filter used in our method was able to identify all the genes shown. The functional pathways associated with these genes are shown in Supplementary Table 2.

Table 2(a)

| Gene name | Normal vs. low grade tumor | Normal vs. high grade tumor | Low vs. high grade tumor |
|---|---|---|---|
| | Student's t-test p-value | | |
| AKT1 <br> v-akt murine thymoma viral oncogene homolog 1 | **0.001** | **0.001** | 0.770 |
| RAP1GAP <br> RAP1 GTPase activating protein | **0.001** | **0.000** | 0.251 |
| APBA2BP <br> N-terminal EF-hand calcium binding protein 3 | **0.001** | **0.001** | 0.626 |
| ARID1A <br> AT rich interactive domain 1A (SWI-like) | **0.001** | **0.001** | **0.047** |
| CACNA2D2 <br> calcium channel, voltage-dependent, alpha 2/delta subunit 2 | **0.002** | **0.022** | 0.143 |
| ATG4B <br> ATG4 autophagy related 4 homolog B (S. cerevisiae) | **0.002** | **0.001** | 0.638 |
| SPINT1 <br> serine peptidase inhibitor, Kunitz type 1 | **0.002** | **0.001** | 0.465 |
| GMPPA <br> GDP-mannose pyrophosphorylase A | **0.003** | **0.009** | 0.169 |
| CAMK2B <br> calcium/calmodulin-dependent protein kinase II beta | **0.003** | **0.000** | 0.999 |
| AMZ2 <br> archaelysin family metallopeptidase 2 | **0.003** | **0.002** | 0.516 |
| TTLL12 <br> tubulin tyrosine ligase-like family, member 12 | **0.004** | **0.000** | 0.195 |
| MSH5 <br> mutS homolog 5 (E. coli) | **0.005** | **0.001** | 0.169 |
| FGFR2 <br> fibroblast growth factor receptor 2 | **0.008** | 0.114 | 0.161 |
| REPIN1 <br> replication initiator 1 | **0.013** | **0.002** | **0.047** |
| LSS <br> lanosterol synthase (2,3-oxidosqualene-lanosterol cyclase) | **0.014** | 0.159 | **0.023** |
| TSPAN4 <br> tetraspanin 4 | **0.016** | **0.038** | 0.266 |
| KLF10 <br> Kruppel-like factor 10 | **0.020** | **0.011** | 0.371 |

| | | | |
|---|---|---|---|
| PYCR1<br>sirtuin (silent mating type information regulation 2 homolog) 7 (S. cerevisiae) | **0.023** | **0.005** | 0.586 |
| CTSH<br>cathepsin H | **0.039** | **0.007** | 0.526 |
| MGLL<br>monoglyceride lipase | 0.050 | 0.208 | 0.112 |
| CENTG2<br>ArfGAP with GTPase domain, ankyrin repeat and PH domain 1 | 0.051 | **0.007** | 0.175 |
| TEAD3<br>TEA domain family member 3 | 0.052 | 0.135 | 0.148 |
| IGF1R<br>insulin-like growth factor 1 receptor | 0.053 | 0.050 | 0.823 |
| MYL9<br>myosin, light chain 9, regulatory | 0.065 | 0.542 | 0.075 |
| GABRE<br>gamma-aminobutyric acid (GABA) A receptor, epsilon | 0.073 | 0.061 | 0.918 |
| SIM2<br>single-minded homolog 2 (Drosophila) | 0.083 | **0.001** | 0.186 |
| NRP1<br>neuropilin 1 | 0.096 | 0.189 | 0.656 |
| PAQR4<br>progestin and adipoQ receptor family member IV | 0.211 | 0.944 | 0.052 |
| PER2<br>period homolog 2 (Drosophila) | 0.250 | 0.532 | 0.525 |
| ECGF1<br>sphingosine-1-phosphate receptor 1 | 0.272 | 0.086 | 0.710 |
| FZD10<br>frizzled homolog 10 (Drosophila) | 0.310 | **0.049** | 0.404 |
| LMNA<br>lamin A/C | 0.406 | 0.085 | 0.427 |
| PAOX<br>polyamine oxidase (exo-N4-amino) | 0.422 | 0.470 | 0.858 |
| RASSF7<br>Ras association (RalGDS/AF-6) domain family (N-terminal) member 7 | 0.439 | **0.031** | 0.065 |
| SECTM1 secreted and transmembrane 1 | 0.513 | **0.030** | 0.068 |
| DUSP8<br>dual specificity phosphatase 8 | 0.666 | 0.542 | 0.966 |
| FOSL2<br>FOS-like antigen 2 | 0.776 | 0.809 | 0.605 |
| CRYBA2<br>crystallin, beta A2 | 0.787 | 0.934 | 0.663 |

Table 2(b)

| Gene name | Normal vs. low grade tumor | Normal vs. high grade tumor | Low vs. high grade tumor |
|---|---|---|---|
| | Student's t-test p-value | | |
| EFS<br>embryonal Fyn-associated substrate | **0.009** | 0.147 | **0.000** |
| NBL1<br>neuroblastoma, suppression of tumorigenicity 1 | **0.026** | 0.179 | **0.000** |

| | | | |
|---|---|---|---|
| GPC1<br>glypican 1 | **0.028** | 0.868 | **0.003** |
| SCRIB scribbled homolog (Drosophila) | **0.035** | **0.007** | **0.005** |
| EML3<br>echinoderm microtubule associated protein like 3 | **0.001** | **0.000** | 0.025 |
| IRAK1<br>interleukin-1 receptor-associated kinase 1 | 0.392 | **0.049** | **0.100** |
| SS18L1<br>synovial sarcoma translocation gene on<br>chromosome 18-like 1 | **0.010** | **0.003** | 0.130 |
| DLK2<br>delta-like 2 homolog (Drosophila) | 0.840 | 0.179 | 0.171 |
| BCOR<br>BCL6 co-repressor | 0.089 | **0.018** | 0.209 |
| C20orf149<br>pancreatic progenitor cell differentiation and<br>proliferation factor homolog (zebrafish) | 0.074 | 0.173 | 0.223 |
| IGF2<br>insulin-like growth factor 2 (somatomedin A);<br>insulin; INS-IGF2 readthrough transcript | 0.076 | 0.317 | 0.247 |
| CACNA1H<br>calcium channel, voltage-dependent, T type, alpha<br>1H subunit | **0.037** | 0.172 | 0.249 |
| IRX5<br>iroquois homeobox 5 | 0.076 | 0.220 | 0.300 |
| MDK<br>Mesomelic dysplasia, Kantaputra type | 0.128 | 0.216 | 0.465 |
| SALL1<br>sal-like 1 (Drosophila) | 0.697 | 0.623 | 0.511 |
| FGFR3<br>fibroblast growth factor receptor 3 | **0.047** | 0.103 | 0.548 |
| INHBB<br>inhibin, beta B | 0.938 | 0.588 | 0.585 |
| BCL11A<br>B-cell CLL/lymphoma 11A (zinc finger protein) | 0.791 | 0.618 | 0.796 |
| GATA3<br>GATA binding protein 3 | 0.617 | 0.468 | 0.801 |
| MFAP2<br>microfibrillar-associated protein 2 | **0.007** | **0.002** | 0.909 |
| SH3BP4<br>SH3-domain binding protein 4 | 0.167 | 0.165 | 0.989 |

## 4. Discussion

Cancers are diseases of dysregulated pathways. Disease initiation is triggered by multiple changes in the wiring of the homeostatic network of normal tissue and disease progression caused by further changes which allow the cancer to evade regulatory pathways and drugs, spread outside its tissue of origin and establish in other organs. In spite of this well known model for cancer and the knowledge that genes do not act independently, most laboratory studies as well as studies of high throughput cancer data try to identify individual genes to understand disease initiation, progression and metastasis. In this paper we develop a complementary approach to infer the network of co-regulated genes which is dysregulated when normal tissue transforms to low grade cancer and when the low grade cancer progresses to high grade disease.

By using co-regulation of genes in microarray data to define a preliminary network, reducing the noise using strict statistical tests and retaining only biologically significant hubs using transcription factor information, our method successfully identified robust hubs in the gene networks of normal prostate and low/high grade prostate cancer. We consider our method as complementary and orthogonal to single gene analysis or analysis from pathways identified based on aggregates of single gene based results. The method we propose is general and can be used to analyze in a host of other contexts (such as protein-protein interaction networks, etc). It can also be extended to different data modalities (copy number variation data, single nucleotide polymorphisms, microRNA expression etc.). With additional data and other data modalities it should be able to address more complex biological questions. Changing the measure we use (co-regulation of genes) to identify genes with high centrality measure (genes with highest number of shortest paths going through it) would suggest how the pathway changes might affect tissue function. It is also possible to define measures to identify modules or sub-networks characteristic of disease or progression. Other extensions involve comparing groups of samples to identify similarities and differences. Cell line and animal studies and data are often used to model human tumors. Our method should be useful on such data to quantify the similarities and differences between the phenotypes in these studies to generate hypotheses for further testing.

To assess the significance of co-regulation identified we used the p-value of an empirically derived null distribution instead of using a threshold on the correlation coefficient. The advantages of this are: (a) threshold selection is often arbitrary and is imposed globally for all genes pairs, while in fact each gene should have its own threshold (b) selecting p-value threshold in this way is independent of the sample size (c) comparing p-values across dataset is now meaningful.

Using only data driven networks, we cannot distinguish between correlations which are just associations from those that are likely to be functionally relevant connections. When we used only correlation networks, we identified FGF2 as a major hub (with ~1000 connections). On further analysis we found that the reason for this is probably because FGF2 is upstream of the FGF signaling pathways in prostate cancer, and the correlations that we observe in the data are mainly associations. To distinguish between associations and actual connections we added the transcription factor network as a biological filter. This increased both the robustness of the underlying network and also identified potentially relevant (possibly mechanistic) connections.

Our analysis identified STAT5a as a major hub in the disease initiation and progression network of prostate cancer. Signal transducer and activator of transcription 5a and 5b (STAT5a/b) belong to the seven-member STAT gene family of transcription factors [26, 12]. STAT5a and STAT5b, encoded by two separated genes at chromosome 17q21, are latent cytoplasmic proteins that act as both cytoplasmic signaling proteins and nuclear transcription factors. Activations of STAT5a/b are mediated by phosphorylation of a conserved tyrosine residue (Y694 for STAT5a and Y699 for STAT5b) in the carboxy-terminal domain by a tyrosine kinase typically of the JAK protein family [20, 21]. The phosphorylated STAT5a and STAT5b dimerize, translocate into the nucleus and bind to specific STAT5 response elements of target gene promoters [8].

STAT5 activation is known to be strongly associated with high histological grade of prostate cancer [16, 17], and STAT5 activation in primary prostate cancer predicts development of castration-resistant recurrent prostate cancer [17]. STAT5 critically regulates growth and viability of human prostate cancer cells in culture and prostate cancer xenograft tumor growth in nude mice [1, 5, 6, 9]. Specifically, adenoviral expression of a dominant-negative (DN) mutant of STAT5, antisense oligonucleotide or siRNA inhibition of STAT5 all induce massive and rapid apoptotic death of human prostate cancer cells in culture [1]. In addition, inhibition of STAT5 blocked human prostate cancer xenograft tumor growth (both subcutaneous and orthotopic) in nude mice and down-regulated BclX$_L$ and Cyclin-D1 protein levels in prostate cancer cells [6]. Nuclear STAT5a/b is over-expressed in castration-resistant clinical prostate cancers [25, 26], and STAT5a/b transcriptionally synergizes with androgen receptor [25]. Given that STAT5a/b and AR are both anti-apoptotic and growth-promoting transcription factors in prostate cancer cells and expressed at high levels in castration-resistant prostate cancers, induction of AR transcriptional activity by STAT5a/b in the presence of low levels of androgens may contribute to castration-resistant growth of prostate cancer. AR, in turn, by promoting transcriptional activity of STAT5a/b, may critically support viability of prostate cancer cells in growth conditions where prostate cancer cells would normally undergo apoptosis. In summary, there is accumulating evidence supporting a key role for STAT5 in prostate cancer progression.

We also find strong evidence for cross-talk between the STAT5a and STAT3 signaling pathways. It is known that STAT5a is required for the survival of prostate cells, while STAT3 is involved in metastasis [9]. Our data supports this claim – the expression of STAT5a is high in normal adjacent prostate tissue (which can be considered to be premalignant samples). However, once the tumor progresses, STAT3 expression is high (involved in metastasis) showing that STAT5a and STAT3 play key but complementary roles in tumor initiation and progression.

**References**

[1]  Ahonen, T.J., Xie, J., LeBaron, M.J., Zhu, J., Nurmi, M., Alanen, K., Rui, H., and Nevalainen, M.T., Inhibition of transcription factor Stat5 induces cell death of human prostate cancer cells, *J Biol Chem*, 278:27287-92, 2003.

[2]  Ambs, S., Prueitt, R.L., Yi, M., Hudson, R.S., Howe, T.M., Petrocca, F., Wallace, T.A., Liu, C.G., Volinia, S., Calin, G.A., Yfantis, H.G., Stephens, R.M., and Croce, C.M., Genomic profiling of microRNA and messenger RNA reveals deregulated microRNA expression in prostate cancer *Cancer Res.*, 68:6162-70, 2008.

[3]  Butte, A.J., Tamayo, P., Slonim, D., Golub, T.R., and I.S., K., Discovering functional relationships between RNA expression and chemotherapeutic susceptibility using relevance networks, *Proc Natl Acad Sci USA*, 97:12182-86, 2000.

[4]  Chuang, H.-Y., Lee, E., Liu, Y.-T., Lee, D., and Ideker, T., Network-based classification of breast cancer metastasis, *Molecular Systems Biology*, 3:140, 2007.

[5]  Dagvadorj, A., Collins, S., Jomain, J.B., Abdulghani, J., Karras, J., Zellweger, T., Li, H., Nurmi, M., Alanen, K., Mirtti, T., Visakorpi, T., Bubendorf, L.,

Goffin, V., and Nevalainen, M.T., Autocrine prolactin promotes prostate cancer cell growth via Janus kinase-2-signal transducer and activator of transcription-5a/b signaling pathway, *Endocrinology*, 148:3089-101, 2007.

[6]     Dagvadorj, A., Kirken, R.A., Leiby, B., Karras, J., and Nevalainen, M.T., Transcription factor signal transducer and activator of transcription 5 promotes growth of human prostate cancer cells in vivo, *Clin Cancer Res*, 14:1317-24, 2008.

[7]     Dai, M., Wang, P., Boyd, A.D., Kostov, G., Athey, B., Jones, E.G., Bunney, W.E., Myers, R.M., Speed, T.P., Akil, H., Watson, S.J., and Meng, F., Evolving gene/transcript definitions significantly alter the interpretation of GeneChip data, *Nucl. Acids Res.*, 33:e175, 2005.

[8]     Gouilleux, F., Wakao, H., Mundt, M., and Groner, B., Prolactin induces phosphorylation of Tyr694 of Stat5 (MGF), a prerequisite for DNA binding and induction of transcription, *Embo J*, 13:4361-9., 1994.

[9]     Gu, L., Dagvadorj, A., Lutz, J., Leiby, B., Bonucelli, G., Lisanti, M.P., Addya, S., Fortina, P., Dasgupta, A., Hyslop, T., Rhimm, J., Bubendorf, L., and Nevalainen, M.T., Transcription factor Stat3 stimulates metastatic behavior of human prostate cancer cells in vivo, while Stat5 has a preferential role in the promotion of prostate cancer cell viability and tumor growth, *Am J of Pathol*, 176:1959-72, 2009.

[10]    Holloway, D.T., Kon, M., and Delisi, C., Machine learning for regulatory analysis and transcription factor target prediction in yeast, *Syst Synth Biol.*, 1:25-46, 2007.

[11]    Hu, Z., Hung, J.H., Wang, Y., Chang, Y.C., Huang, C.L., Huyck, M., and DeLisi, C., VisANT 3.5: multi-scale network visualization, analysis and inference based on the gene ontology, *Nucl. Acids Res.*, 37:W115-W121, 2009.

[12]    Ihle, J.N., The Stat family in cytokine signaling, *Curr Opin Cell Biol*, 13:211-7., 2001.

[13]    Irizarry, R.A., Hobbs, B., Collin, F., Beazer-Barclay, Y.D., Antonellis, K.J., Scherf, U., and Speed, T.P., Exploration, normalization, and summaries of high density oligonucleotide array probe level data, *Biostat.*, 4:249-64, 2003.

[14]    Langfelder, P., and Horvath, S., WGCNA: an R package for weighted correlation network analysis, *BMC Bioinformatics*, 9:559, 2008.

[15]    Li, A., and Horvath, S., Network Neighborhood Analysis with the multi-node topological overlap measure, *Bioinformatics*, 2:142, 2006.

[16]    Li, H., Ahonen, T.J., Alanen, K., Xie, J., LeBaron, M.J., Pretlow, T.G., Ealley, E.L., Zhang, Y., Nurmi, M., Singh, B., Martikainen, P.M., and Nevalainen, M.T., Activation of signal transducer and activator of transcription 5 in human prostate cancer is associated with high histological grade, *Cancer Res*, 64:4774-82, 2004.

[17]    Li, H., Zhang, Y., Glass, A., Zellweger, T., Gehan, E., Bubendorf, L., Gelmann, E.P., and Nevalainen, M.T., Activation of signal transducer and activator of transcription-5 in prostate cancer predicts early recurrence, *Clin Cancer Res*, 11:5863-8, 2005.

[18]     Oldham, M., Horvath, S., and Geschwind, D., Conservation and evolution of gene coexpression networks in human and chimpanzee brains, *Proc Natl Acad Sci USA*:17973-78, 2006.

[19]     Pagel, P., Kovac, S., Oesterheld, M., Brauner, B., Dunger-Kaltenbach, I., Frishman, G., Montrone, C., Mark, P., Stümpflen, V., Mewes, H., Ruepp, A., and Frishman, D., The MIPS mammalian protein-protein interaction database, *Bioinformatics*, 21:832-834, 2005.

[20]     Rui, H., Djeu, J.Y., Evans, G.A., Kelly, P.A., and Farrar, W.L., Prolactin receptor triggering. Evidence for rapid tyrosine kinase activation, *J Biol Chem*, 267:24076-81., 1992.

[21]     Rui, H., Kirken, R.A., and Farrar, W.L., Activation of receptor-associated tyrosine kinase JAK2 by prolactin, *J Biol Chem*, 269:5364-8., 1994.

[22]     Sandberg, R., and Larsson, O., Improved precision and accuracy for microarrays using updated probe set definitions, *BMC Bioinformatics* 8:48, 2007.

[23]     Steuer, R., On the analysis and interpretation of correlations in metabolomic data, *Brief Bioinform.*, 151:151-158, 2006.

[24]     Talloen, W., Clevert, D.-A., Hochreiter, S., Amaratunga, D., Bijnens, L., Kass, S., and H.W.H., G., I/NI-calls for the exclusion of non-informative genes: a highly effective filtering tool for microarray data, *Bioinformatics*, 23:2897-2902, 2007.

[25]     Tan, S.H., Dagvadorj, A., Shen, F., Gu, L., Liao, Z., Abdulghani, J., Zhang, Y., Gelmann, E.P., Zellweger, T., Culig, Z., Visakorpi, T., Bubendorf, L., Kirken, R.A., Karras, J., and Nevalainen, M.T., Transcription factor Stat5 synergizes with androgen receptor in prostate cancer cells, *Cancer Res*, 68:236-48, 2008.

[26]     Tan, S.H., and Nevalainen, M.T., Signal transducer and activator of transcription 5A/B in prostate and breast cancers, *Endocr Relat Cancer*, 15:367-90, 2008.

[27]     Taylor, I.W., Linding, R., Warde-Farley, D., Liu, Y., Pesquita, C., Faria, D., Bull, S., Pawson, T., Morris, Q., and Wrana, J.L., Dynamic modularity in protein interaction networks predicts breast cancer outcome, *Nat. Biotechnol.*, 27:199-204, 2009.

[28]     Wallace, T.A., Prueitt, R.L., Yi, M., Howe, T.M., Gillespie, J.W., Yfantis, H.G., Stephens, R.M., Caporaso, N.E., Loffredo, C.A., and Ambs, S., Tumor immunobiological differences in prostate cancer between African-American and European-American men, *Cancer Res.*, 68:927-36, 2008.

[29]     Wilson, D., Charoensawan, V., Kummerfeld, S., and Teichmann, S., DBD--taxonomically broad transcription factor predictions: new content and functionality, *Nucl. Acids Res.*, 36:D88-D92, 2008.

[30]     Yeung, M.K., Tegnér, J., and Collins, J.J., Reverse engineering gene networks using singular value decomposition and robust regression, *Proc Natl Acad Sci U S A*, 99:6163-8, 2002.

# ANALYZING GENE COEXPRESSION DATA BY AN EVOLUTIONARY MODEL

MORITZ SCHÜTTE[1]
schuette@mpimp-golm.mpg.de

MAREK MUTWIL[1]
mutwil@mpimp-golm.mpg.de

STAFFAN PERSSON[1]
persson@mpimp-golm.mpg.de

OLIVER EBENHÖH[2,3]
ebenhoeh@abdn.ac.uk

[1] *Max Planck Institute of Molecular Plant Physiology, Am Mühlenberg 1, 14476 Potsdam–Golm, Germany*
[2] *Institute for Complex Systems and Mathematical Biology, University of Aberdeen, Aberdeen, AB24 3UE, UK*
[3] *Institute of Medical Sciences, Foresterhill, University of Aberdeen, Aberdeen, AB25 2ZD, UK*

Coexpressed genes are tentatively translated into proteins that are involved in similar biological functions. Here, we constructed gene coexpression networks from collected microarray data of the organisms *Arabidopsis thaliana*, *Saccharomyces cerevisiae*, and *Escherichia coli*. Their degree distributions show the common property of an overrepresentation of highly connected nodes followed by a sudden truncation. In order to analyze this behavior, we present an evolutionary model simulating the genetic evolution. This model assumes that new genes emerge by duplication from a small initial set of primordial genes. Our model does not include the removal of unused genes but selective pressure is indirectly taken into account by preferentially duplicating the old genes. Thus, gene duplication represents the emergence of a new gene and its successful establishment. After a duplication event, all genes are slightly but iteratively mutated, thus altering their expression patterns. Our model is capable of reproducing global properties of the investigated coexpression networks. We show that our model reflects the mean inter-node distances and especially the characteristic humps in the degree distribution that, in the biological examples, result from functionally related genes.

*Keywords*: gene; coexpression; model; evolution; network.

## 1. Introduction

The increasing amount and easy accessibility of microarray gene expression data [3, 9] allows for systematic comparison of gene expression patterns on an organism scale under a wide variety of conditions [24]. Based on this data, coexpression networks are often constructed with the goal to identify clusters of similarly expressed genes which, it is assumed, are functionally related [18]. In the construction of the coexpression networks, genes are usually represented as nodes of a graph which are connected by edges if the similarity of the respective expression patterns lies above a predefined threshold. This approach is useful to support hypotheses about the functions of unknown genes. For example, Mutwil et al. [14] reported that this approach was successfully applied and several genes were correctly predicted to result in an embryo-lethal phenotype.

The study of graph properties of biological networks has a long tradition and encompasses diverse networks such as protein-protein interaction networks, transcriptional regulatory networks, signaling networks, and metabolic networks [5, 7, 8, 13]. Interestingly, many of the studied examples exhibit characteristic features that distinguish them from random network structures. The most widely discussed properties are the scale-freeness [2, 11], which means that the degree distributions follow a power law of the form $d(n) \propto n^{-\gamma}$, and the observation that they are organized in a small-world structure, which means that the average minimal path length between nodes is shorter than expected for random networks [1, 22].

Here, we observe that the coexpression networks of *Arabidopsis thaliana*, *Saccharomyces cerevisiae*, and *Escherichia coli* (hereafter called *A. thaliana*, *S. cerevisiae*, and *E. coli*) roughly follow a power-law degree distribution. However, for high degrees, there is a characteristic overrepresentation of highly coexpressed genes followed by a sharp truncation. To investigate these properties we introduce a model to mimic the growth of the coexpression network. For biological networks growth models are of special interest as they potentially allow to draw conclusions about the evolutionary pressures that have shaped these networks and thus to obtain hints about their design principles [21]. The genome's growth is mainly determined by gene duplications which account for about ninety percent of all eukaryotic genes [6, 15, 19]. Whereas existing growth models rather utilize graph theoretical methods [4, 17, 19, 20, 25] which basically follow preferential attachment [2] for the selection of duplicated nodes, we introduce a model that is based on numerical vectors as nodes in the network and connect them if their correlation values lie above a threshold. In this way, we simulate the emergence of coexpression patterns based on gene duplications and mutations, starting from a small number of initial, primordial, genes. Our simulation results support the view that the characteristic degree distribution of coexpression networks largely results from the functional and homological relatedness of the highly connected genes.

## 2. Data Sets

We downloaded microarray data for three different organisms, *A. thaliana*, *S. cerevisiae*, and *E. coli*. For *A. thaliana* we used the Affymetrix ATH1 array. Then, 1,428 ATH1 microarray data sets were collected from TAIR (http://www.arabidopsis.org/). For *S. cerevisiae* and *E. coli*, we took the Affymetrix Yeast Genome S98 and Affymetrix Ecoli_ASv2 GeneChips. The microarry data was downloaded from Gene Expression Omnibus (http://www.ncbi.nlm.nih.gov/geo/). Data of all organisms were RMA normalized and quality controlled by several steps. First, for Cel files that have shown either artifacts on RMA residual plots or deviated from the majority of the box plots of positive matches by visual inspection, data is removed. Then, those experiments that are mutually very similar were removed by dropping microarrays that have Pearson correlation coefficient higher than 0.95 with more than three other microarrays to reduce a bias in the experimental conditions [14]. From the data we constructed coexpression networks for different Pearson correlation coefficient cut-offs (called Pearson cut-off in Tab. 1), see Tab. 1 and Fig. 1.

Analyzing the data, the degree distributions roughly follow a power-law behavior but we observe an overrepresentation of highly connected nodes followed by a sharp edge, see Fig. 1. To get a biological interpretation, we exemplarily investigated

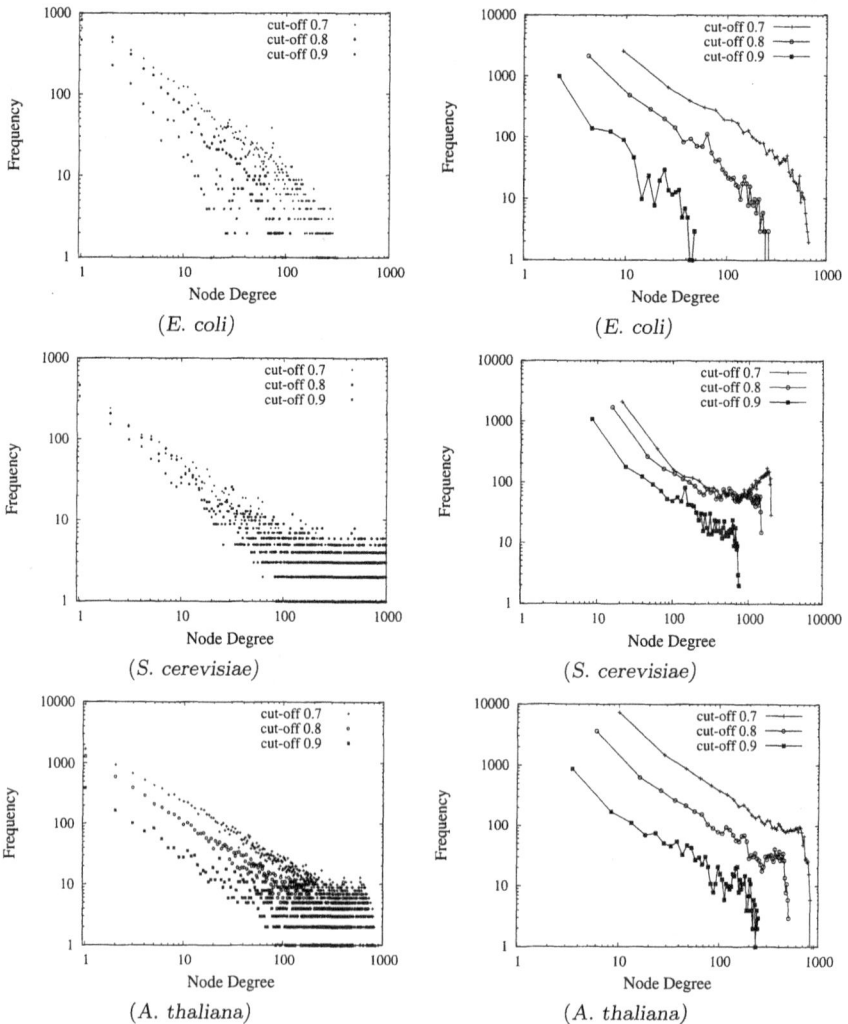

Fig. 1.    Degree distribution of data sets for different cut-offs (0.7, 0.8, 0.9). Left side: pure degree distribution, right: in a binned version. The binned distributions of *A. thaliana* and *S. cerevisiae* clearly show the same property of a concentration of highly connected nodes followed by a sharp edge.

Table 1. Network characteristics of the collected coexpression networks.

| Pearson cut-off | total number of genes | experiments | coexpressed pairs (edges) | connected (isolated) nodes | average path length |
|---|---|---|---|---|---|
| | | | *A. thaliana* | | |
| 0.7 | 22810 | 351 | 882836 | 15318 (7492) | 5.5 ± 1.7 |
| 0.8 | 22810 | 351 | 231881 | 7178 (15632) | 9.7 ± 4.4 |
| 0.9 | 22810 | 351 | 39119 | 2045 (20765) | 2.9 ± 1.6 |
| | | | *S. cerevisiae* | | |
| 0.7 | 9335 | 789 | 2607992 | 7052 (2283) | 4.2 ± 2.2 |
| 0.8 | 9335 | 789 | 1075760 | 5145 (4190) | 5.9 ± 3.7 |
| 0.9 | 9335 | 789 | 177226 | 2610 (6725) | 2.7 ± 1.3 |
| | | | *E. coli* | | |
| 0.7 | 7311 | 244 | 99188 | 5606 (1705) | 5.6 ± 2.0 |
| 0.8 | 7311 | 244 | 18983 | 3425 (3886) | 10.3 ± 5.2 |
| 0.9 | 7311 | 244 | 3299 | 1312 (5999) | 2.3 ± 1.3 |

genes of these nodes. In the case of *S. cerevisiae* with cut-off 0.9, we find significantly many genes that translate into proteins of the large and small ribosomal subunits. Of the total number of 2610 genes 221 are associated with ribosomes but 11 of the 23 most connected ($8\% \rightarrow 48\%$, p-value $p < 10^{-6}$). For *E. coli* 19 of the 22 most connected genes code for flagella proteins of the bacterium's chemotaxis movement, compared to 38 flagella of 1312 total genes ($3\% \rightarrow 86\%$, $p < 10^{-15}$). In *A. thaliana*, several genes of the protein family PF00069 (http://pfam.sanger.ac.uk/), protein kinases, are overrepresented, 43 of 2045 to 10 of 51 ($2\% \rightarrow 20\%$, $p < 10^{-6}$). These results indicate that highly connected genes, hubs, are also mutually highly connected [7, 10, 27] which to some extent is contrary to the idea of distinct functional modules [16]. To support this hypothesis, we calculate the clustering coefficient [23]. For a node $i$ it is defined as $c_i = 2n/k_i \cdot (k_i - 1)$ where n is the actual number of edges between the neighbors of $i$ and $k_i \cdot (k_i - 1)/2$ is the maximum possible number of edges between these. Complete connection leads to $c_i = 1$ and no clustering to $c_i = 0$. We obtain for the sets of high degree nodes listed above mean clustering coefficients $\bar{c}_{Yeast} = 0.49$, $\bar{c}_{E.coli} = 0.80$, and $\bar{c}_{Ara.} = 0.58$, and mean clustering coefficients only within the subnetwork of these highly connected nodes $\bar{c}_{Yeast}^{(high)} = 1$, $\bar{c}_{E.coli}^{(high)} = 0.93$, and $\bar{c}_{Ara.}^{(high)} = 0.96$.

## 3. Model

The aim of our model is to simulate the evolution of the organism's gene expression profile. We start from a small number of given initial genes. Then at every iteration step, we duplicate with high probability one of these initial genes. This is motivated from the known results of preferential attachment models [2], in which earlier nodes more likely become hubs. The chosen selectivity criterion mimics that established genes are robust in their expression patterns while their duplicates are redundant shortly after duplication but gain a new function by divergence. Due to the strong selectivity towards duplicating established genes, we can exclude gene loss as an explicit process [12, 26]. After duplication, all current genes undergo a slight random mutation. Such mutations might change the coexpression pattern of a gene. Iteratively, gene duplication and subsequent mutations are repeated, until the genome reaches a predefined size.

Technically, our model is designed along the proximity to experimental data, see Fig. 2 as a 2D illustration. Experiments are usually run under a variety of different conditions like stress in temperature or nutrient supply. Hence, the data consists of vectors where every entry belongs to a certain experimental condition. Therefore, we also represent genes as D-dimensional unit vectors. Randomly, we produce a set of $I$ initial vectors. Then at every step in the process we duplicate one of the vectors with strong selectivity towards established genes. This is implemented by randomly choosing a candidate gene $g$ for duplication according to a Fermi-Dirac distribution

$$P(g, A) = \left(1 + \exp\left\{\frac{g - I}{f(A)}\right\}\right)^{-1}, \tag{1}$$

where genes are indexed by their order of appearance ($g \leq I$ standing for the initial genes), $A$ is the current genome size, and $f(A)$ a function of $A$, here chosen as $f(A) = c \cdot A$, where $c$ is an adjustable parameter. In the limit $f(A) \to 0$, Eq. (1) is equal to a Heaviside step function with the step at $I$, for higher values the edge softens and approaches a Boltzmann distribution. With this particular choice of distribution, every gene in the genome that has itself emerged by a duplication event is selected for duplication with only a small probability, whereas in most of the cases one of the initial genes is selected.

After every duplication, the second step is the mutation of all current genes. This is done by the addition of a normally distributed random number with zero mean and a small variance $\sigma_{mut}$ to every entry of a vector. The size of the variance $\sigma_{mut}$ is chosen quite small but still by chance and the total number of mutations it is possible to obtain a drastic change in a vector. After each mutation step all vectors are normalized to unity again. Completing the last duplication step, we construct a network form the vectors. For every vector pair, we calculate the corresponding entry in the correlation matrix using the standard definition of the Pearson correlation coefficient and connect them if they exceed a given threshold. In summary, five parameters control the model behavior: dimension $D$, mutation variance $\sigma_{mut}$, correlation threshold $\theta$, the number of initial vectors $I$, and the constant $c$.

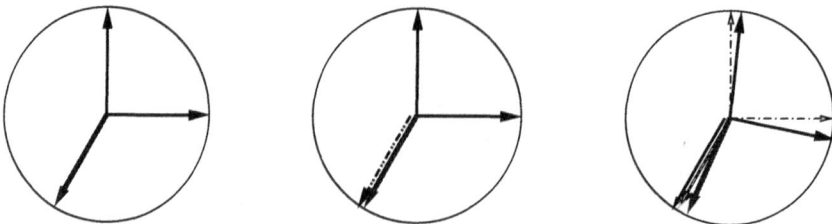

Fig. 2.  2D scheme of the attachment procedure. From left to right: Three current genes, the thick gray one is chosen to be duplicated, then all are mutated along the circle. Right image: Final situation with now four genes (dashed ones represent previous direction).

## 4. Results

The model behavior for different parameter combinations is summarized in Fig. 3 and Tab. 2. For small dimensions the behavior is drastically different from the experimental data (see Fig. 3a). For $D = 3$ and $D = 5$ the distributions miss the typical slope for small degrees but rather appear highly connected over a broad range. This becomes apparent in the average path length (Tab. 2), which is very short for low dimensions. This indicates that a critical number of experimental

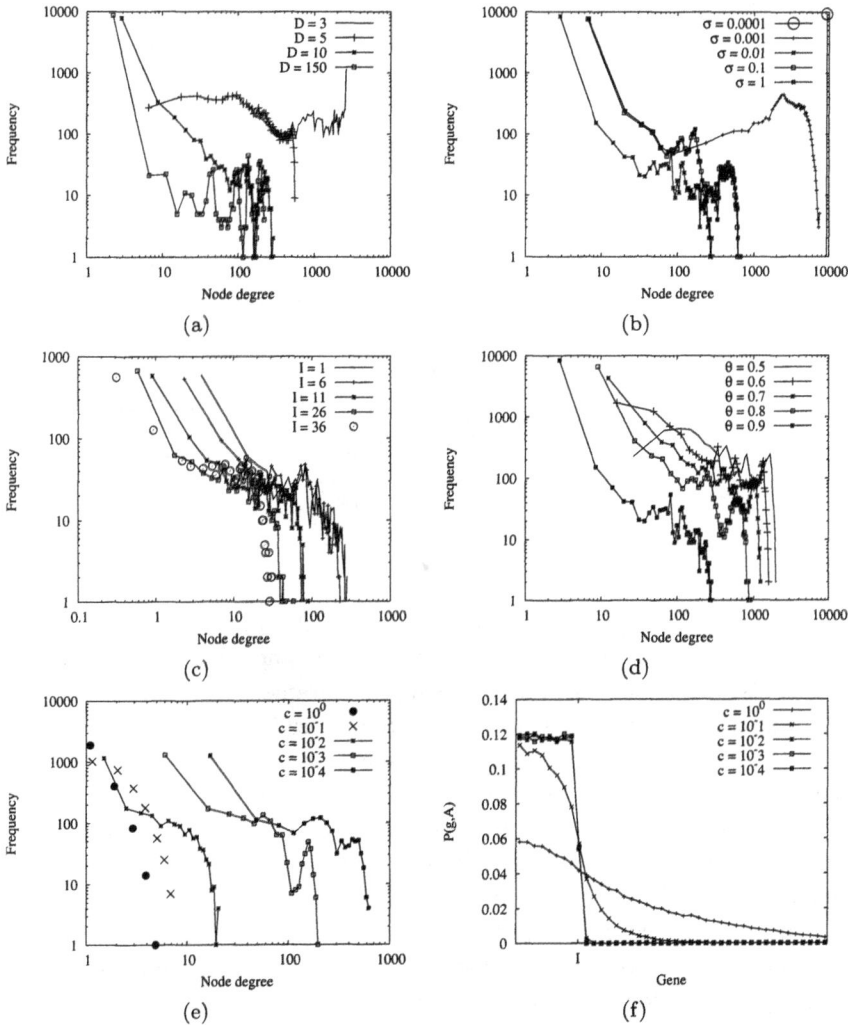

Fig. 3. Degree distribution of simulated data depending on the different parameters. The reference set of fixed parameters is: $I = 1$, $\theta = 0.9$, $D = 16$, $\sigma_{mut} = 0.01$, $c = 10^{-4}$. (a)–(e) variation of Dimension, mutation rate $\sigma_{mut}$, number of initial nodes, Pearson correlation cut-off to create the network, parameter $c$ of Fermi distribution. (f) Sketch of the Fermi distribution for different $c$.

Table 2.   Effect of different parameter values on simulated coexpression networks with 9335 nodes as in *S. cerevisiae*. The reference set is the same as in Fig. 3 and its resulting values are listed under dimension $D = 16$.

| varied parameter | coexpressed pairs (edges) | connected (isolated) nodes | average path length |
|---|---|---|---|
| | Dimension $D$ | | |
| 3 | 7549584 | 9335 (0) | $1.7 \pm 0.5$ |
| 5 | 859392 | 9335 (0) | $1.9 \pm 0.2$ |
| 10 | 53982 | 4217 (5118) | $3.1 \pm 0.5$ |
| 16 | 39697 | 1531 (7804) | $6.3 \pm 2.0$ |
| 40 | 42869 | 927 (8408) | $4.7 \pm 1.8$ |
| 150 | 35413 | 638 (8697) | $4.6 \pm 1.7$ |
| | Mutation rate $\sigma_{mut}$ | | |
| 0.0001 | 43566445 | 9335 (0) | $1 \pm 0$ |
| 0.001 | 14724849 | 9335 (0) | $1.6 \pm 0.5$ |
| 0.1 | 163742 | 3340 (5995) | $6.9 \pm 2.9$ |
| 1 | 168799 | 3378 (5957) | $6.8 \pm 2.9$ |
| | Initial nodes $I$ | | |
| 6 | 42355 | 1907 (7428) | $5.4 \pm 1.6$ |
| 11 | 15367 | 1534 (7801) | $6.6 \pm 2.1$ |
| 26 | 7426 | 1265 (8070) | $7.4 \pm 2.4$ |
| 36 | 7137 | 1293 (8042) | $7.2 \pm 2.1$ |
| | Pearson threshold $\theta$ | | |
| 0.5 | 3360623 | 9335 (0) | $1.9 \pm 0.2$ |
| 0.6 | 1908730 | 9333 (2) | $2.0 \pm 0.2$ |
| 0.7 | 944298 | 8989 (346) | $2.4 \pm 0.5$ |
| 0.8 | 337840 | 5728 (3607) | $3.3 \pm 0.7$ |
| | Fermi parameter $c$ | | |
| $10^{-3}$ | 57146 | 5354 (3981) | $6.9 \pm 2.4$ |
| $10^{-2}$ | 8489 | 4452 (4883) | $13.2 \pm 4.3$ |
| $10^{-1}$ | 3317 | 3682 (5653) | $29.2 \pm 10.4$ |
| $10^{0}$ | 1508 | 2400 (6935) | $10.2 \pm 7.6$ |

repetitions under different conditions is necessary to obtain reliable results. A very low mutation rate leaves all vectors almost unchanged and leads to one large spike (see see Fig. 3b for $\sigma_{mut} = 0.0001$). Increasing the parameter this spike melts apart but still a broader maximum of highly connected nodes is observable. The number of initial nodes causes only a shift in the curves leaving the actual shape unchanged (Fig. 3c). This behavior can be understood as we mainly see a superposition of similar but smaller networks. A similar observation as for the dimensions can be made for the cut-off of the Pearson correlation coefficient, Fig. 3d. For $\theta >= 0.7$ the distributions are similar except for the scale. However, lower cut-offs do change the shape qualitatively as they seem to allow influence of random relations between nodes. By this analysis, we conclude that the different parameters cannot be tuned independently. Dimension and cut-off have a similar impact on how strictly nodes are connected. The dimension also determines the overall effect of mutations: For a certain mutation strength, an original and a mutated duplicated gene are more likely to be connected for a smaller dimensionality of the vector.

Figures 3e and f depict the influence of the selective pressure. The distributions for high values of $c$ ($c = 0.1$ and $c = 1$), which correspond to a uniform probability of picking a gene, lead to a degree distribution in which highly connected nodes are not overrepresented. The lower values ($c = 10^{-3}$ and $c = 10^{-4}$) produce qualitatively similar curves, which are shifted to higher degrees for lower values of $c$, while in both cases high degree nodes are overrepresented. Fig. 3f illustrates the softening of the Fermi edge depending on the parameter $c$. There is a rather strong change from $c = 10^{-1}$ to $c = 10^{-2}$ which leads to an enormous increase in probability for values higher than the edge position.

In order to demonstrate that the model is capable of reproducing important features of the correlation networks determined from experimental microarray data, we manually fit the model parameters to the data (see Fig. 4). The mutual dependence of parameters allows to reduce the dimensionality in comparison with the data sets by adjusting mutation strength or threshold.

## 5. Conclusion

We presented an evolutionary model based on numerical representations of gene coexpression data that can explain some observed properties of the coexpression networks of *A. thaliana*, *S. cerevisiae*, and *E. coli*. These networks contain a group

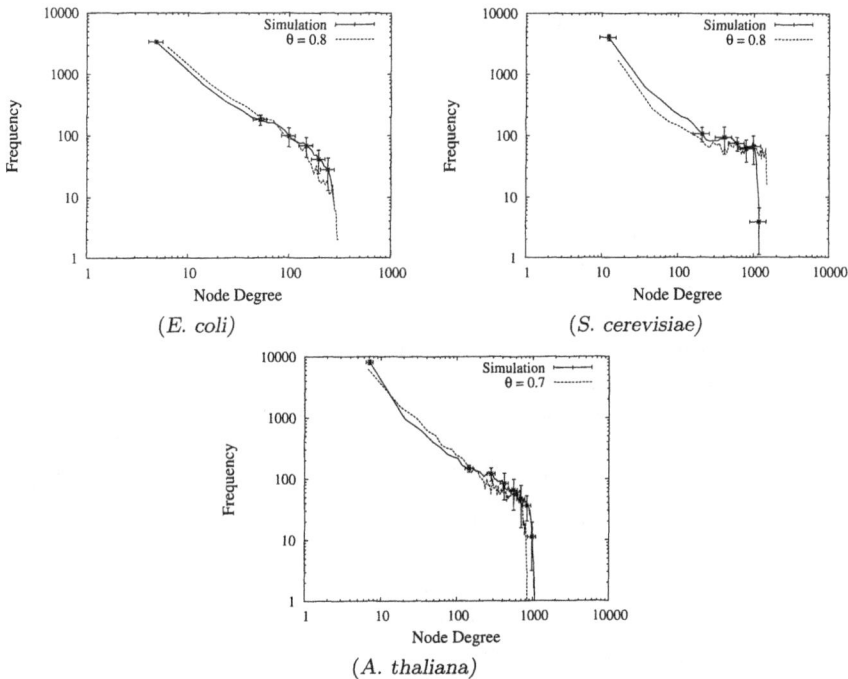

Fig. 4.   Degree distribution of simulations manually fitted to real data. Parameters chosen: *E. coli* 0.8 cut-off: $c = 10^{-3}$, $D = 14$, $I = 2$, $\sigma_{mut} = 0.05$, $\theta = 0.8$; *S. cerevisiae* 0.8: $c = 10^{-4}$, $D = 14$, $I = 1$, $\sigma_{mut} = 0.05$, $\theta = 0.8$; *A. thaliana* 0.7: $D = 14$, $I = 4$, $\sigma = 0.01$, $\theta = 0.88$, $c = 10^{-4}$. With error bars resulting from several simulations with different random initialization.

of highly coexpressed genes that by data analysis code for proteins with very similar function, flagella in E. coli and ribosome in *S. cerevisiae*, or of the same protein family, PF00069 protein kinases in *A. thaliana*. Due to the capability of the model to reflect characteristic features of the experimental data, in particular the over-representation of highly coexpressed genes, it allows to assess which evolutionary parameters are critical for these features to emerge. A robust observation from our modelling results is that the characteristic overrepresentation of highly connected genes can only be reproduced under high selective pressures towards duplicating established genes. This finding is consistent with the notion of preferential attachment which assumes that those genes that are already highly coexpressed with other genes have a higher chance to establish a new gene by duplication. Further model results demonstrate the necessity to take great care when interpreting coexpression network properties. The importance of the dimensionality of the expression vectors, representing the number of different experimental conditions, shows that the possibility cannot be ruled out that the inclusion of even more data will lead to completely new network properties. Similarly, the choice of the threshold value used to construct a coexpression network must be critically assessed as a too low or a too high value may result in dramatically different network characteristics.

This work was supported by the International Research Training Group *Genomics and Systems Biology of Molecular Networks* IRTG 1360 (MS), and the Scottish Universities Life Science Alliance SULSA (OE).

# References

[1] Arita, M. The metabolic world of escherichia coli is not small, *Proc Natl Acad Sci U S A*, 101(6):1543–1547, 2004.

[2] Barabasi, A.-L. and Albert, R. Emergence of scaling in random networks, *Science*, 286(5439):509–512, 1999.

[3] Barrett, T, Troup, D. B., Wilhite, S. E., Ledoux, P., Rudnev, D., Evangelista, C., Kim, I. F., Soboleva, A., Tomashevsky, M., Marshall, K. A., Phillippy, K. H., Sherman, P. M., Muertter, R. N., and Edgar, R. Ncbi geo: archive for high-throughput functional genomic data, *Nucleic Acids Res*, 37(Database issue):D885–D890, 2009.

[4] Bhan, A., Galas, D. J., and Dewey, T. G. A duplication growth model of gene expression networks, *Bioinformatics*, 18(11):1486–1493, 2002.

[5] Dokholyan, N. V., Shakhnovich, B., and Shakhnovich, E. I. Expanding protein universe and its origin from the biological big bang, *Proc Natl Acad Sci U S A*, 99(22):14132–14136, 2002.

[6] Gough, J., Karplus, K., Hughey, R., and Chothia, C. Assignment of homology to genome sequences using a library of hidden markov models that represent all proteins of known structure, *J Mol Biol*, 313(4):903–919, 2001.

[7] Jeong, H., Mason, S. P., Barabási, A. L., and Oltvai, Z. N. Lethality and centrality in protein networks, *Nature*, 411(6833):41–42, 2001.

[8] Kartal, Ö. and Ebenhöh, O. Ground state robustness as an evolutionary design principle in signaling networks, *PLoS One*, 4(12):e8001, 2009.

[9] Lee, H. K., Hsu, A. K., Sajdak, J., Qin, J., and Pavlidis, P. Coexpression analysis of human genes across many microarray data sets, *Genome Res*, 14(6):1085–1094, 2004.

[10] Lee, R. E. C. and Megeney, L. A. The yeast kinome displays scale free topology with functional hub clusters, *BMC Bioinformatics*, 6:271, 2005.

[11] Lima-Mendez, G. and van Helden, J. The powerful law of the power law and other myths in network biology, *Mol Biosyst*, 5(12):1482–1493, 2009.

[12] Lynch, M. and Conery, J. S. The evolutionary fate and consequences of duplicate genes, *Science*, 290(5494):1151–1155, 2000.

[13] Milo, R., Shen-Orr, S., Itzkovitz, S., Kashtan, N., Chklovskii, D., and Alon, U. Network motifs: simple building blocks of complex networks, *Science*, 298(5594):824–827, 2002.

[14] Mutwil, M., Usadel, B., Schütte, M., Loraine, A., Ebenhöh, O., and Persson, S. Assembly of an interactive correlation network for the arabidopsis genome using a novel heuristic clustering algorithm, *Plant Physiol*, 152(1):29–43, 2010.

[15] Ohno, S. *Evolution by gene duplication*, Springer-Verlag, 1970.

[16] Ravasz, E. Detecting hierarchical modularity in biological networks, *Methods Mol Biol*, 541:145–160, 2009.

[17] Rzhetsky, A. and Gomez, S. M. Birth of scale-free molecular networks and the number of distinct dna and protein domains per genome, *Bioinformatics*, 17(10):988–996, 2001.

[18] Tatusov, R. L., Koonin, E. V., and Lipman, D. J. A genomic perspective on protein families, *Science*, 278(5338):631–637, 1997.

[19] Teichmann, S. A. and Babu, M. M. Gene regulatory network growth by duplication, *Nat Genet*, 36(5):492–496, 2004.

[20] van Noort, V., Snel, B., and Huynen, M. A. The yeast coexpression network has a small-world, scale-free architecture and can be explained by a simple model, *EMBO Rep*, 5(3):280–284, 2004.

[21] Wagner, A. Robustness and evolvability: a paradox resolved, *Proc Biol Sci*, 275(1630):91–100, 2008.

[22] Wagner, A. and Fell, D. A. The small world inside large metabolic networks, *Proc Biol Sci*, 268(1478):1803–1810, 2001.

[23] Watts, D. J. and Strogatz, S. H. Collective dynamics of 'small-world' networks, *Nature*, 393(6684):440–442, 1998.

[24] Zampieri, M., Soranzo, N., Bianchini, D., and Altafini, C. Origin of co-expression patterns in e. coli and s. cerevisiae emerging from reverse engineering algorithms, *PLoS One*, 3(8):e2981, 2008.

[25] Zhang, Z., Luo, Z. W., Kishino, H., and Kearsey, M. J. Divergence pattern of duplicate genes in protein-protein interactions follows the power law, *Mol Biol Evol*, 22(3):501–505, 2005.

[26] Zhu, J., Sanborn, J. Z., Diekhans, M., Lowe, C. B., Pringle, T. H., and Haussler, D. Comparative genomics search for losses of long-established genes on the human lineage, *PLoS Comput Biol*, 3(12):e247, 2007.

[27] Zotenko, E., Mestre, J., O'Leary, D. P., and Przytycka, T. M. Why do hubs in the yeast protein interaction network tend to be essential: reexamining the connection between the network topology and essentiality, *PLoS Comput Biol*, 4(8):e1000140, 2008.

# COLLOCATION-BASED SPARSE ESTIMATION
# FOR CONSTRUCTING DYNAMIC GENE NETWORKS

TEPPEI SHIMAMURA[1]
shima@ims.u-tokyo.ac.jp
MAI YAMAUCHI[2]
cyowako@ims.u-tokyo.ac.jp
YOSHINORI TAMADA[1]
tamada@ims.u-tokyo.ac.jp

SEIYA IMOTO[1]
imoto@ims.u-tokyo.ac.jp
RUI YAMAGUCHI[1]
ruiy@ims.u-tokyo.ac.jp
NORIKO GOTOH[2]
ngotoh@ims.u-tokyo.ac.jp

MASAO NAGASAKI[1]
masao@ims.u-tokyo.ac.jp
ANDRÉ FUJITA[3]
andrefujita@riken.jp
SATORU MIYANO[1]
miyano@ims.u-tokyo.ac.jp

[1] *Human Genome Center, Institute of Medical Science, University of Tokyo, 4-6-1 Shirokanedai, Minato-ku, Tokyo 108-8639, Japan*
[2] *Division of Systems Biomedical Technology, Institute of Medical Science, University of Tokyo, 4-6-1 Shirokanedai, Minato-ku, Tokyo 108-8639, Japan*
[3] *RIKEN, 2-1 Hirosawa, Wako, Saitama 351-0198, Japan*

One of the open problems in systems biology is to infer dynamic gene networks describing the underlying biological process with mathematical, statistical and computational methods. The first-order difference equation-based models such as dynamic Bayesian networks and vector autoregressive models were used to infer time-lagged relationships between genes from time-series microarray data. However, two primary problems greatly reduce the effectiveness of current approaches. The first problem is the tacit assumption that time lag is stationary. The second is the inseparability between measurement noise and process noise (unmeasured disturbances that pass through time process).

To address these problems, we propose a stochastic differential equation model for inferring continuous-time dynamic gene networks under the situation in which both of the process noise and the observation noise exist. We present a collocation-based sparse estimation for simultaneous parameter estimation and model selection in the model. The collocation-based approach requires considerably less computational effort than traditional methods in ordinary stochastic differential equation models. We also incorporate various biological knowledge easily to refine the estimation accuracy with the proposed method. The results using simulated data and real time-series expression data of human primary small airway epithelial cells demonstrate that the proposed approach outperforms competing approaches and can provide significant genes influenced by gefitinib.

*Keywords*: stochastic differential equation model; elastic net; functional data analysis; time-series microarray data; transcriptional regulatory networks.

## 1. Background

Transcriptional gene regulation is a fundamental dynamic process controlled by transcription factors (TFs) which bind to specific DNA sequences and activate or repress their target genes. This process can be represented as a network where nodes represent genes and edges describe interactions between TFs and their targets. One of the main challenges in systems biology is to infer such a network describing the underlying biological process from time-series gene expression data using mathematical, statistical and computational methods. A number of approaches were used to model and reverse-engineer gene regulatory networks from time-series data based on dynamic Bayesian networks [1–4] and vector autoregressive models [5–7].

However, these existing algorithms for inferring gene regulatory networks have neglected two aspects of the problem which is critical for formulating the dynamics of gene regulation approximately. The first aspect is the time interval among the lagged temporal variables. As of April 2010, to the best of our knowledge, time-series datasets with equally-spaced intervals are rarely seen in gene expression omnibus (GEO) [8] and ArrayExpress [9]. Nonetheless, the difference equation-based models including dynamic Bayesian networks and vector autoregressive models usually assume that gene expressions are measured at consecutive sampling points separated by equal intervals of time. This discrepancy might introduce a systematic bias into the model. The second is the inseparability between two noises, process noise (unmeasured disturbances that pass through the process) and measurement noise. The difference equation models can be considered as approximations of the stochastic differential equation model where only the process noise exists by replacing the time derivative with the first-forward difference, which will be shown in the latter section. If measurement errors from time-series microarray data are large, the use of these noise-free models will lead to a lower power for identifying time-lagged relationships between genes.

In order to address these problems, we propose a stochastic differential equation (SDE) model for inferring continuous-time dynamic gene networks from time-series data under the situation in which the process noise and the measurement noise separately exist. This methodology is related to principal differential analysis (PDA) [10, 11] in which discrete measurements of gene expressions are fitted empirically using penalized smoothing splines. This enables us to obtain time-derivative curves. The time-derivative information is then substituted into the SDEs, converting the parameter estimation problem of the dynamic system into a much simpler optimization problem. We also present a new estimation approach called collocation-based sparse estimation for producing sparse estimators for unknown parameters in the model. The proposed approach can easily incoorpolate prior biological knowledge to parameter estimation for refining the estimation accuracy. We also evaluate the performance of the proposed method through synthetic data. We also applied the proposed method to real time-series microarray data of normal human small airway epithelial cells (SAECs) with epidermal growth factor (EGF) stimulation and with both EGF stimulation and treatment of gefitinib, an EGF receptor inhibitor (EGFR) in order to examine gene expression dynamics induced by EGF stimulation and a specific inhibitor of EGFR tyrosine kinase.

## 2. Methods

### 2.1. Continuous-time dynamic gene networks

Let $p$ be the number of a set of genes whose expression levels are measured at time series under a given biological condition. Let $m$ represent the number of a subset of these $p$ genes that are transcription factors (TFs). Without loss of generality, we assume that the first $m$ genes encode for TFs.

For illustrating a dynamic system of gene regulation, we consider the following continuous-time stochastic dynamic model for

$$\frac{dX_k(t)}{dt} = \beta_{0,k}^* + \sum_{j=1}^{m} \frac{\beta_{j,k}}{1 + \exp\{-(X_j(t) - \mu_j)/r_j\}} - d_k X_k(t) + \eta_k(t) \qquad (1)$$

$$= \beta_{0,k} + \sum_{j=1}^{m} \frac{\beta_{j,k}}{1 + \exp\{-(X_j(t) - \mu_j)/r_j\}} + \eta_k(t), \qquad (2)$$

$$Y_k(t) = X_k(t) + \varepsilon_k(t), \qquad (3)$$

where $X_k(t)$ represents a transcript abundance of the $k$-th gene at time $t$, $Y_k(t)$ indicates its noisy observation, $\beta_{0,k}^*$ and $\beta_{j,k}$ are the baseline level of the $k$-th gene and regulatory capability from the $j$-th TF to the $k$-th target gene, $d_k$ is the self-degradation rate, $\mu_j$ and $r_j$ indicate the mean and standard deviation of the $j$-th TF, and $\beta_{0,k} = \beta_{0,k}^* - d_k X_k(t)$. $\eta_k(t)$ is a continuous zero-mean stationary white-noise process with variance $\xi_k^2$ and $\varepsilon_k(t)$ is a zero-mean uncorrelated random variable with variance $\sigma_k^2$. The assumed model is a first-order stochastic differential equation model which extends the first-order differential equation model [12] to the situation in which the process noise $\eta_k(t)$ exists separately from the measurement noise $\varepsilon_k(t)$.

We describe a connection between the proposed model (2) and the first-order dynamic Bayesian networks. Let $t \in \{t_1, \ldots, t_n\}$ denote the sampling time points with equally-spaced intervals. We assume that the measurement time interval is lower than the gene evolution characteristic time. Replacing the time derivative in the model (2) with the first-forward difference $dX_j(t)/dt = (X_j(t_{i+1}) - X_j(t_i))/(t_{i+1} - t_i)$ yields

$$X_k(t_{i+1}) = (t_{i+1} - t_i) \left\{ \beta_{0,k} + \sum_{j=1}^{m} \frac{\beta_{j,k}}{1 + \exp\{-(X_j(t_i) - \mu_j)/r_j\}} + \eta_k(t) \right\} + X_k(t_i).$$

$$\qquad (4)$$

Under the assumption that the transcription rate of the $j$-th TF with a logistic function in the model (2) can be approximated by the simple transcript abundance of the $j$-th TF, the model (4) can be described as:

$$X_k(t_{i+1}) \approx (t_{i+1} - t_i) \left\{ \beta_{0,k} + \sum_{j=1}^{m} X_j(t_i)\beta_{j,k} + \eta_k(t) \right\} + X_k(t_i)$$

$$= \tilde{\beta}_{0,k} + \sum_{j=1}^{m} X_j(t_i)\tilde{\beta}_{j,k} + \tilde{\eta}_k(t_i), \qquad (5)$$

where $\tilde{\beta}_{0,k} = (t_{i+1} - t_i)\beta_{0,k}$,

$$\tilde{\beta}_{j,k} = \begin{cases} 1 + (t_{i+1} - t_i)\beta_{j,k} & \text{if } j = k \\ (t_{i+1} - t_i)\beta_{j,k} & \text{otherwise} \end{cases} \quad \text{and} \quad \tilde{\eta}_k(t_i) = (t_{i+1} - t_i)\eta_k(t_i).$$

The model (5) can be seen as the first-order linear dynamic Bayesian networks or the first-order vector autoregressive model with a diagonal noise matrix [6, 7]. It should be noticed that the difference $(t_{i+1} - t_i)$ influences $\tilde{\beta}_{j,k}$ in the model (5). If the discrepancy between the assumption of the model (5) and real data is large, it is inappropriate to use the model (5). We also note that only the assumption of the model (5) cannot be separated the process noise $\tilde{\eta}(t_i)$ from the measurement noise, which leads to a misleading result, that is, causes a decrease of identifying true relationships between TFs and their target genes if the measurement noise is large.

In this article, we do not estimate the degradation rates $d_k$ $(k = 1, \ldots, p)$ from observations and focus on parameter estimation in the model (2) under the assumption (3). The inferred regulation capabilities $\hat{\beta}_{j,k}$ provide a network topology between TFs and their target genes and can be described as a bipartite graph where parent nodes represent TFs and child nodes indicate their target genes. We expect most genes to be regulated by only a small subset of the total TFs, and thus assume that a large number of regulation capabilities will be zero, that is, the inferred regulation capabilities will be sparse.

## 2.2. Collocation-based sparse estimation

Suppose that we have $n$ noisy observations $y_k(t_i)$ $(i = 1, \ldots, n)$ for each $k$-th gene at time step $t \in \{t_1, \ldots, t_n\}$. Notice that these observations need not be placed at equal intervals. Our goal is to estimate unknown parameters $(\mu_1, \ldots, \mu_m, r_1, \ldots, r_m, \beta_{0,1}, \ldots, \beta_{0,p}, \beta_{1,1} \ldots, \beta_{m,p}, \xi_1^2, \ldots, \xi_p^2,$ and $\sigma_1^2, \ldots, \sigma_p^2)$ from observations. Especially, sparse estimators of the regulation capabilities $\beta_{1,1}, \ldots, \beta_{m,p}$ for providing TF-gene relationships are of primary interest. In this section, we propose collocation-based sparse estimation for estimating these parameters.

Our approach is a kind of collocation-based methods [10, 11, 13–15] in which the solutions for the stochastic differential equation in the model (2), say $x_j(t)$, $t \in \{t_1, \ldots, t_n\}$, can be approximated by a linear expansion of B-splines $\phi_{l,k}(t)$

$(l = 1, \ldots, q)$ as follows:

$$y_k(t) = x_k(t) + \varepsilon_k(t) = \sum_{l=1}^{q} c_{l,k} \phi_{l,k}(t) + \varepsilon_k(t), \tag{6}$$

where $c_{l,k}$ $(l = 1, \ldots, q)$ are coefficients. The number of basis functions, $q$, should be chosen to ensure enough the flexibility of capturing the variation in the approximated function $x_k(t)$ and its derivatives. Let $\boldsymbol{c}_k = (c_{1,k}, \ldots, c_{q,k})'$ represent the coefficient vector and $\boldsymbol{\phi}_k(t) = (\phi_{1,k}(t), \ldots, \phi_{q,k}(t))'$ denote the basis function vector where the superscript $'$ indicates the transpose. Once $\boldsymbol{c}_k$ is estimated, the empirical curve $\tilde{x}_k(t)$ and its one-order derivative can be represented by:

$$\tilde{x}_k(t) = \sum_{l=1}^{q} \hat{c}_{l,k} \phi_{l,k}(t) = \boldsymbol{\phi}_k'(t) \hat{\boldsymbol{c}}_k, \tag{7}$$

$$\frac{d}{dt} \tilde{x}_k(t) = \frac{d}{dt} \left( \sum_{l=1}^{q} \hat{c}_{l,k} \phi_{l,k}(t) \right) = \dot{\boldsymbol{\phi}}_k'(t) \hat{\boldsymbol{c}}_k, \tag{8}$$

where $\dot{\boldsymbol{\phi}}_k(t)$ indicates the first-order derivative vector of the basis functions. Next we describe the detail of the collocation-based sparse estimation.

We start by introducing the following notations:

$$\boldsymbol{y}_k = \begin{pmatrix} y_k(t_1) \\ y_k(t_2) \\ \vdots \\ y_k(t_n) \end{pmatrix} \quad \text{and} \quad \boldsymbol{\Phi}_k = \begin{pmatrix} \phi_{1,k}(t_1) & \phi_{2,k}(t_1) & \cdots & \phi_{q,k}(t_1) \\ \phi_{1,k}(t_2) & \phi_{2,k}(t_2) & \cdots & \phi_{q,k}(t_2) \\ \vdots & \vdots & \ddots & \vdots \\ \phi_{1,k}(t_n) & \phi_{2,k}(t_n) & \cdots & \phi_{q,k}(t_n) \end{pmatrix}. \tag{9}$$

The first step is to estimate the empirical curve $\tilde{x}_k(t)$ and its derivative by selecting the spline coefficient vector $\boldsymbol{c}_k$ that minimizes the following object function:

$$\hat{\boldsymbol{c}}_k = \arg\min_{\boldsymbol{c}_k} \left\{ \| \boldsymbol{y}_k - \boldsymbol{\Phi}_k \boldsymbol{c}_k \|_2^2 + \gamma_k \int [\sum_{l=1}^{m} c_{l,k} \frac{d^2}{dt^2} \phi_{l,k}(t)]^2 dt \right\}, \tag{10}$$

where $\gamma_k$ is a regularization parameter which influences the trade-off between the data fitting function and the smoothness penalty function. It is important to choose an appropriate regularization parameter since poor spline fits give misleading derivation information, which results in inaccurate parameter estimation from the next step. If the measurement-noise variance $\sigma_k^2$ is available, we try to choose $\gamma_k$ that minimizes the following loss function as follows:

$$\hat{\gamma}_k = \arg\min_{\gamma_k} (\sigma_k^2 - \hat{\sigma}_k^2)^2, \tag{11}$$

where

$$\hat{\sigma}_k^2 = \frac{1}{n} \| \boldsymbol{y}_k - \boldsymbol{\Phi}_k \hat{\boldsymbol{c}}_k \|_2^2. \tag{12}$$

If $\sigma_k^2$ is unavailable, we choose $\gamma_k$ which minimizes generalized cross-validation measure GCV [16]. For details, see Ramsay and Silverman [10].

With the data expressed in functional form, the next step is to estimate the unknown parameters in the model (2). We first estimate $\mu_j$ and $r_j$ for each TF by the sample mean and sample standard deviation of $\phi'_j(t)\hat{c}_j$:

$$\hat{\mu}_j = \frac{1}{n}\sum_{i=1}^{n}\phi'_j(t_i)\hat{c}_j, \tag{13}$$

$$\hat{r}_j = \sqrt{\frac{1}{n-1}\sum_{i=1}^{n}(\phi'_j(t_i)\hat{c}_j - \hat{\mu}_j)^2}. \tag{14}$$

Substituting $c_1, \ldots, c_p$, $\mu_1, \ldots, \mu_m$, and $r_1, \ldots, r_m$ by these estimators, we obtain the following matrix:

$$z_k = \begin{pmatrix} \dot{\phi}'_k(t_1)\hat{c}_k \\ \dot{\phi}'_k(t_2)\hat{c}_k \\ \vdots \\ \dot{\phi}'_k(t_n)\hat{c}_k \end{pmatrix}, \quad S = \begin{pmatrix} S_1(\phi'_1(t_1)\hat{c}_1) & S_2(\phi'_2(t_1)\hat{c}_2) & \cdots & S_m(\phi'_m(t_1)\hat{c}_m) \\ S_1(\phi'_1(t_2)\hat{c}_1) & S_2(\phi'_2(t_2)\hat{c}_2) & \cdots & S_m(\phi'_m(t_2)\hat{c}_m) \\ \vdots & \vdots & \ddots & \vdots \\ S_1(\phi'_1(t_n)\hat{c}_1) & S_2(\phi'_2(t_n)\hat{c}_2) & \cdots & S_m(\phi'_m(t_n)\hat{c}_m) \end{pmatrix},$$

where

$$S_j(\phi'_j(t)\hat{c}_j) = \frac{1}{1 + \exp\{-(\phi'_j(t)\hat{c}_j - \hat{\mu}_j)/\hat{r}_j\}}.$$

In the previous studies [10, 11, 14, 15], unknown parameters in the differential equation model are estimated by minimizing the residual sum of squares or maximizing the likelihood function with respect to these parameters for the fixed $B$-spline coefficient vectors. However, when the unknown parameters are much larger than the number of sampling time points, these existing approaches often produce unstable estimators. Furthermore, our interest is a sparse estimator of $\beta_{j,k}$. Therefore, we need to further investigate whether $\beta_{j,k}$ is zero or non-zero, which is a structure learning problem. In many cases, this task is computationally hard [17].

For simultaneous parameter selection and model selection for $\beta_{j,k}$ ($j = 1, \ldots, m$), for each $k$-th target gene, we minimize the following penalized residual sum of squares:

$$\begin{aligned} \text{PRSS}(\beta_{0,k}, \boldsymbol{\beta}_k) &= \frac{1}{2n}\sum_{i=1}^{n}\left(\frac{d}{dt}\tilde{x}_k(t_i) - \beta_{0,k} - \sum_{j=1}^{m}S_j(\tilde{x}_j(t_i))\beta_{j,k}\right)^2 \\ &\quad + \lambda_k\sum_{j=1}^{m}\left\{\frac{(1-\alpha)w_{j,k}}{2}(\beta_{j,k})^2 + \alpha w_{j,k}|\beta_{j,k}|\right\} \\ &= \frac{1}{2n}\|z_k - \beta_{0,k}\mathbf{1}_n - S\boldsymbol{\beta}_k\|_2^2 + \lambda_k P(\boldsymbol{\beta}_k), \tag{15} \end{aligned}$$

where

$$P(\boldsymbol{\beta}_k) = \sum_{j=1}^{m}\left\{\frac{(1-\alpha)w_{j,k}}{2}(\beta_{j,k})^2 + \alpha w_{j,k}|\beta_{j,k}|\right\}$$

is an weighted version of the elastic net penalty [7, 20, 21], $\boldsymbol{\beta}_k = (\beta_{1,k}, \ldots, \beta_{m,k})'$ is the coefficient vector, $w_{1,k}, \ldots, w_{m,k}$ are variable importance weights, $\alpha$ is a mixture parameter between an $L_1$-penalty and an $L_2$-penalty, $\lambda_k$ is a regularization parameter, and $\mathbf{1}_n$ is one vector with length $n$. The minimization of (15) gives a point-wise estimate of $\beta_{j,k}$ evaluated at observed time points. The $L_1$-penalty called the lasso penalty produces a sparse estimator, that is, produces the estimator where some coefficients are zero [18]. While the $L_2$-penalty called the ridge penalty stabilizes the estimator and encourages the grouping effect [19]. $P(\boldsymbol{\beta}_k)$ is the mixture penalty between the lasso penalty ($\alpha = 0$) and the ridge penalty ($\alpha = 1$). The variable importance weights $w_{j,k}$ ($j = 1, \ldots, m$) enhance the sparseness of the estimator and reduces the number of false positives drastically if it is chosen properly [7, 20, 21]. The $w_{j,k}$ also allows for incorporating prior information elicited from the known biological networks which will be discussed in the next section. At a result, the estimators are given by

$$\hat{\boldsymbol{\beta}}_k = \arg\min_{\boldsymbol{\beta}_k} \{\text{PRSS}(\beta_{0,k}, \boldsymbol{\beta}_k)\},$$

$$\hat{\beta}_{0,k} = \bar{z}_k - \bar{\boldsymbol{s}}'\hat{\boldsymbol{\beta}}_k,$$

$$\hat{\xi}_k^2 = \frac{\|\boldsymbol{z}_k - \hat{\beta}_{0,k}\mathbf{1}_n - \boldsymbol{S}\hat{\boldsymbol{\beta}}_k\|_2^2}{n},$$

where $\bar{z}_k$ and $\bar{\boldsymbol{s}}$ are the mean of $\boldsymbol{y}_k$ and the mean vector for each column of $\boldsymbol{S}$, respectively. The optimization problem of minimizing the loss function (15) can be solved by the coordinate-wise descent algorithm [22]. The coordinate-wise descent algorithm sequentially calculates $\hat{\boldsymbol{\beta}}_k$ for given values of $\lambda_k$ and can handle large $n$ and $p$ problems. The R package `glmnet` for the coordinate-wise descent algorithm is available from the Comprehensive R Archive Network (CRAN) [23].

## 2.3. Incorporating various biological knowledge

In this section, we describe some examples of biological knowledge for gene regulatory networks and how to incorporate them to parameter estimation of $\beta_{j,k}$. The suitable prior knowledge about network topology can refine the estimation accuracy of gene networks. Imoto et al. [24] provided a general framework for combining microarray data and biological knowledge aimed at estimating a gene network with Bayesian networks.

As mentioned in the previous section, we can incorporate prior knowledge for network topologies to parameter estimation of $\beta_{j,k}$ through minimizing the loss function (15) by using the variable importance weights. In this article, we consider two designs for constructing $w_{j,k}$. The first weights are defined by

$$w_{j,k} = \begin{cases} \rho, & \text{if prior knowledge of the } j\text{-th TF and the } k\text{-th gene exists} \\ 1, & \text{if no prior knowledge of the } j\text{-th TF and the } k\text{-th gene exists} \end{cases}, \quad (16)$$

where $\rho$ is a non-zero and non-negative value smaller than one. The $\beta_{j,k}$ with small value of $\rho$ tends to be estimated at nonzero more than the $\beta_{j,k}$ with $\rho$ equal to one in the problem of minimizing (15). While the second weights are given by

$$
w_{j,k} = \begin{cases} 1, & \text{if prior knowledge of the } j\text{-th TF and the } k\text{-th gene exists} \\ \infty, & \text{if no prior knowledge of the } j\text{-th TF and the } k\text{-th gene exists} \end{cases} \tag{17}
$$

The infinite weight enforces the estimator to be zero in the problem of minimizing (15). The difference between the weights (16) and (17) is that the $\beta_{j,k}$ without prior knowledge is always estimated at zero under the weight (17) while it is not always estimated at zero under the weights (16). Which weights are used depends on the purpose of data analysis. The design of the weights (16) is suitable for discovering unknown TF-gene interactions from observations as possible as approaching to the known biological networks. While the design of the weights (17) is intended for finding which TF-gene interactions are activated out of known TF-gene interactions.

To construct these weights, we utilize two biological knowledges: Motif information and Protein-Protein Interaction (PPI) information.

*Motif information*

A transcription factor (TF) regulates the transcriptions of its target genes by binding to a specific regulatory motif in transcription factor binding site (TFBS) that is located in the upstream regions of the genes. Thus, if genes have a consensus motifs of the $j$-th TF, it is natural that we might consider these genes as children of the $j$-th TF. In our analysis, we use a public database DBTSS [25] for finding transcription starting site for each gene and a commercial database TRANSFAC [26] for searching TFBSs in the 1,000 base-pairs upstream regions and 1,000 base-pairs downstream regions for each gene with MATCH [27].

*PPI information*

Proteins frequently bind together in pairs or complexes to take part in biological processes. Recently, the large number of protein-protein interactions (PPI) were identified and these PPI information are available from several public databases [29–33]. These protein-protein interactions include different types of interactions; regulatory interactions between a transcription factor and its candidate targets, interactions within cellular signaling cascades, and biochemical interactions among orthologous proteins such as ribosomal proteins or the ATPase. In the first case, co-expression might indicate a regulatory relationship. In the second case, co-expression has been reported but its mechanism has been still poorly understood. In the third case, co-expression might represent co-participation in a complex. Although we cannot distinguish between them, these connections are utilized as prior knowledge. We use the integrated PPI data set publicly available in Genome Network Platform (GNP) [28] (released on March 27, 2009). GNP collected the PPI data from the public PPI databases of BIND [29], BioGrid [30], HPRD [31], IntAct [32], and MINT [33]. It also contains their own experimental data by yeast two hybrid experiments and by nano LC/MS/MS experiments. In total, GNP PPI data consists of 53,417 non-redundant PPIs.

As other available sources, we can also introduce the network information directly from some databases such as KEGG [34], TRANSPATH [35] and Ingenuity Pathways Analysis [36].

When constructing the weights (16) from both motif information and PPI information, we define $\rho$ separately, that is, $\rho_{\text{motif}}$ and $\rho_{\text{PPI}}$. If prior knowledge between the $j$-th TF and the $k$-th gene exists in both motif information and PPI information, we define $\rho$ by $\rho = \min\{\rho_{\text{motif}}, \rho_{\text{PPI}}\}$. How much prior knowledge is incorporated to parameter estimation with the weights (16), that is, how to choose $\rho$, is a crucial problem which will be discussed in the next section.

### 2.4. Model selection

In parameter estimation with the collocation-based sparse estimation, it is important to choose suitable tuning parameters $\lambda_k$, $\alpha$ and $\rho$ since they determine on the estimation accuracy based on the model and the model sparsity (how many parameters are estimated at zero). The selection of these parameters is considered as a model selection problem in regression. In the previous study [7], we derived an unbiased estimator for the degrees of freedom in the weighted elastic net and proposed to choose tuning parameters which minimize a modified version of corrected Akaike information criterion (AICc) [37, 38]. Following this idea, we choose $\lambda_k$, $\alpha$ and $w_{j,k}$ (depends on $\rho$) by minimizing the following AICc criterion:

$$
\text{AICc}_k = n + n \log \left( \frac{\|z_k - \hat{\beta}_0 1_n - S\hat{\beta}_k\|_2^2}{n} \right) + \frac{2n(\text{tr}[\tilde{S}(\tilde{S}'\tilde{S} + n\lambda_k \tilde{D}_k)^{-1}\tilde{S}] + 1)}{n - \text{tr}[\tilde{S}(\tilde{S}'\tilde{S} + n\lambda_k \tilde{D}_k)^{-1}\tilde{S}] - 2},
\tag{18}
$$

where $\tilde{S}$ is the sub matrix whose column vector corresponds to the nonzero elements of $\hat{\beta}_k$, and $\tilde{D} = \text{diag}((1-\alpha)w_{j,k})_{j \in \mathcal{N}_k}$ with $\mathcal{N}_k = \{j; \hat{\beta}_{j,k} \neq 0\}$.

### 2.5. Computational algorithm

The algorithm for estimating the stochastic differential equation model with the collocation-based sparse estimation is summarized as:

(1) Set subsets of grid values for $\gamma_k$, $\rho$, $\alpha$ and $\lambda_k$ denoted by $\mathcal{G}$, $\mathcal{R}$, $\mathcal{A}$ and $\mathcal{L}$.

(2) For each variable $k = 1, \ldots, p$

    (a) For each value $\gamma_k \in \mathcal{G}$, estimate the spline coefficient vector $\hat{c}_k(\lambda_k)$ by the minimizing problem (10).

    (b) Select $\hat{c}_k$ with $\hat{\gamma}_k$ by the criterion (11) or by minimizing the GCV criterion [10, 16].

(3) For each variable $j = 1, \ldots, m$, estimate the mean $\hat{\mu}_j$ and the variance $\hat{r}_j$ from (13) and (14).

(4) Using $\rho$ selected from $\mathcal{R}$, construct the variable weights $w_{j,k}$ ($j = 1, \ldots, m$ and $k = 1, \ldots, p$) from (16) or (17).

(5) Select $\alpha$ from $\mathcal{A}$.

(6) For each variable $k = 1, \ldots, p$

   (a) For each value $\lambda_k \in \mathcal{L}$ and given $\rho$ and $\alpha$, estimate the vector of the regulation capabilities $\hat{\beta}_k(\rho, \alpha, \lambda_k)$ by minimizing the penalized loss function (15).

   (b) Select $\hat{\beta}_k$ with $\hat{\lambda}_k$ which provides the minimum AICc criterion (18), denoted by $\mathrm{minAICc}_k(\rho, \alpha) = \min_{\lambda_k} \mathrm{AICc}_k(\rho, \alpha, \lambda_k) = \mathrm{AICc}_k(\rho, \alpha, \hat{\lambda}_k)$.

(7) Select the regulation capabilities, say $\hat{\beta}_1(\hat{\rho}, \hat{\alpha}, \hat{\lambda}_1), \ldots, \hat{\beta}_p(\hat{\rho}, \hat{\alpha}, \hat{\lambda}_p)$, with $\hat{\rho}$ and $\hat{\alpha}$ by the following criterion

$$(\hat{\rho}, \hat{\alpha}) = \arg\min_{\rho, \alpha} \sum_{k=1}^{p} \mathrm{minAICc}_k(\rho, \alpha). \tag{19}$$

## 3. Numerical experiments

We evaluated the performance of the proposed model on both simulated data and experimental data. Using simulated data, we show that our proposed method can indeed identify significant edges against a perfectly known simulated networks in a computer model of gene regulation. Using experimental data, we also show that we can find new biological findings.

### 3.1. *Simulation results*

As a simulated network in the computer model, we considered a dynamic model (2) of 40 TFs and their 200 target genes where each gene is regulated by 3 TFs. The procedure of data generation was as follows:

(1) We generated transcription capabilities between TFs and their targets from a uniform distribution $U[-1, 1]$.

(2) To simulate 40 true transcript abundances of TFs, 40 average smooth functions were produced based on real time-series expression data.

(3) Initial values of transcription abundances of 200 target genes were generated from a uniform distribution $U[-1, 1]$.

(4) Using the simulated transcription abundances of 40 TFs, 200 transcription abundances of 200 target genes were generated by the Euler approximation [39].

(5) Finally, we constructed simulated data by adding Gaussian error with standard deviation $0.05 \times (\max(X_k(t_i)) - \min(X_k(t_i)))$ to each generated transcription abundance $X_k(t_i)$ at 25 unequally-spaced time points.

The performances of the graphical modeling approaches were monitored using the rate of true and false positives in receiver operator characteristics (ROC) curves. The true positive rate and false positive rate are defined as $\mathrm{TPR} = \mathrm{TP}/(\mathrm{TP} + \mathrm{FN})$

and FPR $=$ FP/(TN $+$ FP) where TP, FP, TN and FN are the numbers of true positives, false positives, true negatives, and false negatives, respectively. The ROC curve depicts the true-positive rate as a function of the false-negative rate. As an alternative for evaluating the performance, precision-recall (PR) curves are also considered. Let the precision quantity Precision $=$ TP/(TP $+$ FP) measure the fraction of true edges among the identified edges and the recall quantity Recall $=$ TP/(TP $+$ FN) also know as true positive rate (TPR), denote the fraction of true edges that are correctly inferred. The PR curve is a diagram which plots the precision versus the recall. As a competing method, a vector autoregressive (VAR) model with James-Stein Shrinkage [5] was used.

Fig. 1 shows the average ROC curve and the average PR curve for the inferred networks with the proposed stochastic differential equation (SDE) model and the VAR model through 100 simulations, respectively. Here the solid line represents the result of the SDE model and the dash line describes that of the VAR model. In the ROC curve, left upper area indicates better performance. It could be seen that the SDE model outperforms the VAR model in the ROC curves. The PR curve also provides further evidence that the proposed SDE model have a better performance than the VAR model.

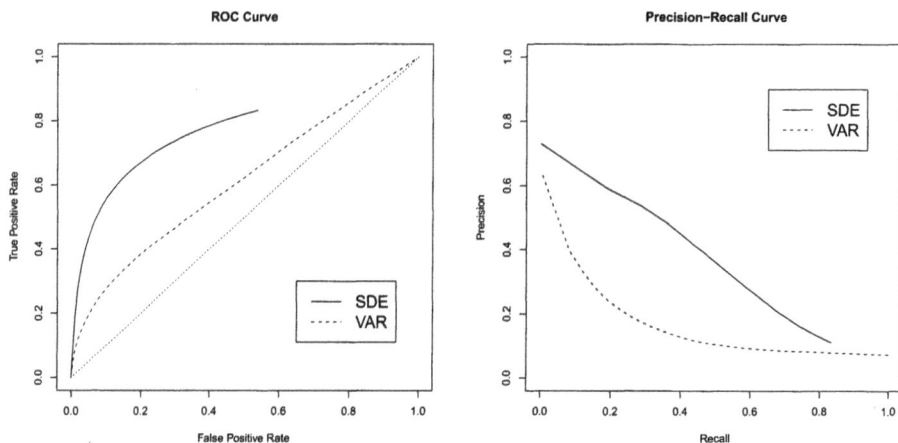

Fig. 1.   Comparison of ROC curves and precision-recall curves on the inferred networks with the proposed SDE model and the VAR model

## 3.2. *Application to experimental data*

We applied the proposed method to gene expression datasets of human primary small airway epithelial cells (SAECs) under two biological conditions: stimulation with epidermal growth factor (EGF), and both stimulation with EGF and treatment with gefitinib. EGF is a ligand for EGF receptor (EGFR/ErbB1), a cell-surface receptor with an intrinsic tyrosine kinase activity, and EGFR and its downstream targets ERK and Akt are phosphorylated upon stimulation with EGF. Gefitinib is an inhibitor for EGF receptor (EGFR) tyrosine kinase developed by AstraZeneca for treatment of refractory non-small cell lung cancer (NSCLC) and other types of cancers. This study intended to examine gene expression dynamics induced by EGF stimulation and a specific inhibitor for EGFR tyrosine kinase in normal lung cells. These datasets will be available soon after the publication of the main study results. The number of time points in each condition was 19 (0, 0.5, 1, 2, 4, 6, 9, 12, 15, 18, 21, 24, 27, 30, 33, 36, 39, 43, and 48 hours).

We first selected one probe for each gene that had the largest number of Present (P) flags which indicate fairly good quality of data acquisition, and then obtained 19267 probes. Next we focused on 1500 probe-sets (mapped to 1500 unique genes) from published literatures and analysis with Ingenuity Pathways Analysis [36]. At a result, data matrices of 1500 probe-sets by 19 time points under two conditions (EGF stimulation, and EGF and gefitinib stimulation) were used in the network inference phase. In this analysis, 100 equally spaced knots were used to approximate $X_k(t)$ $(k = 1, \ldots, 1500)$ by cubic $B$-splines. The regularization parameters $\gamma_k$ $(k = 1, \ldots, p)$ are selected by the GCV criterion. We also selected $\alpha$ from the set of grid values, precisely $\{0.1, 0.2, \ldots, 0.8, 0.9\}$. The sequence of 1000 values of $\lambda_k$ are generated from $\lambda_{\max}$ to $\lambda_{\min}$ on the log scale where $\lambda_{\max} = \max_j |s'_j z_k|/n\alpha$ with $s_j$ being the $j$-th column vector of $S$ and $\lambda_{\min} = 0.001\lambda_{\max}$. The optimal values of these parameters are chosen by the criterion (18). We used the stochastic differential equation model with prior based on the variable importance weights (17).

To compare the two inferred gene networks stimulated by EGF, and both EGF and gefitinib from characteristics of graph structure, we defined two measures, TF-interactivity and target-interactivity, defined as $\sum_{k=1}^{p} |\hat{\beta}_{j,k}^{**}|$ and $\sum_{j=1}^{m} |\hat{\beta}_{j,k}^{**}|$ for each TF $(j = 1, \ldots, m)$ and each target gene $(k = 1, \ldots, p)$, respectively, where $\hat{\beta}_{j,k}^{**}$ is the standardized coefficient estimated by the proposed method. Here the TF-interactivity of the $j$-th TF indicates how much the $j$-th TF influences target genes and the target-interactivity of the $k$-th target gene shows how much the $k$-th target genes are influenced by TFs. All the TFs and all the target genes were then ranked according to the difference of the TF-interactivity and target-activity between the two conditions, respectively. Table 1 indicates top five TFs with respect to the increased and decreased TF-interactivity. Table 2 also shows top five target genes with respect to the increased and decreased target-interactivity. These ranked genes were considered as important genes in the regulatory networks influenced by gefitinib which might be targets or off-targets of gefitinib.

Table 1.    Ranking of TFs with respect to increased and decreased TF-interactivity

| TF | Increased TF-Interactivity | | TF | Decreased TF-Interactivity | |
|---|---|---|---|---|---|
| | EGF | EGF and gefitinib | | EGF | EGF and gefitinib |
| MYC | 114.16 | 529.96 | TP53 | 30.19 | 16.00 |
| SMAD3 | 42.42 | 171.39 | ETS2 | 31.05 | 22.09 |
| NFKB2 | 88.66 | 186.70 | OVOL1 | 12.69 | 3.76 |
| EGR1 | 107.70 | 205.42 | HDAC4 | 19.04 | 12.21 |
| NFKB1 | 77.94 | 172.95 | KHDRBS1 | 8.60 | 1.79 |

Table 2.    Ranking of target genes with respect to increased and decreased target-interactivity

| Target | Increased Target-Interactivity | | Target | Decreased Target-Interactivity | |
|---|---|---|---|---|---|
| | EGF | EGF and gefitinib | | EGF | EGF and gefitinib |
| ATP7B | 0.00 | 63.78 | CDR2 | 33.18 | 0.00 |
| NFKBIZ | 0.00 | 62.23 | SYNJ2 | 44.41 | 17.83 |
| FOS | 36.82 | 79.35 | NEDD4 | 24.87 | 0.00 |
| GLI3 | 9.41 | 49.37 | DOCK9 | 24.61 | 0.00 |
| SMAD7 | 4.71 | 42.40 | PKIG | 24.60 | 0.00 |

## 4.  Conclusion

We addressed the two primary problems for current methods for inferring dynamic gene networks from time-series microarray data: the tacit assumption that time lag is stationary and the inseparability between the process noise and the measurement noise. We proposed a stochastic differential equation model under the situation in which the two noises separately exist to solve these problems. The proposed estimation method called the collocation-based sparse estimation achieved simultaneous parameter estimation and model selection for unknown parameters in the model and enabled us to infer large-scale gene networks from time-series data. Through simulated data and real data analysis of SAEC time-series data, we found that the proposed method outperformed the vector autoregressive model and could provide candidate genes influenced by EGF and gefitinib.

Future works of this study are two-fold. The first future work is the comparison of the proposed method with the state space model based on Kallman Filtering approach [40] which is another solution for the inseparability problem between the process noise and the measurement noise. The second future work is to replace the smoothness penalty in the penalized loss function (8) by a model-based penalty [14, 41]. Poyton et al. [11] showed that the smoothness penalty in the first step sometimes cannot capture the detailed dynamic behavior of the system and the model-based penalty could solve this problem. This replacement would further provide higher sensitivity and higher accuracy in inferring gene regulatory networks with the proposed method.

## Acknowledgements

The computational resource was provided by the Super Computer System, Human Genome Center, Institute of Medical Science, University of Tokyo.

## References

[1] Murphy, K. P. and Mian, S., Modelling gene expression data using dynamic Bayesian networks, *Tech. Rep., MIT Artificial Intelligence Laboratory*, 1999.

[2] Perrin, B. E., Ralaivola, L., Mazurie, A., Bottani, S., Mallet, J., and D'Alche-Buc, F., Gene networks inference using dynamic Bayesian networks, *Bioinformatics*, 19(Suppl 2): II138–II148, 2003.

[3] Kim, S., Imoto, S., and Miyano, S., Dynamic Bayesian network and nonparametric regression for nonlinear modeling of gene networks from time series gene expression data, *Biosystems*, 75(1-3): 57–65, 2004.

[4] Zou, M. and Conzen, S. D., A new dynamic Bayesian network (DBN) approach for identifying gene regulatory networks from time course microarray data, *Bioinformatics*, 21(1): 71–79, 2005.

[5] Opgen-Rhein, R. and Strimmer, K., Learning causal networks from systems biology time course data: an effective model selection procedure for the vector autoregressive process, *BMC Bioinformatics*, 8(Suppl. 2): S3, 2007.

[6] Fujita, A., Sato, J. R., Garay-Malpartida, H. M., Yamaguchi, R., Miyano, S., Sogayar, M. C., and Ferreira, C. E., Modeling gene expression regulatory networks with the sparse vector autoregressive model, *BMC Sys. Biol.*, 1: 39, 2007.

[7] Shimamura, T., Imoto, S., Yamaguchi, R., Fujita, A., Nagasaki, M., and Miyano, S., Recursive regularization for inferring gene networks from time-course gene expression profiles, *BMC Sys. Biol.*, 3: 41, 2009.

[8] GEO: http://www.ncbi.nlm.nih.gov/geo/

[9] ArrayExpress: http://www.ebi.ac.uk/microarray-as/ae/

[10] Ramsay, J. O. and Silverman, B. W., *Functional Data Analysis*, 2nd edn., Springer, New York, 2005.

[11] Poyton, A. A., Varziri, M. S., McAuley, K. B., McLellan, P. J., and Ramsay, J. O., Parameter estimation in continuous-time dynamic models using principal differential analysis, *Comp. Chem. Eng.*, 30(4): 698–708, 2006.

[12] Chen, H. C., Lee, H. C., Lin, T. Y., Li, W. H., and Chen, B. S., Quantitative characterization of the transcriptional regulatory network in the yeast cell cycle, *Bioinformatics*, 20(12): 1914–1927, 2004.

[13] Varah, J. M., A spline least squares method for numerical parameter estimation in differential equations, *SIAM J. Scient. Comput.*, 3(1): 28–46, 1982.

[14] Ramsay, J. O., Hooker, G., Campbell, D., and Cao, J., Parameter estimation for differential equations: a generalized smoothing approach, *J. R. Statist. Soc. B*, 69(5): 741–796, 2007.

[15] Cao, J. and Zhao, H., Estimating dynamic models for gene regulation networks, *Bioinformatics*, 24(14): 1619–1624, 2008.

[16] Craven, P. and Wahba, G., Smoothing noisy data with spline functions: Estimating the correct degree of smoothing by the method of generalized cross validation, *Numer. Math.*, 31(4): 377–403, 1979.

[17] Chickering, D. M., Heckerman, D., and Meek, C., Large-sample learning of Bayesian networks is NP-hard, *J. Mach. Learn. Res.*, 5: 1287–1330, 2004.

[18] Tibshirani, R., Regression shrinkage and selection via the lasso, *J. R. Statist. Soc. B*, 58(1): 267–288, 1996.

[19] Zou, H. and Hastie, T., Regularization and variable selection via the elastic net, *J. R. Statist. Soc. B*, 67(2): 301–320, 2005.

[20] Zou, H. and Zhang, H. H., On the adaptive elastic-net with a diverging number of parameters, *Ann. Stat.*, 37(4): 1733–1751, 2009.

[21] Shimamura, T., Imoto, S., Yamaguchi, R., Nagasaki, M., and Miyano, S., Inferring dynamic gene networks under varying conditions for transcriptomic network comparison, *Bioinformatics*, 26(8): 1064–1072, 2010.

[22] Friedman, J., Hastie, T., Hoefling, H., and Tibshirani, R., Pathwise coordinate optimization, *Ann. App. Stat.*, 1(2): 302–332, 2007.

[23] Comprehensive R Archieve Network: `http://cran.r-project.org/`

[24] Imoto, S., Higuchi, T., Goto, T., Tashiro, K., Kuhara, S., and Miyano, S., Combining microarrays and biological knowledge for estimating gene networks via Bayesian networks, *Proc. 2nd Computational Systems Bioinformatics*, 104–113, 2003.

[25] Suzuki, Y., Yamashita, R., Nakai, K., and Sugano, S., DBTSS: DataBase of human Transcriptional Start Sites and full-length cDNAs, *Nucleic Acids Res.*, 30: 328-331, 2002.

[26] Matys, V., Kel-Margoulis, O. V., Fricke, E., Liebich, I., Land, S., Barre-Dirrie, A., Reuter, I., Chekmenev, D., Krull, M., and Hornischer, K., *et al.*, TRANSFAC and its module TRANSCompel: transcriptional gene regulation in eukaryotes, *Nucleic Acids Res.*, 34: 108–110, 2006.

[27] Chekmenev, D. S., Haid, C., and Kel, A. E., P-Match: transcription factor binding site search by combining patterns and weight matrices, *Nucleic Acids Res.*, 33: 432–437, 2005.

[28] GNP: `http://genomenetwork.nig.ac.jp/`

[29] BIND: `http://www.bind.ca/`

[30] BioGrid: `http://www.thebiogrid.org/`

[31] HPRD: `http://www.hprd.org/`

[32] IntAct: `http://www.ebi.ac.uk/intact/`

[33] MINT: `http://minto.bio.uniroma2.it/mint/`

[34] KEGG: `http://www.genome.jp/kegg/`

[35] Krull, M., Voss, N., Choi, C., Pistor, S., Potapov, A., and Wingender, E., TRANSPATH: an integrated database on signal transduction and a tool for array analysis, *Nucleic Acids Res.*, 31(1): 97–100, 2003.

[36] Ingenuity Pathways Analysis: `http://www.ingenuity.com/`

[37] Sugiura, N., Further analysis of the data by Akaike's information criterion and the finite corrections, *Comm. Statist. A*, 7(1): 13–26, 1978.

[38] Hurvich, C. M. and Tsai, C. L., Bias of the corrected AIC criterion for underfitted regression and time series model, *Biometrika*, 78(3): 499–509, 1991.

[39] Stefano, M., *Simulation and Inference for Stochastic Differential Equations*, Springer Series in Statistics, 2008.

[40] Kojima, K., Yamaguchi, R., Imoto, S., Yamauchi, M., Nagasaki, M., Yoshida, R., Shimamura, T., Ueno, K., Higuchi, T., Gotoh, N., and Miyano, S., Sparse representation of VAR models with sparse learning for dynamic gene networks, *Genome Informatics*, 22: 56–68, 2009.

[41] Heckman, N. E. and Ramsay, J. O., Penalized regression with model-based penalties, *Can. J. Stat.*, 28(2): 241–258, 2000.

# DIFFERENT GROUPS OF METABOLIC GENES CLUSTER AROUND EARLY AND LATE FIRING ORIGINS OF REPLICATION IN BUDDING YEAST

THOMAS W. SPIESSER

thomas.spiesser@biologie.hu-berlin.de

EDDA KLIPP

edda.klipp@hu-berlin.de

*Theoretical Biophysics, Humboldt University of Berlin, Invalidenstr. 42, 10115 Berlin, Germany.*

DNA replication is a fundamental process that is tightly regulated during the cell cycle. In budding yeast it starts from multiple origins of replication and proceeds in a timely fashion according to a reproducible temporal program until the entire DNA is replicated exactly once per cell cycle. In this program an origin seems to have an inherent firing probability at a specific time in S-phase that is conserved over the population. However, what exactly determines the origin initiation time remains obscure. In this work, we analyze the gene content that clusters around replication origins following the assumption that inherent origin properties that determine staggered initiation times could potentially be mirrored in the close origin proximity. We perform a Gene Ontology term enrichment test and find that metabolic genes are significantly over-represented in the regions that are close to the starting points of DNA replication. Furthermore, functional analysis also reveals that catabolic genes cluster around early firing origins, whereas anabolic genes can rather be found in the proximity of late firing origins of replication. We speculate that, in budding yeast, gene function around replication origins correlates with their intrinsic probability to initiate DNA replication at a given point in S-phase.

*Keywords*: DNA replication; GO term enrichment; origins of replication; metabolic genes; S-phase.

## 1. Introduction

DNA replication is one of the most fundamental processes in life. It ensures genomic integrity for dividing cells, which is why it occurs in a highly regulated fashion once, and only once per cell cycle. DNA replication initiates from multiple discrete sites that are scattered across the genome, called origins of replication. In the budding yeast, *Saccharomyces cerevisiae*, all replication origins share a common feature, an about 200 base pair sequence, called Autonomously Replicating Sequence (ARS) [19]. Within this region, an eleven base pair sequence, the so called ARS consensus sequence (ACS) is specifically recognized by the origin recognition complex [29]. A sequence match to the ACS is essential, although the presence of this element alone does not define origin function *per se* [4, 21]. Origins initiate DNA replication throughout the Synthesis-phase (S-phase) of the cell cycle. Although, most origins fire near mid S-phase, it has been argued that there are chromosomal regions that can be classified into early and late replicating domains [18, 21, 35]. Early origins initiate the replication in the first half of the S-phase (early domains), and late origins in the second half (late domains). Various studies have mapped altogether 735 origins to the budding yeast genome and for a large part of these origins the time of initiation has been measured as well [8, 21, 24, 33–35].

DNA replication initiates when components of the pre-replicative complex, a multi protein complex that assembles on the replication origins during the Gap1-phase ($G_1$-phase), get phosphorylated by the cyclin-dependent protein kinase (CDK) Cdc28 (for details see [32]). CDK triggered phosphorylation eventually leads to the unwinding and separation of the two DNA strands by the putative replicative helicase minichromosome maintenance complex (Mcm2-7) and successively to the replication by the polymerases [10, 16, 31].

Cdc28 is active throughout S-phase, however the corresponding positive regulatory subunits, the cyclins, change. In the first half of S-phase, Clb6 is expressed and bound to the kinase, ensuring its activity. Clb6 gets degraded near mid S-phase and the cyclin Clb5 binds Cdc28 [14]. Both complexes (Cdc28-Clb6 and Cdc28-Clb5) can activate replication origins [6, 15, 26]. McCune and colleagues have studied DNA replication in a $\triangle clb5$ environment and demonstrated that only for a subset of origins the initiation time is altered in this condition [18]. They labeled regions in the genome that showed altered replication kinetics in the $\triangle clb5$ mutant as Clb5-dependent-regions (CDRs) and those unaffected as non-Clb5-dependent-regions (non-CDRs). That exactly the latter half of the S-phase is affected in the $\triangle clb5$ mutant has also been investigated by mathematical modeling [3, 28], supporting the distinction of early and late replicating domains.

It has been argued that origins contain an intrinsic relative firing probability, where the ones with a relatively higher probability are more likely to fire early in S-phase, whereas the ones with a relatively lower probability are rather unlikely to do so [25]. However, the regulatory event which eventually determines the timing of origin initiation remains obscure. Several mechanisms have been proposed that could potentially account for variations in the relative origin firing probability. Chromatin status, transcription and the number of Mcm2-7 molecules loaded onto the DNA are amongst them [25]. It has been observed that heterochromatin replicates late [13], which is consistent with a view in which chromatin density delimits the accessibility of replication origins. Indeed, in budding yeast the chromosomal context influences the origin firing time [9, 11]. Consistently, a correlation between larger, transcriptionally active regions and early replication has been observed in *Drosophila* [17], supporting the idea that an open chromatin structure facilitates origin activation and thus earlier firing. Altogether, the time of replication initiation is potentially governed by a combination of factors that act within the genomic origin domain. Thus, the activation time might be mirrored to some extent in properties of the imminent origin vicinity, and genomic analysis could shed some light onto how origin sequences have evolved and how this affects the replication program [23].

A direct implication of the genomic separation into early and late domains is that the replication of genes in the vicinity of origins is also confined to early and late domains. In this work, we investigate the functional relationship of genes that co-localize with the origins of replication assuming that inherent origin properties that cause the observed staggered replication patterns could potentially be present as well in the wider origin domains. We analyze the gene function in origin proximity using a Gene Ontology term enrichment test and find that metabolic genes are significantly over-represented. Furthermore, the analysis of genes that are localized in CDRs and non-CDRs also shows that genes with catabolic function cluster around early firing origins, whereas genes with anabolic function are rather found in the vicinity of late firing origins. We hypothesise that gene function is but one of many factors that mirrors intrinsic origin properties in budding yeast such as the relative firing probability.

## 2. Materials and Methods

Information about origins of replication were taken from the OriDB [20]. In this work we have considered all origins that are currently listed (735: confirmed, likely and dubious). Information regarding genomic features of budding yeast were obtained from the *Saccharomyces* Genome Database (SGD) [5] in form of the downloadable SGD_feature.tab and the chromosome_length.tab files [27]. We have identified all verified open reading frames (ORFs) that are located in the vicinity of the origins of replication (target gene set). Herein, vicinity is defined as the region that spans 2kb up- and downstream of the medial position of the origin (i.e. 4kb region). A gene is positively identified if the 3' end, the 5' end or the whole gene lies within or stretches over the whole region, as illustrated in Fig. 1.

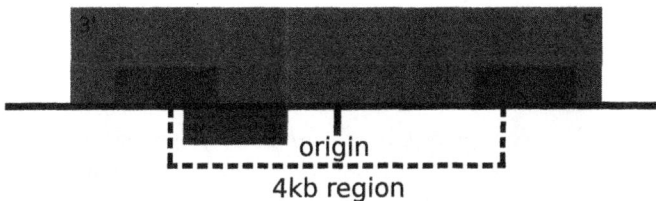

Fig. 1. Scheme of the gene-origin association criterion. The medial position of an origin was chosen to define a 4 kb region around the replication origin on the genome. A gene is associated with this region if the 3' end, the 5' end or the whole gene lies within (shown in grey) or stretches over (lightgrey) the whole region.

Using all verified ORFs of the budding yeast as a reference set, we performed a functional analysis of the target gene set. The analysis is based on the association of Gene Ontology (GO) terms [2] to genes and has been performed using GOstats [7]. GOstats is a package of the $R$ statistics environment [22] and is available from the Bioconductor project [12]. We have tested for over-representation of GO terms in our target gene set by applying a conditional hypergeometric test [7], with a $p$-value ($p$)

cutoff of 0.01. The conditioning of the commonly used hypergeometric test corrects for the problem of the hierarchical structure of GO. GO terms usually inherit the annotations from more specific descendants. This often leads to classification of directly related GO terms that have a high degree of gene overlap as significant at a specific $p$-value cutoff. The conditioning, implemented by Falcon and Gentleman [7] solves this problem by removal of all genes that are annotated at significant children from the gene list of the parent, an approach similar to that proposed by Alexa et al. [1].

For comparison, we also performed the conditional hypergeometric test on 1000 gene target sets identified on the basis of 735 random locations. A random number generator has been used to randomize the positions of the origins on the chromosomes. However, the origin position change is only allowed within the appropriate chromosome, thus, the positions of the origins change but the number per chromosome remains the same. The new positions were sampled from a uniform distribution with density:

$$f(x) = \frac{1}{(max - min)} \tag{1}$$

for $min \leq x \leq max$. For every chromosome we used $min = 1$ and $max = length\,of\,the\,chromosome$.

Then, we approximated the density distribution function of the $p$-value distribution, resulting from the 1000 tests using a non-parametric estimator:

$$\hat{f}_h(x) = \frac{1}{nh} \sum_{i=1}^{n} K\left(\frac{x - x_i}{h}\right) \tag{2}$$

with $x_i, i \in (1, n)$ being the samples of the random variable, a Gaussian kernel $K$ (mean $= 0$, variance $= 1$) and bandwidth $h$ that is automatically chosen, thus:

$$K\left(\frac{x - x_i}{h}\right) = \frac{1}{\sqrt{2\pi}} e^{-\frac{(x-x_i)^2}{2h^2}} \tag{3}$$

Furthermore, we calculated the empirical cumulative distribution function (ECDF) from the distribution of the $p$-values. Generally, the ECDF is used to calculate the probability to obtain a certain value (or smaller) under a given distribution and has the following form:

$$Fn(x) = \frac{1}{n} \sum_{i=1}^{n} I(X_i \leq x), \tag{4}$$

where $I(A)$ is the indicator function (or characteristic function) of an event $A$.

Finally, we have divided the origins of replication into two different clusters. We classify the origins according to the time at which they initiate DNA replication in a given S-phase of the cell cycle. Different studies have identified initiation times for a large number of the origins, nonetheless the information available remains incomplete [24, 35]. Therefore, we have used a whole genome study by McCune and colleagues [18] to classify whether an origin of replication lies within a CDR or non-CDR. This procedure allowed for the separation of the origins into early and late firing origins. We have tested for GO term over-representation of genes associated with both clusters.

All tasks were implemented and analyzed with the programming language *Python* [30] and the *R* statistics environment [22]. *Rpy*, a high level *Python*-module for managing the lookup of *R* objects, has been used for the internal communication between *Python* and *R*.

## 3. Results

Table 1 shows significant hits of the GO term enrichment analysis for genes that are located in the vicinity of origins of replication. In this paper, we used all verified ORFs of *Saccharomyces cerevisiae* (4844) as a reference set (gene universe) for the conditional hypergeometric test and analyzed 1388 genes located in the immediate origin local area. 21 terms have been identified using a $p$-value cutoff of 0.01 and an origin vicinity margin of 4kb.

We have further subcategorized the enriched GO terms. The first and largest of the subgroups represents metabolic processes. 10 out of the 12 most significant hits fall into that category, e.g. alcohol catabolic process ($p \approx 2.15422e - 06$) or thiamin biosynthetic process ($p \approx 0.00077$). Directly related to metabolic processes is the category transport of metabolites. This group represents functional enriched genes of carbohydrate, hexose and glycerol transport annotations. The third group contains genes of cell cycle processes and development, e.g. synaptonemal complex assembly ($p \approx 0.00879$) and the fourth group RNA processing type of genes. Response to toxin ($p \approx 4.61234e - 06$) could not be assorted to any of the categories.

For comparison we have performed 1000 GO term enrichment analysis for genes located near origins with random (uniformly distributed) positions. Exemplarily, we present the result of one random test, where we identified 11 terms with a $p$-values below 0.01 (Tab. 2). Five of the terms concern various types of regulation, e.g. positive regulation of organelle organization ($p \approx 0.00106$) or positive regulation of of glucose metabolic process ($p \approx 0.00829$). Three terms assort to metabolic processes and the last three terms concern response to copper ion, flocculation and mitochondrial genome maintenance. The $p$-values are generally higher than the ones determined using the original origin positions. The density distribution of the $p$-values (1000 tests) and the associated ECDF are displayed in Fig. 2. The density distribution shows an almost bimodal shape with peaks near 0.003 and 0.008, whereas the ECDF increases nearly linearly (solid line). The ECDF obtained using the $p$-values from the original origin positions (points) is also shown. It increases in the first half in a saturated curve-like manner and then converges into linear growth in the latter half.

We divided the origins into clusters of early and late replication to study whether different groups of genes are replicated at distinguishable times in S-phase. Tables

Table 1.   GO term enrichment analysis results for 1388 genes associated with origins of replication. GO terms, count of genes in target set, count of genes in reference set, $p$-values (rounded up) and GOBPIDs are shown for significantly enriched terms using a 0.01 $p$-value cutoff.

| Term | count target set | count reference set | $p$-value | GOBPID |
|---|---|---|---|---|
| alcohol catabolic process | 40 | 73 | 2.15422e-06 | GO:0046164 |
| hexose catabolic process | 34 | 60 | 4.61234e-06 | GO:0019320 |
| response to toxin | 24 | 38 | 9.44885e-06 | GO:0009636 |
| monocarboxylic acid metabolic process | 68 | 151 | 1.00416e-05 | GO:0032787 |
| monosaccharide metabolic process | 63 | 142 | 3.75132e-05 | GO:0005996 |
| carbohydrate catabolic process | 42 | 89 | 0.00014 | GO:0016052 |
| gluconeogenesis | 19 | 31 | 0.00015 | GO:0006094 |
| glycolysis | 20 | 34 | 0.00022 | GO:0006096 |
| thiamin biosynthetic process | 13 | 20 | 0.00077 | GO:0009228 |
| thiamin and derivative metabolic process | 14 | 23 | 0.00125 | GO:0042723 |
| carbohydrate transport | 19 | 36 | 0.00187 | GO:0008643 |
| carboxylic acid catabolic process | 26 | 57 | 0.00456 | GO:0046395 |
| endonucleolytic cleavage to generate mature 5'-end of SSU-rRNA from (SSU-rRNA, 5.8S rRNA, LSU-rRNA) | 15 | 28 | 0.00473 | GO:0000472 |
| endonucleolytic cleavage in 5'-ETS of tricistronic rRNA transcript (SSU-rRNA, 5.8S rRNA, LSU-rRNA) | 14 | 26 | 0.00594 | GO:0000480 |
| hexose transport | 14 | 26 | 0.00594 | GO:0008645 |
| cellular developmental process | 132 | 384 | 0.00636 | GO:0048869 |
| glycerol transport | 4 | 4 | 0.00672 | GO:0015793 |
| cell wall organization | 89 | 248 | 0.00672 | GO:0007047 |
| ncRNA 5'-end processing | 15 | 29 | 0.00725 | GO:0034471 |
| reproduction of a single-celled organism | 79 | 218 | 0.00787 | GO:0032505 |
| synaptonemal complex assembly | 6 | 8 | 0.00879 | GO:0007130 |

Table 2.   Exemplary GO term enrichment analysis results for genes associated with 735 random locations on the genome. GO terms, count of genes in target set, count of genes in reference set, $p$-values (rounded up) and GOBPIDs are shown for significantly enriched terms using a 0.01 $p$-value cutoff.

| Term | count target set | count reference set | $p$-value | GOBPID |
|---|---|---|---|---|
| positive regulation of organelle organization | 9 | 12 | 0.00106 | GO:0010638 |
| regulation of gene-specific transcription | 17 | 31 | 0.00168 | GO:0032583 |
| cellular amino acid derivative biosynthetic process | 14 | 25 | 0.00332 | GO:0042398 |
| positive regulation of specific transcription from RNA polymerase II promoter | 10 | 16 | 0.00449 | GO:0010552 |
| hydrogen peroxide metabolic process | 4 | 4 | 0.00643 | GO:0042743 |
| response to copper ion | 4 | 4 | 0.00643 | GO:0046688 |
| flocculation | 7 | 10 | 0.00748 | GO:0000128 |
| mitochondrial genome maintenance | 16 | 32 | 0.00759 | GO:0000002 |
| positive regulation of glucose metabolic process | 6 | 8 | 0.00829 | GO:0010907 |
| positive regulation of carbohydrate metabolic process | 6 | 8 | 0.00829 | GO:0045913 |
| amine catabolic process | 21 | 46 | 0.00894 | GO:0009310 |

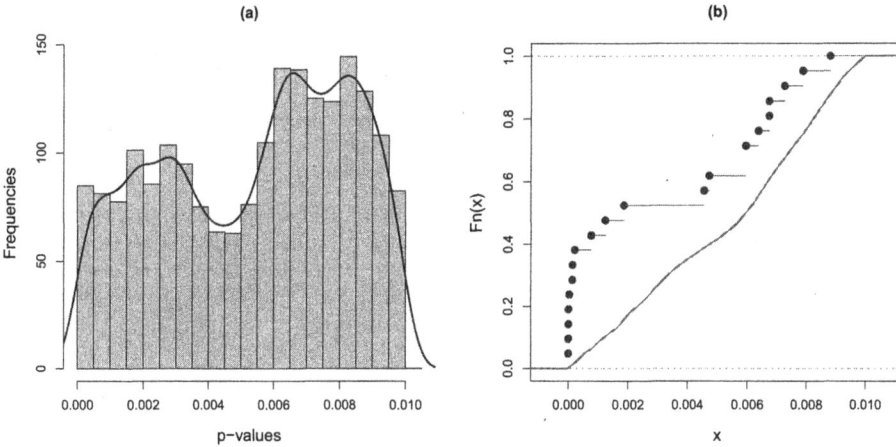

Fig. 2. Distribution of *p*-values obtained from 1000 enrichment tests using random locations. (a) Frequencies are shown as histogram, approximated density distribution is shown as black line. (b) Empirical cumulative distribution function is shown for random location *p*-values (solid line) and for the *p*-values obtained from testing the original positions of replication origins (points).

3 and 4 show the results for the GO term enrichment analysis for genes in early (non-CDR) and late (CDR) replicating regions, respectively. We found 16 enriched GO terms for 558 genes that are associated with early firing origins. Remarkably, more than half of them (9 out of 16) are related to catabolic processes, e.g. organic acid catabolic process ($p \approx 0.00049$) or aromatic compound catabolic process ($p \approx 0.0079$). Two are associated with metabolic or biosynthetic processes, two with RNA processing, two with DNA packing and one with organelle inheritance ($p \approx 0.0059$). Concerning the 773 genes that are localized close to late firing origins, we found 30 enriched GO terms (Tab. 4). 14 terms are related to various kinds of metabolic processes, e.g. vitamin metabolic process ($p \approx 7e - 05$) or gluconeogenesis ($p \approx 0.00623$), 9 terms represent genes that we classified as cell cycle and development related, as e.g. developmental process ($p \approx 0.00076$) or meiosis I ($p \approx 0.00908$), 6 terms concern genes of compartmentalization, e.g. cell wall organization ($p \approx 0.00192$) or spore wall biogenesis ($p \approx 0.00235$) and one term represents genes that are involved in the transport of glycerol ($p \approx 0.00064$).

## 4. Discussion

In this work, we performed a functional analysis of genes that we found to be positioned close to origins of replication. A conditional hypergeometric test was used to cluster functionally related genes according to their GO terms and to determine significant over-representation. We found that genes related to metabolic processes were most prominently over-represented amongst the genes that were tested (10 out of the 12 best hits, see Tab. 1). We calculated *p*-values that could be expected by

Table 3.   GO term enrichment analysis results for 558 genes associated with early firing origins of replication (positioned in non-CDRs). GO terms, count of genes in target set, count of genes in reference set, *p*-values (rounded up) and GOBPIDs are shown for significantly enriched terms using a 0.01 *p*-value cutoff.

| Term | count target set | count reference set | *p*-value | GOBPID |
|---|---|---|---|---|
| organic acid catabolic process | 16 | 57 | 0.00049 | GO:0016054 |
| nucleosome assembly | 10 | 29 | 0.00099 | GO:0006334 |
| nitrogen compound catabolic process | 16 | 61 | 0.0011 | GO:0044270 |
| threonine catabolic process | 3 | 3 | 0.00152 | GO:0006567 |
| maturation of 5.8S rRNA from tricistronic rRNA transcript (SSU-rRNA, 5.8S rRNA, LSU-rRNA) | 17 | 69 | 0.00165 | GO:0000466 |
| DNA packaging | 17 | 70 | 0.00195 | GO:0006323 |
| pyruvate metabolic process | 12 | 42 | 0.00206 | GO:0006090 |
| cellular amino acid catabolic process | 12 | 42 | 0.00206 | GO:0009063 |
| monosaccharide catabolic process | 16 | 68 | 0.00369 | GO:0046365 |
| allantoin catabolic process | 4 | 7 | 0.00458 | GO:0000256 |
| glucose catabolic process to ethanol | 4 | 7 | 0.00458 | GO:0019655 |
| organelle inheritance | 15 | 65 | 0.0059 | GO:0048308 |
| endonucleolytic cleavage in 5'-ETS of tricistronic rRNA transcript (SSU-rRNA, 5.8S rRNA, LSU-rRNA) | 8 | 26 | 0.00697 | GO:0000480 |
| aromatic compound catabolic process | 5 | 12 | 0.0079 | GO:0019439 |
| catabolic process | 121 | 866 | 0.00835 | GO:0009056 |
| alcohol biosynthetic process | 17 | 81 | 0.00952 | GO:0046165 |

chance, using the results of 1000 tests with randomized positions and the probabilities of the *p*-values obtained from the original test. The probabilities to obtain the *p*-values of the first 8 hits are around 1%. This means that the odds to obtain such an association by chance lie around 1%.

In addition, the gene target set has been split to test whether different groups of genes cluster around early and late firing replication origins. Genome-wide data concerning the dependency of replication times on Clb5 was used to classify the genes to either lie in early or late replicating domains. Fig. 3 shows all genes that have been identified to be in the vicinity of origins, using a vicinity margin of 4kb on a genome scale, where replication origins, CDRs, non-CDRs and inconclusive regions are indicated as well. Since origins, as well as genes, occupy a certain terrain on the genome, it seems apparent that a gene could generally be classified to belong to more than one origin region. Theoretically, the two origins could lie on the border of a CDR and a non-CDR, so the gene in question could, in that particular case, not unambiguously be assigned to be located in an early or late firing domain. In order to test for this special case, we investigated how many genes allocate to more than one origin. We found this to be true for 107 genes. Consequently, we further tested how many of them could potentially fall into both (CDR and non-CDR) regions and detected this to be the case for only three genes (YLR081W, YMR246W and YER136W). Hence, the three ambiguous genes have not been considered in the

Table 4. GO term enrichment analysis results for 773 genes associated with late firing origins of replication (positioned in CDRs). GO terms, count of genes in target set, count of genes in reference set, $p$-values (rounded up) and GOBPIDs are shown for significantly enriched terms using a 0.01 $p$-value cutoff.

| Term | count target set | count reference set | $p$-value | GOBPID |
|---|---|---|---|---|
| vitamin metabolic process | 32 | 102 | 7e-05 | GO:0006766 |
| thiamin and derivative metabolic process | 11 | 23 | 0.00034 | GO:0042723 |
| thiamin biosynthetic process | 10 | 20 | 0.0004 | GO:0009228 |
| glycerol transport | 4 | 4 | 0.00064 | GO:0015793 |
| developmental process | 96 | 447 | 0.00076 | GO:0032502 |
| hexose catabolic process | 19 | 60 | 0.00176 | GO:0019320 |
| cell wall organization | 57 | 248 | 0.00192 | GO:0007047 |
| cell differentiation | 55 | 239 | 0.00223 | GO:0030154 |
| ascospore wall assembly | 15 | 44 | 0.00235 | GO:0030476 |
| spore wall biogenesis | 15 | 44 | 0.00235 | GO:0070590 |
| cell wall assembly | 15 | 44 | 0.00235 | GO:0070726 |
| fungal-type cell wall biogenesis | 17 | 54 | 0.00324 | GO:0009272 |
| regulation of cell division | 6 | 11 | 0.00363 | GO:0051302 |
| reproduction of a single-celled organism | 50 | 218 | 0.00374 | GO:0032505 |
| alcohol catabolic process | 21 | 73 | 0.00386 | GO:0046164 |
| medium-chain fatty acid biosynthetic process | 3 | 3 | 0.00405 | GO:0051792 |
| water-soluble vitamin biosynthetic process | 18 | 60 | 0.00445 | GO:0042364 |
| reproductive process | 42 | 178 | 0.00446 | GO:0022414 |
| pentose-phosphate shunt | 7 | 15 | 0.00508 | GO:0006098 |
| pyridine nucleotide metabolic process | 16 | 52 | 0.00545 | GO:0019362 |
| gluconeogenesis | 11 | 31 | 0.00623 | GO:0006094 |
| M phase of meiotic cell cycle | 44 | 192 | 0.00644 | GO:0051327 |
| monocarboxylic acid metabolic process | 36 | 151 | 0.00686 | GO:0032787 |
| coenzyme metabolic process | 35 | 146 | 0.00697 | GO:0006732 |
| ascospore formation | 28 | 111 | 0.00733 | GO:0030437 |
| sexual sporulation | 28 | 111 | 0.00733 | GO:0034293 |
| premeiotic DNA synthesis | 4 | 6 | 0.00737 | GO:0006279 |
| NADP metabolic process | 9 | 24 | 0.0087 | GO:0006739 |
| alkaloid metabolic process | 15 | 50 | 0.00905 | GO:0009820 |
| meiosis I | 22 | 83 | 0.00908 | GO:0007127 |

analysis. Furthermore, it has to be mentioned that one gene (YAR020C) lies within a region that was termed 'Inconclusive' by McCune et al. [18] and is therefore neither classified as early nor late. Thus, we did not consider YAR020C in the analysis of the early and late replicating domains either. Additionally, the analysis of McCune et al. [18] does not give any information regarding the first and last 12 kb of every chromosome, which gives a total of 57 genes (including the ones mentioned above) that have not been considered in the analysis of early and late replicating domains.

We found 16 GO terms for functionally enriched genes close to early and 30 terms for genes close to late replicating origins. Genes related to metabolic processes also dominate the GO terms in both domains when separated. However, it seems that metabolic genes that cluster around early origins mostly concern catabolic reactions

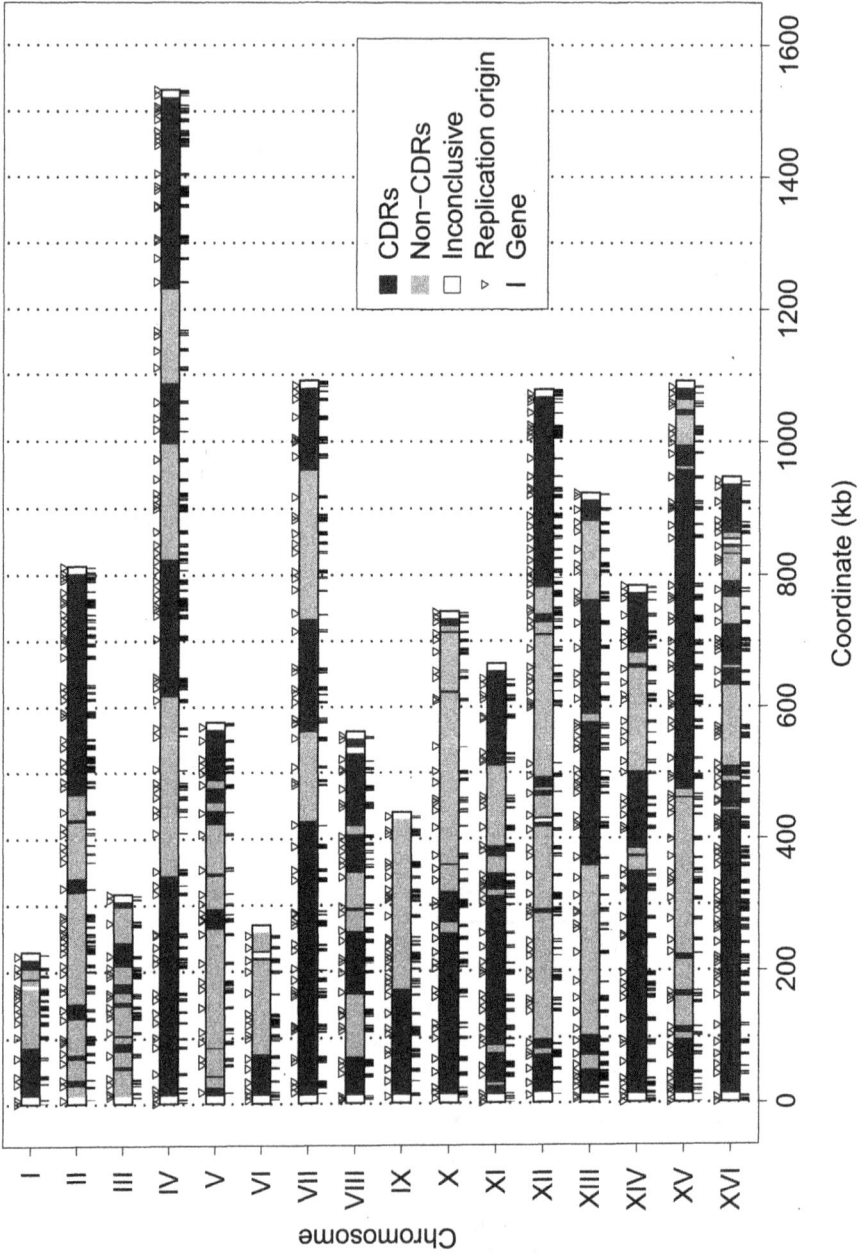

Fig. 3.   Chromosomal location of replication origins (triangles) and associated genes in origin vicinity (black lines). Furthermore, CDRs (lightgrey), non-CDRs (darkgrey) and Inconclusive regions (white) are shown, as identified by McCune et al. [18].

(9 out of 11). Intrigued by this, we investigated the metabolic genes around late origins in more detail as well. 14 terms relate to metabolic processes, where 7 of them cannot be distinguished on first sight (e.g. vitamin metabolic process), 5 concern anabolic reactions (e.g. thiamin biosynthetic process) and two of them catabolic ones (e.g. hexose catabolic process). Therefore, we investigated the structure of the GO tree around the 7 indistinguishable metabolic terms and the genes in the gene target set that relate to them. We found that, e.g. vitamin metabolic process (74 genes) has the following four children: regulation (2 genes), water-soluble vitamin metabolic process (70 genes), biosynthetic (64 genes) and catabolic (1 gene) in budding yeast. A closer look into our gene set told us that the catabolic gene is not part of our gene set. Furthermore, since we applied the hypergeometric test with the conditional correction and water-soluble vitamin biosynthetic process (a child of water-soluble vitamin metabolic process) is a significant term and, therefore, taken out of the set when testing the vitamin metabolic process, it follows that the majority of genes to be tested must be out of the 64 annotated biosynthetic genes. Thus, we conclude that the vast majority of vitamin metabolic process genes actually concerns anabolic reactions in the target set, since no catalytic ones could be found and 64 out of 74 are anabolic. The same procedure has been applied for the other 6 indistinguishable metabolic terms. It became apparent that also for thiamin (vitamin B1) and derivative metabolic process, pyridine nucleotide metabolic process, NADP metabolic process and alkaloid metabolic process no catalytic genes were in the gene target set. Regarding monocarboxylic acid metabolic process and coenzyme metabolic process we could not fully determine the single contributions of our gene target set due to complexity of the gene composition concerning those GO terms. A more sophisticated method needs to be developed in the future to investigate those nondistinctive terms. Nonetheless, it seems that, in budding yeast, catabolic genes cluster around early and anabolic genes around late origins of replication.

We speculate that this phenomenon might be the results of an evolutionary optimization designed to cope with the increasing costs during cell division. The early replication of catabolic genes results in early duplicates of those genes, which increases their transcriptional capacity and thus potentially their mRNA levels as well (Fig. 4). Consequently, cells that double their catabolic genes in early S-phase can benefit much longer from a potentially hightend catabolic capacity. In this particular case, the genomic position might function as a modifier of gene expression.

## 5. Conclusions

In conclusion, we found that especially metabolic genes are localized close to replication origins. Probabilities for such highly significant over-representations have been calculated using the probability distribution that could be obtained by random location tests. Under the assumption that certain origin properties, such as probabilities for early or late initiation, could potentially be mirrored in the origin environment, we separately tested genes in early and late firing domains according

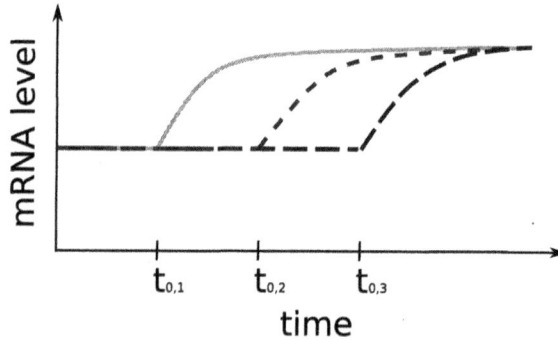

Fig. 4.    Schematic of mRNA level during S-phase. Given that a gene is transcribed with rate $k_1$ and the resulting mRNA is degraded with rate $k_2$, the mRNA level is constant after a while. If a particular gene gets replicated early in S-phase ($t_{0,1}$) the gene itself and its copy can be transcribed until the cell finally divides. This doubles rate $k_1$, which results in increased mRNA levels. Accordingly, the mRNA levels are affected later, and thus shorter, for genes that are replicated late in S-phase ($t_{0,3}$). Therefore, the location of a gene can influence its expression during S-phase.

to functional over-representation. Indeed, apart from chromatin status and correspondingly transcriptional activity, two factors that are most closely connected with origin activation time *per se*, also the gene function around origins seems to reflect some basic property of DNA replication. That is to say that metabolic genes near early firing origins mostly concern catabolic reactions and the majority of the metabolic genes near late firing origins are responsible for anabolic processes. It is tempting to speculate that origins and gene sequences in their close proximity might have evolved through e.g. duplication events, optimizing energy allocation and conserving inherent properties of a particular genomic region along the way.

## Acknowledgments

This work was supported by grants of the German Research Council (DFG, International Research Training Group, GRK 1360) and the European Commission (EU project Unicellsys HEALTH-2007-201142) to EK.

## References

[1] Alexa A., Rahnenführer J., and Lengauer T., Improved scoring of functional groups from gene expression data by decorrelating go graph structure, *Bioinformatics*, 22(13):1600–1607, 2006.

[2] Ashburner M., Ball C.A., Blake J.A., Botstein D., Butler H., Cherry J.M., Davis A.P., Dolinski K., Dwight S.S., Eppig J.T., Harris M.A., Hill D.P., Issel-Tarver L., Kasarskis A., Lewis S., Matese J.C., Richardson J.E., Ringwald M., Rubin G.M., and Sherlock G., Gene ontology: tool for the unification of biology. the gene ontology consortium, *Nat Genet*, 25(1):25–29, 2000.

[3] Barberis M., Spiesser T.W., and Klipp E., Replication origins and timing of temporal replication in budding yeast: how to solve the conundrum, *Curr Genomics*, 11(3):199–211, 2010.

[4] Breier A.M., Chatterji S., and Cozzarelli N.R., Prediction of saccharomyces cerevisiae replication origins, *Genome Biol*, 5(4):R22, 2004.

[5] Cherry J.M., Ball C., Weng S., Juvik G., Schmidt R., Adler C., Dunn B., Dwight S., Riles L., Mortimer R.K., and Botstein D., Genetic and physical maps of saccharomyces cerevisiae, *Nature*, 387(6632 Suppl):67–73, 1997.

[6] Epstein C.B. and Cross F.R., Clb5: a novel b cyclin from budding yeast with a role in s phase, *Genes Dev*, 6(9):1695–1706, 1992.

[7] Falcon S. and Gentleman R., Using gostats to test gene lists for go term association, *Bioinformatics*, 23(2):257–258, 2007.

[8] Feng W., Collingwood D., Boeck M.E., Fox L.A., Alvino G.M., Fangman W.L., Raghuraman M.K., and Brewer B.J., Genomic mapping of single-stranded dna in hydroxyurea-challenged yeasts identifies origins of replication, *Nat Cell Biol*, 8(2):148–155, 2006.

[9] Ferguson B.M. and Fangman W.L., A position effect on the time of replication origin activation in yeast, *Cell*, 68(2):333–339, 1992.

[10] Forsburg S.L., Eukaryotic mcm proteins: beyond replication initiation, *Microbiol Mol Biol Rev*, 68(1):109–131, 2004.

[11] Friedman K.L., Diller J.D., Ferguson B.M., Nyland S.V., Brewer B.J., and Fangman W.L., Multiple determinants controlling activation of yeast replication origins late in s phase, *Genes Dev*, 10(13):1595–1607, 1996.

[12] Gentleman R.C., Carey V.J., Bates D.M., etal., Bioconductor: Open software development for computational biology and bioinformatics, *Genome Biology*, 5:R80, 2004.

[13] Goren A. and Cedar H., Replicating by the clock, *Nat Rev Mol Cell Biol*, 4(1):25–32, 2003.

[14] Jackson L.P., Reed S.I., and Haase S.B., Distinct mechanisms control the stability of the related s-phase cyclins clb5 and clb6, *Mol Cell Biol*, 26(6):2456–2466, 2006.

[15] Kühne C. and Linder P., A new pair of b-type cyclins from saccharomyces cerevisiae that function early in the cell cycle, *EMBO J*, 12(9):3437–3447, 1993.

[16] Labib K. and Diffley J.F., Is the mcm2-7 complex the eukaryotic dna replication fork helicase? *Curr Opin Genet Dev*, 11(1):64–70, 2001.

[17] MacAlpine D.M., Rodríguez H.K., and Bell S.P., Coordination of replication and transcription along a drosophila chromosome, *Genes Dev*, 18(24):3094–3105, 2004.

[18] McCune H.J., Danielson L.S., Alvino G.M., Collingwood D., Delrow J.J., Fangman W.L., Brewer B.J., and Raghuraman M.K., The temporal program of chromosome replication: genomewide replication in clb5Delta saccharomyces cerevisiae, *Genetics*, 180(4):1833–1847, 2008.

[19] Newlon C.S. and Theis J.F., The structure and function of yeast ars elements, *Curr Opin Genet Dev*, 3(5):752–758, 1993.

[20] Nieduszynski C.A., Hiraga S., Ak P., Benham C.J., and Donaldson A.D., Oridb: a dna replication origin database, *Nucleic Acids Res*, 35(Database issue):D40–D46, 2007.

[21] Nieduszynski C.A., Knox Y., and Donaldson A.D., Genome-wide identification of replication origins in yeast by comparative genomics, *Genes Dev*, 20(14):1874–1879, 2006.

[22] R-Development-Core-Team, *R: A language and environment for statistical computing*. R Foundation for Statistical Computing, Vienna, Austria, 2008, ISBN 3-900051-07-0.

[23] Raghuraman M.K. and Brewer B.J., Molecular analysis of the replication program in unicellular model organisms, *Chromosome Res*, 18(1):19–34, 2010.

[24] Raghuraman M.K., Winzeler E.A., Collingwood D., Hunt S., Wodicka L., Conway A., Lockhart D.J., Davis R.W., Brewer B.J., and Fangman W.L,. Replication dynamics of the yeast genome, *Science*, 294(5540):115–121, 2001.

[25] Rhind N., Yang S. C.-H., and Bechhoefer J., Reconciling stochastic origin firing with defined replication timing, *Chromosome Res*, 18(1):35–43, 2010.

[26] Schwob E. and Nasmyth K., Clb5 and clb6, a new pair of b cyclins involved in dna replication in saccharomyces cerevisiae, *Genes Dev*, 7(7A):1160–1175, 1993.

[27] SGDproject, "saccharomyces genome database" http://downloads.yeastgenome.org/chromosomal_feature/03.03.2010.

[28] Spiesser T.W., Klipp E., and Barberis M., A model for the spatiotemporal organization of dna replication in saccharomyces cerevisiae, *Mol Genet Genomics*, 282(1):25–35, 2009.

[29] Theis J.F. and Newlon C.S., The ars309 chromosomal replicator of saccharomyces cerevisiae depends on an exceptional ars consensus sequence, *Proc Natl Acad Sci U S A*, 94(20):10786–10791, 1997.

[30] Guido van Rossum, *Python reference manual*, CWI (Centre for Mathematics and Computer Science), Amsterdam, The Netherlands, The Netherlands, 1995.

[31] Walter J. and Newport J., Initiation of eukaryotic dna replication: origin unwinding and sequential chromatin association of cdc45, rpa, and dna polymerase alpha, *Mol Cell*, 5(4):617–627, 2000.

[32] Weinreich M., Palacios DeBeer M. A., and Fox C.A., The activities of eukaryotic replication origins in chromatin, *Biochim Biophys Acta*, 1677(1-3):142–157, 2004.

[33] Wyrick J.J., Aparicio J.G., Chen T., Barnett J.D., Jennings E.G., Young R.A., Bell S.P., and Aparicio O.M., Genome-wide distribution of orc and mcm proteins in s. cerevisiae: high-resolution mapping of replication origins, *Science*, 294(5550):2357–2360, 2001.

[34] Xu W., Aparicio J.G., Aparicio O.M., and Tavaré S., Genome-wide mapping of orc and mcm2p binding sites on tiling arrays and identification of essential ars consensus sequences in s. cerevisiae, *BMC Genomics*, 7:276, 2006.

[35] Yabuki N., Terashima H., and Kitada K., Mapping of early firing origins on a replication profile of budding yeast, *Genes Cells*, 7(8):781–789, 2002.

# INTEGER PROGRAMMING-BASED METHOD FOR COMPLETING SIGNALING PATHWAYS AND ITS APPLICATION TO ANALYSIS OF COLORECTAL CANCER

TAKEYUKI TAMURA[1]
tamura@kuicr.kyoto-u.ac.jp

YOSHIHIRO YAMANISHI[2,3,4]
yoshihiro.yamanishi@ensmp.fr

MAO TANABE[1]
mao@scl.kyoto-u.ac.jp

SUSUMU GOTO[1]
goto@kuicr.kyoto-u.ac.jp

MINORU KANEHISA[1]
kanehisa@kuicr.kyoto-u.ac.jp

KATSUHISA HORIMOTO[5]
k.horimoto@aist.go.jp

TATSUYA AKUTSU[1]
takutsu@kuicr.kyoto-u.ac.jp

[1]*Bioinformatics Center, Institute for Chemical Research, Kyoto University, Uji, Kyoto 611-0011, Japan*
[2]*Mines ParisTech, Centre for Computational Biology, 35 rue Saint-Honore, F-77305 Fontainebleau Cedex, France*
[3]*Institut Curie, F-75248, Paris, France*
[4]*INSERM U900, F-75248, Paris, France*
[5]*Computational Biology Research Center, 2-42 Aomi, Koto-ku, Tokyo 135-0064, Japan*

Signaling pathways are often represented by networks where each node corresponds to a protein and each edge corresponds to a relationship between nodes such as activation, inhibition and binding. However, such signaling pathways in a cell may be affected by genetic and epigenetic alteration. Some edges may be deleted and some edges may be newly added. The current knowledge about known signaling pathways is available on some public databases, but most of the signaling pathways including changes upon the cell state alterations remain largely unknown. In this paper, we develop an integer programming-based method for inferring such changes by using gene expression data. We test our method on its ability to reconstruct the pathway of colorectal cancer in the KEGG database.

*Keywords*: signaling pathway; integer programming; missing interaction.

## 1. Introduction

Signaling pathways represent relationships among proteins in cells such as activation, inhibition and binding. Existing information on signaling pathways is available on some public databases [3, 8, 16]. Although most of their contents are based on facts confirmed by *in vivo* and *in vitro* experiments, it may also be possible to obtain new knowledge by *in silico* experiments. For example, when genetic and epigenetic alteration occurs, signaling pathways in cells are affected. However degree of such change in pathways can be considered not so large since many functions of cells remain unchanged even when they suffer cancers. Therefore, it may be possible to determine differences between signaling pathways of normal cells and those of cancer cells *in silico* utilizing existing knowledge and data.

Inferring biological networks is an important topic in bioinformatics and systems biology. In the problem of inferring genetic networks, a function along with input genes that regulates each gene should be the output when a series of gene expression profiles are given. By using a series of gene expression profiles, signaling pathways can also be derived as networks [15]. However, it is not necessary to infer whole networks for our aim. Since the differences between signaling pathways of normal cells and those of cancer cells are not large, it suffices to complete the former utilizing existing other knowledge such as gene expression profiles. Some basic problems for completing networks using observed data are known to be NP-hard [1].

In this paper, we assume a Boolean network model [10] as a model of signaling pathways, where states of nodes are represented by either 0 or 1. When 1 is assigned to a node, it means that the corresponding protein is activated. On the other hand, when 0 is assigned to a node, it means that the corresponding protein is suppressed. Although readers may think that representing states of proteins only by 0 and 1 is a rough modeling, it is reasonable since many existing knowledge on signaling pathways can be represented by Boolean functions.

For example, the family of Mitogen-activated protein kinase (MAPK) is known to activate FBJ murine osteosarcoma virus oncogene (c-Fos) [4]. MAPK1 and MAPK3 are subfamilies of MAPK [12]. Both of MAPK1 and MAPK3 can activate cFOS [4]. Therefore, the condition of activation of c-FOS can be represented by cFOS= MAPK1 ∨ MAPK3, meaning cFOS=1 if either MAPK1 or MAPK3 is 1. In this way, we represent existing knowledge of signaling pathways downloaded from the KEGG database [8] by Boolean models although details are explained in Sec. 2.1.

To detect changes of signaling pathways after cell state alteration, we apply gene expression profiles of cancer cells to signaling pathways representing the state before alteration. Values of gene expression profiles are encoded to 0 or 1 by using thresholds. Details are explained in Sec. 2.2. Since signaling pathways before alteration are represented by Boolean models and states of proteins after alteration are represented by 0 or 1, assigned 0/1 values may sometimes contradict with assigned Boolean functions. The part where such contradictions are detected can be considered as the change by alteration.

For example, if the relationship of cFOS= MAPK1 ∨ MAPK3 is changed by alteration, cFOS=0 may be often observed even when MAPK1=1 or MAPK3=1 holds. However, since it contradicts cFOS= MAPK1 ∨ MAPK3, the change on cFOS= MAPK1 ∨ MAPK3 can be detected.

Of course, data of gene expression profiles and encoding by thresholds often include noises. Moreover, Boolean functions assigned to proteins do not always appropriately represent relationships among proteins. To handle these problems, we develop an integer programming-based method where noises by small number of patients are automatically ignored.

Integer programming (IP) is a very common method in the field of operations research and often applied to complex problems [7]. To apply IP, problems must be formalized to maximize or minimize a given objective function which is a linear function of integer variables and constraints must also be given as linear equations or inequations of integer variables. Details of the proposed method are explained in Sec. 2.3.

To verify the validity of the proposed method, we perform the following computer experiments. Firstly, we represent a signaling pathway of colorectal cancer in the KEGG database as a Boolean model. This data includes information of missing interaction among proteins. Missing interactions are relationships which are observed before alteration, but not observed after alteration. Therefore, if we apply data of gene expression profiles representing states after alteration, contradictions are expected to be detected on the part of missing interactions. For this purpose, we download data of gene expression profiles of colorectal cancer [6] from NCBI.

As a result of computer experiments, contradictions between the signaling pathway and data of gene expression profiles are detected on functions corresponding to beta-catenin (CTNNB1), caspase 9 (CASP9), B-cell leukemia/lymphoma 2 (BCL2) and BCL2-associated agonist of cell death (BAD). Missing interactions to CTNNB1,CASP9 and BCL2 are described in the above KEGG data although a missing interaction to BAD is not. Thus three of the four predicted missing interactions are correct (specificity=75%). Details of the computer experiment and its result are explained in Sec. 3. In Sec. 4, we conclude with future work.

## 2. Method

### 2.1. *Data of signaling pathway*

In the computer experiments of this study, we focus on colorectal cancer of human. The colorectal cancer pathway was obtained from the Human Diseases section of the KEGG PATHWAY database [8]. The pathway can be regarded as a directed graph where each node corresponds to a protein and each edge corresponds to a relationship between proteins. Formally, a network is defined as $N = (V, E)$, where $V = \{v_1, \ldots, v_n\}$ is a set of nodes and $E = \{e_1, \ldots, e_m\}$ is a set of edges. If there is an edge from $v_2$ to $v_1$, $v_2$ is called a *parent* of $v_1$ and $v_1$ is called a *child* of $v_2$.

Protein-protein relationships can be classified into activation, inhibition, missing activation, and missing inhibition, where "missing" means that the protein-protein relationship is present in normal cells, but missing in cancer cells. Moreover, a relationship among a node and its parents can be represented by a combination of "AND", "OR" and "NOT". If $v_1$ is an "AND" node, $v_1$ is 1 only if values of all parents of $v_1$ are 1. For example, in Fig. 1 (a), the relationship among $v_1$, $v_2$, $v_3$ and $v_4$ is denoted by $v_1 = v_2 \wedge v_3 \wedge v_4$ if $v_1$ is an "AND" node. Therefore, $v_1 = 1$ holds only if $v_2 = v_3 = v_4 = 1$ holds. On the other hand, if $v_1$ is an "OR" node, the relationship among $v_1$, $v_2$, $v_3$ and $v_4$ is denoted by $v_1 = v_2 \vee v_3 \vee v_4$. $v_1 = 1$ holds

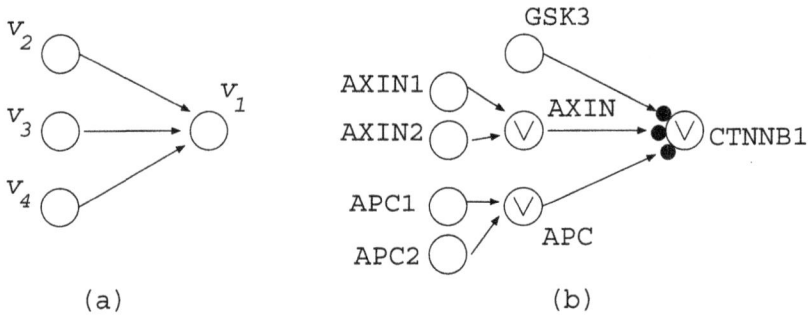

Fig. 1.    (a) The relationship among $v_1, \ldots, v_4$. $v_1$ is a child of $v_2, v_3, v_4$ and $v_2, v_3, v_4$ are parents of $v_1$. If $v_1$ is "AND", $v_1$ is 1 only if $v_2 = v_3 = v_4 = 1$ holds. (b) The relationship among GSK3, AXIN1, AXIN2, APC1, APC2 and CTNNB1.

if one of $v_2$, $v_3$, $v_4$ is 1. Moreover, "NOT" is denoted by overlines. For example, $\overline{v_1}$ is negation of $v_1$ and $v_1 \neq \overline{v_1}$ always holds. Note that every variable takes only 0 or 1. Functions represented by combination of "AND", "OR" and "NOT" are called *Boolean functions*. Thus the signaling pathway of colorectal cancer is represented by a network where each node is assigned a Boolean function.

For example, CTNNB1 is degraded by a beta-catenin destruction complex, which includes Axin, adenomatosis polyposis coli (APC), and glycogen synthase kinase 3 (GSK3). GSK3 phosphorylates beta-catenin, resulting in beta-catenin recognition by an E3 ubiquitin ligase, and subsequent beta-catenin ubiquitination and proteasomal degradation [9, 11, 13]. AXIN has subfamilies AXIN1 and AXIN2. APC has subfamilies APC1 and APC2. Therefore the relationship among CTNNB1, GSK3, AXIN1, AXIN2, APC1 and APC2 can be represented as $\overline{CTNNB1} = GSK3 \wedge (AXIN1 \vee AXIN2) \wedge (APC1 \vee APC2)$. It means that all of GSK3, AXIN and APC are necessary to degrade CTNNB1. Since both AXIN1 and AXIN2 work as AXIN, AXIN is represented as $(AXIN1 \vee AXIN2)$. Similarly APC is represented as $(APC1 \vee APC2)$. By De Morgan's laws, $\overline{CTNNB1} = GSK3 \wedge (AXIN1 \vee AXIN2) \wedge (APC1 \vee APC2)$ is converted into $CTNNB1 = \overline{GSK3} \vee \overline{(AXIN1 \vee AXIN2)} \vee \overline{(APC1 \vee APC2)}$ as shown in Fig. 1 (b). Note that "●" represents a negation in Fig. 1 (b).

The Boolean model of signaling pathway of colorectal cancer constructed from Human Diseases section of the KEGG PATHWAY database [8] of March 2010 contains 57 nodes and 154 edges.

## 2.2. Gene expression data

Gene expression profiles of colorectal cancer patients were obtained from the Gene Expression Omnibus (GEO) [2]. This data set consists of 12 patients and 10 healthy controls. In the original study, the authors performed analysis of normal-appearing colonic mucosa of early onset colorectal cancer patients [6]. Tumor specimens and adjacent grossly normal-appearing tissue were routinely collected and archived from patients undergoing colorectal resection. Healthy controls are those who underwent colonoscopic examination and were found to have no polyps and no known family history or previous colorectal cancer incidence.

To utilize this data, we firstly normalize values as follows. For each gene, let $a_1, \ldots, a_{10}$ and $a_{11}, \ldots, a_{22}$ be values of gene expression profiles of healthy controls and colorectal cancer patients respectively. Let $min$ and $max$ be the minimum and maximum values in $a_1, \ldots, a_{22}$ respectively. Normalized values $a'_1, \ldots, a'_{22}$ are calculated by

$$a'_i = \frac{a_i - min}{max - min},$$

where each $a'_i$ takes values on the range between 0 and 1. Then, $a'_i$ is encoded to 0 or 1 by a threshold. In our experiment, 0.8 was used as the threshold. Note that data of gene expression profiles are not available for all nodes in the signaling pathway.

## 2.3. *Integer programming based method*

In integer programming, every constraint must be represented by a linear equality or inequality of integer variables. To represent Boolean constraints of signaling pathways, a method which we developed in [17] might be applicable. However, it is not straightforward to extend this method so that the completion problem of signaling pathways is formalized as IP. Therefore, in the first part of this section, we summarize how to formalize Boolean constraints as linear equations or inequations.

For "AND" nodes,

$$x_1 = x_2 \wedge x_3 \wedge \cdots \wedge x_k$$

is represented by

$$(x_1 \vee \overline{x_2} \vee \overline{x_3} \vee \cdots \vee \overline{x_k}) \wedge (\overline{x_1} \vee x_2) \wedge (\overline{x_1} \vee x_3) \wedge \cdots \wedge (\overline{x_1} \vee x_k). \qquad (1)$$

Then, (1) is represented by the following linear inequalities:

$$x_1 + \overline{x_2} + \overline{x_3} + \cdots + \overline{x_k} \geq 1 \qquad (2)$$
$$\overline{x_1} + x_2 \geq 1 \qquad (3)$$
$$\overline{x_1} + x_3 \geq 1 \qquad (4)$$
$$\vdots$$
$$\overline{x_1} + x_k \geq 1 \qquad (5)$$

where every variable takes only 0 or 1. On the other hand, for "OR" nodes,

$$x_1 = x_2 \vee x_3 \vee \cdots \vee x_k$$

is represented by

$$(\overline{x_1} \vee x_2 \vee x_3 \vee \cdots \vee x_k) \wedge (x_1 \vee \overline{x_2}) \wedge (x_1 \vee \overline{x_3}) \wedge \cdots \wedge (x_1 \vee \overline{x_k}). \qquad (6)$$

Similar to the case of "AND" nodes, (6) is represented by the following linear inequalities:

$$\overline{x_1} + x_2 + x_3 + \cdots + x_k \geq 1 \tag{7}$$

$$x_1 + \overline{x_2} \geq 1 \tag{8}$$

$$x_1 + \overline{x_3} \geq 1 \tag{9}$$

$$\vdots$$

$$x_1 + \overline{x_k} \geq 1 \tag{10}$$

where every variable takes only 0 or 1. By the above method, constraints of signaling pathways of the Boolean model can be represented by linear inequalities.

To detect missing activations and missing inhibitions, we divide every node $v$ into $v_1$ and $v_2$ as shown in Fig. 2 (a). $v_1$ and $v_2$ are connected by a directed edge $e_1$. The 0/1 value calculated by a Boolean function of $v$, which is determined by values of parents of $v$, is assigned to $v_1$. On the other hand, the 0/1 value encoded from gene expression profiles is assigned to $v_2$. Whichever "AND" or "OR" is assigned to $v_2$, $v_1 = v_2$ must be satisfied since $v_2$ has only one in-edge. However, since the value of $v_1$ is not always the same as the value of $v_2$, a contradiction sometimes occurs. Although this contradiction may come from noises of data, it is reasonable to think that missing activations or missing inhibitions yield the contradiction if we can ignore such noises.

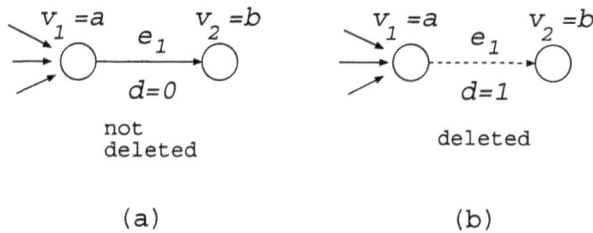

Fig. 2.    Every node $v$ is divided into two nodes $v_1$ and $v_2$. The 0/1 value calculated by the Boolean function corresponding to $v$ is assigned to $v_1$. The 0/1 value encoded from gene expression profiles by a threshold is assigned to $v_2$. To handle a contradiction between $v_1$ and $v_2$, $e_1$ can be deleted if $d = 1$ holds.

However, for IP formalization, contradictions are not allowed. Therefore, every constraint represented by linear equalities or inequalities must be satisfied. To do so, $e_1$ is deleted when $v_1 \neq v_2$ holds in our method. A parameter $d$ is used to represent whether $e_1$ is deleted or not. $d = 0$ means that $e_1$ is not deleted and $d = 1$ means that $e_1$ is deleted. Therefore the relationship among $v_1$, $v_2$ and $d$ are represented as

$$\overline{d}(v_1 v_2 + \overline{v_1} \; \overline{v_2}) + d = 1 \tag{11}$$

where every variable and term take either 0 or 1. $\overline{d}(v_1 v_2 + \overline{v_1} \; \overline{v_2})$ means that $v_1 = v_2 = 1$ (represented by $v_1 v_2$) or $v_1 = v_2 = 0$ (represented by $\overline{v_1} \; \overline{v_2}$) must be satisfied when $d = 0$ (represented by $\bar{d}$). On the other hand, $d$ of (11) means that there is no constraint between $v_1$ and $v_2$ when $d = 1$.

To represent (11) by linear inequalities, (11) is converted into

$$\bar{d}ab + \bar{d}\bar{a}\bar{b} + d \geq 1$$

and further converted into

$$A + B + d \geq 1 \tag{12}$$

where

$$A = \bar{d} \wedge a \wedge b \tag{13}$$
$$B = \bar{d} \wedge \bar{a} \wedge \bar{b}. \tag{14}$$

Since (13) and (14) are not yet linear, (13) is converted into

$$\overline{A} + \bar{d} \geq 1 \tag{15}$$
$$\overline{A} + a \geq 1 \tag{16}$$
$$\overline{A} + b \geq 1 \tag{17}$$
$$A + d + \bar{a} + \bar{b} \geq 1 \tag{18}$$

and (14) is converted into

$$\overline{B} + \bar{d} \geq 1 \tag{19}$$
$$\overline{B} + \bar{a} \geq 1 \tag{20}$$
$$\overline{B} + \bar{b} \geq 1 \tag{21}$$
$$B + d + a + b \geq 1. \tag{22}$$

Thus, whether $e_1$ is deleted is represented by linear inequalities (12) and (15)-(22). Note that every variable takes either 0 or 1.

Since there are 12 patients of colorectal cancer in our dataset, the signaling pathway is duplicated into 12 copies $N_1, \ldots, N_{12}$ as shown in Fig. 3. Note that Fig. 3 describes only one node $v$ although each $N_i$ originally contains 57 nodes and each node is duplicated into two nodes. Then for each pathway 0/1 values of gene expression profiles are applied. Since different parameters are necessary for different copies of the signaling pathway, $d_1, \cdots, d_{12}$ are used to represent whether $e_{i\_1}$ is deleted or not. If $v_{i\_1} \neq v_{i\_2}$, $d_i = 1$ holds and then $e_{i\_1}$ is deleted. Since values of gene expression profiles are different among patients, $d_i = 1$ may hold for some $i$, but $d_i = 0$ holds for other $i$. However, it is reasonable to assume that differences of signaling pathways among patients of colorectal cancer are not very large.

Of course, it is known that some patients of colorectal cancer have relatively different signaling pathways when compared to those of other patients [5]. There are two major types of genomic instability for patients of colorectal cancer, chromosome instability (CIN) and microsatellite instability (MSI). According to the KEGG database, signaling pathways of patients of MSI are partially different from those of CIN. However, since MSI occurs in less than 15% [5] of colorectal cancer, it is

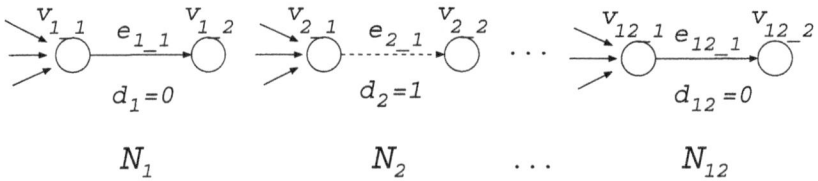

Fig. 3.   $N_i$ corresponds to the network of the $i$-th patient of colorectal cancer. $d_i$ represents whether $e_{i\_1}$ is deleted or not. Note that each $N_i$ contains $2n$ nodes although there are only two nodes for each $N_i$ in this figure.

reasonable to assume that most data of gene expression profile is of CIN patients. Thus, in addition to noises in data of gene expression profile, data from MSI patients are also treated as noises in our method. To appropriately ignore noises, a parameter $D$ determined by $d_1, \ldots, d_{12}$ is used to judge whether a function assigned to $v$ is missing in colorectal cancer.

By adding

$$d_1 + d_2 + \cdots + d_{12} - k \le (12 - k)D \tag{23}$$

as a constraint of IP, contradictions at most $k$ copies are ignored as noises since the objective function of our method is to minimize the sum of "D". Note that "D" is defined on every node of the original network as $D_1, \ldots, D_n$. In our computer experiments, $k = 5$ was used and the objective function was

$$\min \sum_{i=1}^{n} D_i. \tag{24}$$

## 3. Result of Computer Experiment

The computer experiment was done on a PC with a Xeon 3GHz CPU and 8GB RAM under the Linux (version 2.6.24) operating system, where CPLEX (Version 10.1.0) was used as the solver of integer programming.

As a result, in the obtained optimal solution of IP, "D"s assigned on CTNNB1, BAD, CASP9 and BCL2 were 1. It means that contradictions by missing activations or missing inhibitions were detected on functions corresponding to CTNNB1, BAD, CASP9 and BCL2. To verify this prediction result, we compare them with the description on missing activations and missing inhibitions of the KEGG database.

According to the KEGG database, there are missing activations or missing inhibitions on the functions of CTNNB1, CASP9 and BCL2 in colorectal cancer. Therefore, "CTNNB1", "CASP9" and "BCL2" can be considered as correctly predicted missing interactions. Note that both missing activations and missing inhibitions are called as missing interactions. On the other hand, since there is no missing interactions on "BAD" in the KEGG database of colorectal cancer, "BAD" can be considered as incorrectly predicted missing interactions. Although there is a possibility that incorrectly predicted missing interactions are newly found knowledge of missing interactions, we treat them as false prediction results when we estimate the prediction accuracy.

To evaluate the prediction accuracy of our method, the following two measures [14] are used:

$$Sensitivity = \frac{TP}{TP + FN}, \quad Specificity = \frac{TP}{TP + FP}.$$

TP is the number of correctly predicted interactions (true positive). FP is the number of overpredicted interactions (false positive). FN is the number of underpredicted interactions (false negative).

Since TP=3, FN=6 and FP=1 hold in our result, sensitivity is 0.33 and specificity is 0.75. Although specificity is relatively high, sensitivity is not good. However, for our purpose, specificity is much more important than sensitivity since our method may be possible to iterate after adding newly found knowledge into existing knowledge. Note again that FP interactions may be newly found knowledge of missing interactions.

## 4. Concluding Remark

In this paper, we developed an integer programming-based method of inferring changes of signaling pathways after cell state alteration. For this purpose, we modeled signaling pathways of colorectal cancer as a Boolean model, in which every protein-protein interaction is represented by a combination of "AND", "OR" and "NOT", which is called a Boolean function. On the other hand, values of gene expression profiles are encoded to either 0 or 1 by using a threshold. Since two binary values, one is calculated by a Boolean function and the other is encoded from gene expression profiles, are assigned to each node, a contradiction sometimes occurs. When such a contradiction is detected, we regard that the contradiction is yielded by the missing interaction by cell state alteration. To implement the above, every node is divided into two nodes connected by an edge. One node is assigned the value calculated by the Boolean function and the other node is assigned the value encoded from gene expression profiles. The contradiction is represented by whether the edge connecting two nodes is deleted or not.

The topology of signaling pathways of MSI type of colorectal cancer is different from those of CIN type. However, since MSI occurs in less than 15% of patients of colorectal cancer, data by MSI patients are treated as noises in addition to noises by encoding values of gene expression profiles by a threshold. Ignoring appropriately such noises can be realized by a linear inequality.

As a result of the computer experiment, "CTNNB1", "BAD", "CASP9" and "BCL2" were predicted as missing interactions. According to the KEGG database, "CTNNB1", "CASP9" and "BCL2" are correctly predicted missing interactions, but "BAD" is a false prediction. Since 9 missing interactions are included in the network used in the experiment, sensitivity and specificity are 0.33 and 0.75 respectively.

Since specificity is much more important than sensitivity for our purpose, it is seen that our method can efficiently predict missing interactions.

Although the signaling pathway used in our experiment consists of those of colorectal cancer and missing interactions of colorectal cancer, it is preferable to use signaling pathways of normal cells in order to newly predict knowledge of missing interactions and this remains as our future work. Moreover, in the experiment, only one threshold of 0.8 was used for every gene. Using different thresholds for different genes may yield more accurate prediction of missing interactions. Such optimization for thresholds might be realized by linear constraints of IP where the objective function is the number of contradictions. This also remains as a future work.

### Acknowledgement

This work was also supported by the bi-national JSPS/INSERM grant and Japan-France Research Cooperative Program.

### References

[1] Akutsu, T., Tamura, T., and Horimoto, K., Completing networks using observed data, *Proc. 20th International Conference on Algorithmic Learning Theory*, 126-140, 2009.

[2] Barrett, T., Suzek, T.O., Troup, D.B., Wilhite, S.E., Ngau, W.C., Ledoux, P., Rudnev, D., Lash, A.E., Fujibuchi, W., and Edgar, R., NCBI GEO: mining millions of expression profiles–database and tools, *Nucleic Acids Research*, 33:D562-D566, 2005.

[3] Database of Cell Signaling, Science Signaling, http://stke.sciencemag.org/cm/.

[4] Fang, J.Y. and Richardson, B.C., The MAPK signalling pathways and colorectal cancer, *The Lancet Oncology*, 6(5):322-327, 2005.

[5] Grady, W.M., Genomic instability and colon cancer, *Cancer and Metastasis Reviews*, 23:11-27, 2004.

[6] Hong, Y., Ho, K.S., Eu, K.W., and Cheah, P.Y., A susceptibility gene set for early onset colorectal cancer that integrates diverse signaling pathways: implication for tumorigenesis, *Clinical Cancer Research*, 13(4):1107-1114, 2007.

[7] Hromkovic, J., *Algorithmics For Hard Problems*, Springer-Verlag Berlin Heidelberg, 2001.

[8] Kanehisa, M., Goto, S., Furumichi, M., Tanabe, M., and Hirakawa, M., KEGG for representation and analysis of molecular networks involving diseases and drugs. *Nucleic Acids Research*, 38:D355-D360, 2010.

[9] Karim, R.Z., Tse, G.M.K., Putti, T.C., Scolyer, R.A., and Lee, C.S., The significance of the Wnt pathway in the pathology of human cancers, *Pathology*, 36(2):120-128, 2004.

[10] Kauffman, S.A., *The Origins Of Order: Self-organization And Selection In Evolution*, Oxford University Press, New York, 1993.

[11] Komiya, Y. and Habas, R., Wnt signal transduction pathways, *Organogenesis*, 4(2):68-75, 2008.

[12] Kutz, D. and Burg, M., Evolution of osmotic stress signaling via MAP kinase cascades, *The Journal of Experimental Biology* 201:3015-3021, 1998.

[13] MacDonald, B.T., Tamai, K., and He, X., Wnt/beta-catenin signaling: components, mechanisms, and diseases, *Developmental Cell*, 17(1):9-26, 2009.

[14] Mount, D.W., *Bioinformatics, Sequence And Genome Analysis*, Cold Spring Harbor Laboratory Press, 2004.

[15] Sachs, K., Perez, O., Pe'er, D., Lauffenburger D. A., and Nolan., G. P., Causal protein-signaling networks derived from multiparameter single-cell data, *Science*, 308:5721, Apr. 22, 2005.

[16] Schaefer, C.F., Anthony, K., Krupa, S., Buchoff, J., Day, M., Hannay, T., and Buetow, K.H., PID: the pathway interaction database, *Nucleic Acids Research*, 37:D674-D679, 2009.

[17] Tamura, T., Takemoto, K., and Akutsu, T., Finding minimum reaction cuts of metabolic networks under a Boolean model using integer programming and feedback vertex sets, *International Journal of Knowledge Discovery in Bioinformatics*, 1:14-31, 2010.

# G1 AND G2 ARRESTS IN RESPONSE TO OSMOTIC SHOCK ARE ROBUST PROPERTIES OF THE BUDDING YEAST CELL CYCLE

CHRISTIAN WALTERMANN

christian.waltermann@hu-berlin.de

MAX FLOETTMANN

max.floettmann@biologie.hu-berlin.de

EDDA KLIPP

edda.klipp@hu-berlin.de

*Theoretical Biophysics, Humboldt University of Berlin, Invalidenstr. 42, 10115 Berlin, Germany*

Boolean modeling has been successfully applied to the budding yeast cell cycle to demonstrate that both its structure and its timing are robustly designed. However, from these studies few conclusions can be drawn how robust the cell cycle arrest upon osmotic stress and pheromone exposure might be. We therefore implement a compact Boolean model of the *S. cerevisiae* cell cycle including its interfaces with the High Osmolarity Glycerol (HOG) and the pheromone pathways. We show that all initial states of our model robustly converge to a cyclic attractor in the absence of stress inputs whereas pheromone exposure and osmotic stress lead to convergence to singleton states which correspond to G1 and G2 arrest *in silico*. A comparison with random Boolean networks reveals, that cell cycle arrest under osmotic stress is a highly robust property of the yeast cell cycle. We implemented our model using the novel frontend booleannetGUI to the python software booleannet.

*Keywords*: Boolean modeling; budding yeast cell cycle; robustness; osmotic stress; pheromone; cell cycle arrest.

## 1. Introduction

Biological systems often exhibit a high degree of complexity which requires the integration of empirical data into mathematical models for a deeper understanding [18, 19]. Particularly for biochemical systems there is often a lack of parameters for the identification of ordinary differential equation (ODE) models since in many cases, kinetics and parameters have not been reliably determined [22]. In these cases, Boolean models provide a useful alternative framework for description and analysis of the system since their construction is based on logical rules exclusively. The rules underlying the Boolean description are derived from qualitative biochemical and other empircal observations in the system [11]. Here, we present a new compact Boolean model of the budding yeast cell cycle including its interfaces with the High Osmolarity Glycerol (HOG) pathway and the pheromone pathway. Our cell cycle model only takes into account the fundamental regulatory principles and reduces the connected pathways to their downstream MAPKs Hog1p and Fus3p. Fig. 1 and

Tab. 1 provide an overview over the network. Using this approach, we will study the regulation of the cell cycle by HOG and pheromone signaling pathways. Moreover, we will investigate the robustness of the cell cycle and its arrest states by challenging it under constantly and temporarily perturbed conditions *in silico* and compare the results with random Boolean networks [2].

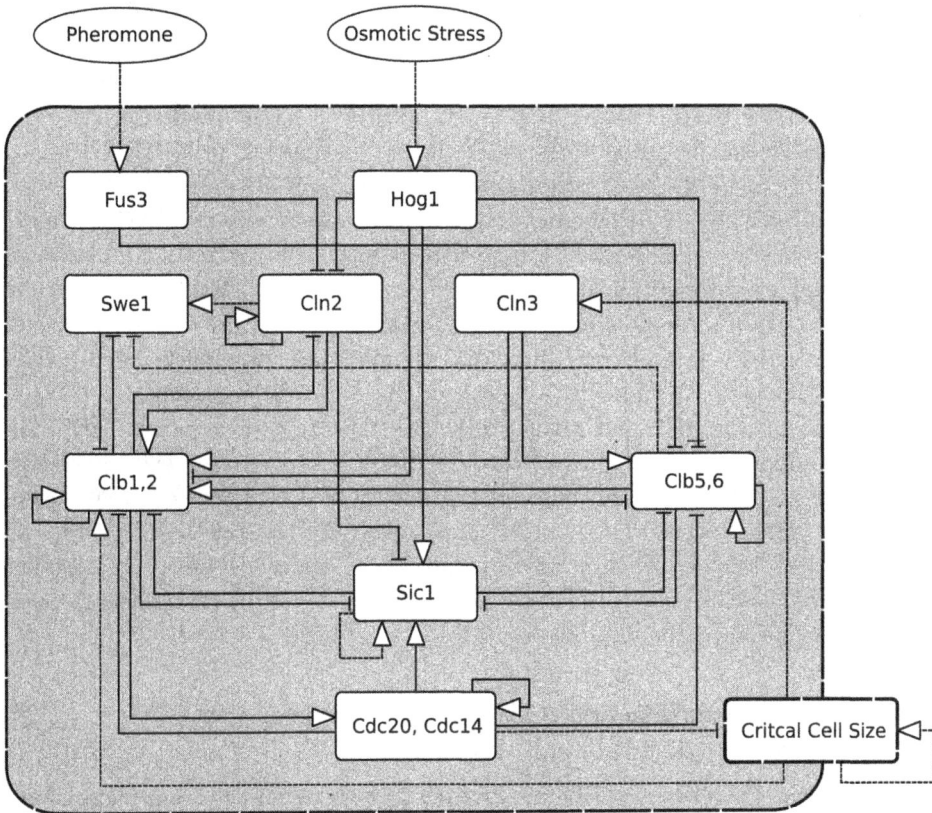

Fig. 1.   Overview over the cell cycle network including the connections to the downstream MAPKs of the pheromone and HOG pathways. Solid lines represent direct and indirect biochemical implications, dotted lines denote other empirical observations such as co-expression and implications from cell division.

## 2. The Model

### 2.1. *Biochemical processes captured in the core cell cycle*

Progression through the budding yeast cell cycle depends on the accumulation of cyclins (CLNs) and B-type cyclins (CLBs) and their binding to the cyclin dependent kinase (CDK) Cdc28p [27]. The cyclin binding lets Cdc28p fulfill different tasks throughout the cell cycle phases. At the beginning of the cell cycle in G1, upon a critical size, a sufficient amount of the cyclin Cln3p is accumulated to activate the transcription factor complexes SBF (Swi4-Swi6) and MBF (Mbp1-Swi6) [33, 38]. Both SBF and MBF activate the transcription of the cyclins *CLN1* and *CLN2* which in turn are able to enhance their own transcription through a positive feedback mechanism [7]. It is suggested that MBF additionally triggers the transcription of *CLB5* [30]. In the model, we combine all the steps from Cln3p accumulation to initiation of *CLN1*, *CLN2*, and *CLB5* expression. Cln1-Cdc28 and Cln2-Cdc28 then phosphorylate the CDK inhibitor (CKI) Sic1p which marks it for degradation [34]. This lifts in particular the inhibition of the B-cyclin-Cdc28 complexes allowing Clb5-Cdc28 to initiate DNA synthesis and entry into the S phase of the cell cycle [9]. The protein kinase Swe1p becomes active during bud formation in late G1 which depends on successful Cln2p accumulation [4, 10]. In its stabilized state Swe1p inhibits Clb2-Cdc28 [15]. If all checkpoints are met, Swe1 becomes unstable in G2 and is degraded [32], allowing Clb2-Cdc28 to perform its role in the G2/M transition. In the model, we correlate the destabilization event of "Swe1" with the ON-state of the species "Clb5". The presence of both Clb5-Cdc28 and Cln2-Cdc28 complexes induces the inactivation of the Cdh1/APC complex [31, 36], which allows Clb2p to rise and transition to mitosis (M phase of the cell cycle) can occur [20]. Clb2-Cdc28 turns off the SBF and MBF transcription factors which cause Cln2p and Clb5p to disappear [3]. Exit from mitosis/cell division is initiated when Clb2-Cdc28 induces the synthesis of Cdc20p allowing the Cdc20/APC complex to become active [37]. The complex degrades Clb2p [37] and causes the phosphatase Cdc14p - an agent for mitotic exit [35] - to enter the scene. In the model we lump together the variables "Cdc20" and "Cdc14" to "Cdc20&Cdc14". Sic1p is stabilized through dephosphorylation by the phosphatase Cdc14p at the end of mitosis [35].

## 2.2. *Model assumptions and links to the pheromone and HOG pathways*

In our Boolean model we describe the division process under the following assumptions and rules:

- The species "CriticalSize" was introduced to be able to track if the cell has reached a sufficient size for cell cycle progression.
- The species "Cdc20&Cdc14" is switched on and remains switched on if the the critical cell size condition is fulfilled and "Clb2" is active. In this case, the variables "CriticalSize" and "Cln3" are set to zero to restart the cell cycle in G1.
- Additionally we ensured that autonomous cell growth is reflected by activating the critical size when it has not been active in the previous step. The critical size also enters the model at the G1/S and G2/M checkpoints and mitotic exit in accordance with biological observations [14, 33]. At the G1 and G2 checkpoints it is a necessary condition for the activation of the species "Clb5" and "Clb2" respectively which represent markers for entry

Table 1. The Boolean model of the cell cycle including its interfaces with the pheromone and the high osmolarity glycerol (HOG) pathway.

| Node | Rules determining node value | Justification and references |
|---|---|---|
| CriticalSize | $\neg (CriticalSize \vee Cdc20\&Cdc14)$ | Rules ensure autonomous growth and reset of CriticalSize to zero on mitotic exit. |
| Cln3 | $CriticalSize \wedge \neg Cdc20\&Cdc14$ $\wedge \neg Fus3$ | Reset of "Cln3" to zero on mitotic exit. Cln3-Cdc28 activity is downregulated by active Fus3p [28] |
| Sic1 | $((Cdc20\&Cdc14 \vee Sic1)$ $\wedge \neg (Clb2 \vee Cln2 \vee Clb5))$ $\vee ((Cdc20 \vee Sic1) \wedge Hog1)$ | Sic1p is stabilized through dephosphorylation by the phosphatase Cdc14p at the end of mitosis [35]. It remains stable in the absense of activators and inhibitors [34]. Cln2-Cdc28, Clb5-Cdc28 [34], and Clb2-Cdc28 [9] phosphorylate Sic1p and mark it for degradation. Active Hog1p can stabalize Sic1p [12]. |
| Cln2 | $(Cln3 \vee Cln2) \wedge \neg Clb2$ $\wedge \neg Fus3 \wedge \neg Hog1$ | *CLN2* expression is upregulated by the SBF transcription factor [7] which in turn is activated by Cln3-Cdc28 [33, 38]. Cln2-Cdc28 can also activate SBF and therefore enhances the transcription of its own gene [7] whereas Clb2-Cdc28 inactivates SBF [3]. Hog1p activity downregulates *CLN2* expression [6]. Pheromone exposure and subsequent Fus3p activation leads to the phosphorylation of Far1p which represses *CLN2* transcription and inhibits Cln2-Cdc28 kinase activity [28]. |
| Clb5 | $Cln3$ $\wedge \neg (Sic1 \vee Clb2 \vee Cdc20\&Cdc14)$ $\wedge \neg Hog1$ | *CLN5* transcription is induced by MBF transcription factor complex [30]. MBF activity is in turn upregulated by Cln3-Cdc28 [38]. Sic1p inhibits Clb5-Cdc28 [9], Clb2-Cdc28 inactivates MBF [3], and Cdc20p induces the degradation of Clb5p [36]. Active Hog1p reduces the expression of *CLB5* [6]. |
| Swe1 | $(Cln2 \wedge \neg (Clb5 \vee Clb2))$ $\vee (Swe1 \wedge Hog1)$ | *SWE1* is coordinately expressed together with *CLN2* [25]. Swe1p is stabilized through Hog1p activity under osmotic shock [8]. The onset of Swe1p degradation has a similar timing as the Clb5p peak in S phase [24]. Additionally, Swe1p degradation is reinforced by Clb2-Cdc28 [15]. |
| Clb2 | $((Clb5 \wedge Cln2) \vee Clb2)$ $\wedge \neg (Sic1 \vee Cdc20 \vee Swe1)$ $\wedge CriticalSize$ | Clb5-Cdc28 and Cln2-Cdc28 inactivate Cdh1p, an activator of anaphase-promoting complex (APC) [31, 36], which in turn degrades Clb2p [20]. Swe1p and Sic1p inhibit Clb2-Cdc28 [15, 29]. Cdc14p activates Cdh1p and thus APC and increases Clb2p degradation [37]. The cell has to have a minimal size to make the transition to mitosis [14]. |
| Cdc20&Cdc14 | $(Clb2 \vee Cdc20) \wedge CriticalSize$ | Clb2-Cdc28 allows Cdc20p and finally Cdc14p - an agent for mitotic exit - to appear [35, 37]. Cdc14p stabilizes Cdc20p thus forming a positive feedback loop [35]. Assumption: A miminal cell size is required to divide. |

into the S and M phase of the cell cylce in our model.

Osmotic stress and subsequent Hog1p activation has three major influences on cell cycle components:

(1) Downregulation of expression of the cyclins *CLN1*, *CLN2*, and *CLB5* mediated by active Hog1p [6]. In G1, an insufficient amount of these cyclins delays the transition to S phase.

(2) Stabilization of the CKI Sic1p during the G1 phase of the cell cycle through phosphorylation by active Hog1p [12]. This inhibits Clb5-Cdc28 from promoting DNA synthesis and causes G1 arrest.

(3) Active Hog1p phosphorylates Hs11p which in turn stabilizes Swe1p [8] which leads to G2 cell cycle arrest.

Pheromone exposure and resulting Fus3 activation has two main effects on cell cycle components which cause arrest in the G1 phase of the cell cycle:

(1) Cln3-Cdc28 activity is inhibited by active Fus3p [16].
(2) Also, Cln1-Cdc28 and Cln2-Cdc28 activity are inhibited by active Fus3p *via* Far1p [28].

Additionally we introduced a rule which ensures that the species "Sic1" remains switched on in the absence of inhibitors and activators in accordance with empirical observations [34].

## 3. Materials and Methods

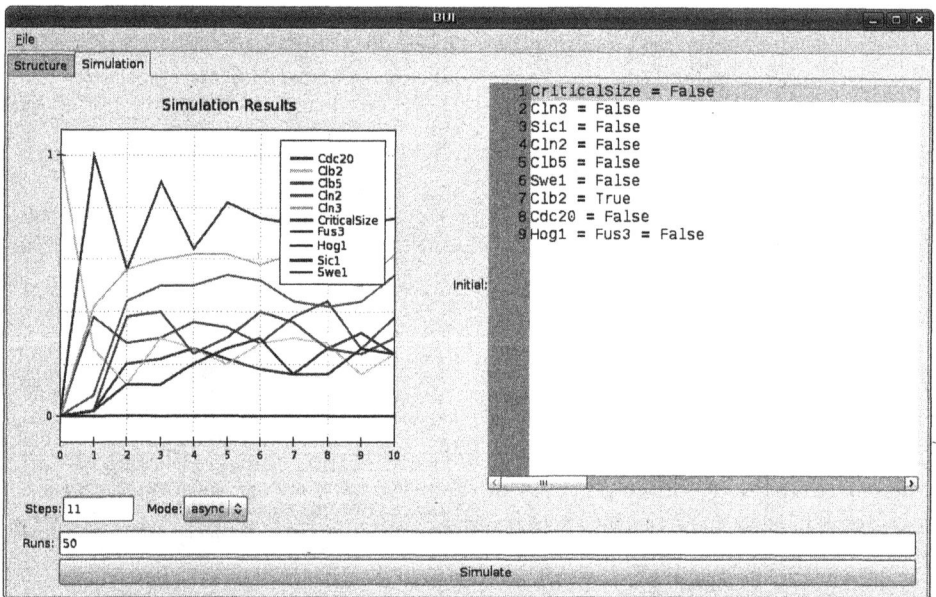

Fig. 2.   Graphical user interface of booleannetGUI which we used to implement the cell cycle model and to simulate its state space under asynchronous and synchronous updating and for PLDE simulations.

Boolean modeling is a discrete deterministic approach where all variables either assume the state "1" for "on" or "0" for "off" [17]. A variable is updated according to logical rules depending on the states of other variables in the Boolean network. We apply both synchronous updating in which the rules for all variables are updated at the same time and asynchrounous simulations, in which one random variable is updated at a time according to the rules which apply to it. As an additional check for robustness against alterations such as mutations in our network, we performed perturbation experiments *in silico* and compared these to random networks with similar structure (details in the *Results* section). The system was implemented and simulated in the synchronous, asynchronous [5], and partially linear differential equation (PLDE) mode [13] using booleannetGUI (Fig. 2), a graphical frontend to the python tool booleannet [1]. The tool is available upon request from the authors. The tool also provides a script for the representation of the state space of a synchronously updated Boolean model. Comparisons with random networks were performed with the R-package boolnet [26]. The statistical significance test implemented in this package computes the percentage of test function results for random networks that are greater than in our original cell cycle network and takes this result as a p-value. The test function itself counts the overall mean percentage of attractors found in the randomly perturbed versions of the network.

## 4. Results

### 4.1. *The model exhibits a robust cyclic attractor*

We simulated our Boolean model under all 1024 initial conditions (10 variables, hence $2^{10} = 1024$ states). For those conditions, in which no osmotic stress and no pheromone was applied and hence the species "Fus3" and "Hog1" are in the zero state, we observe that they converge on a cyclic attractor (Fig. 3A, Tab. 2) progressing through the cell cycle phases. Fig. 3B shows the attractor when simulated as a partially linear differential equation (PLDE). In a second simulation we updated the rules asynchronously and demonstrate that we obtain a complex attractor (Fig. 3C, Fig. 4) encompassing 100 states and 258 possible transitions. We confirmed that the complex attractor contains the core cell cycle states reached under synchronous updating rules and that it is reached independently of the initial configuration. Inspections of repeated simulations of asynchronous updating of the network revealed that the complex attractor is most likely rhythmic. We thus conclude that the rhythmic property of the cell cycle network might be robust towards slight variations of the sequence in which biochemical reactions and other empirically observed events are executed.

Table 2. Progression of states of cell cycle components through the cell cycle in the absence of osmotic stress and pheromone.

| Crit. Size | Cln3 | Sic1 | Cln2 | Clb5 | Swe1 | Clb2 | Cdc20,14 | Hog1 | Fus3 | CC State |
|---|---|---|---|---|---|---|---|---|---|---|
| 1 | 0 | 1 | 0 | 0 | 0 | 0 | 0 | 0 | 0 | G1 |
| 1 | 1 | 1 | 0 | 0 | 0 | 0 | 0 | 0 | 0 | G1 |
| 1 | 1 | 1 | 1 | 0 | 0 | 0 | 0 | 0 | 0 | G1 |
| 1 | 1 | 0 | 1 | 0 | 1 | 0 | 0 | 0 | 0 | G1 |
| 1 | 1 | 0 | 1 | 1 | 1 | 0 | 0 | 0 | 0 | S |
| 1 | 1 | 0 | 1 | 1 | 0 | 0 | 0 | 0 | 0 | G2 |
| 1 | 1 | 0 | 1 | 1 | 0 | 1 | 0 | 0 | 0 | M |
| 1 | 1 | 0 | 0 | 0 | 0 | 1 | 1 | 0 | 0 | M |
| 0 | 0 | 0 | 0 | 0 | 0 | 0 | 1 | 0 | 0 | mitotic exit |

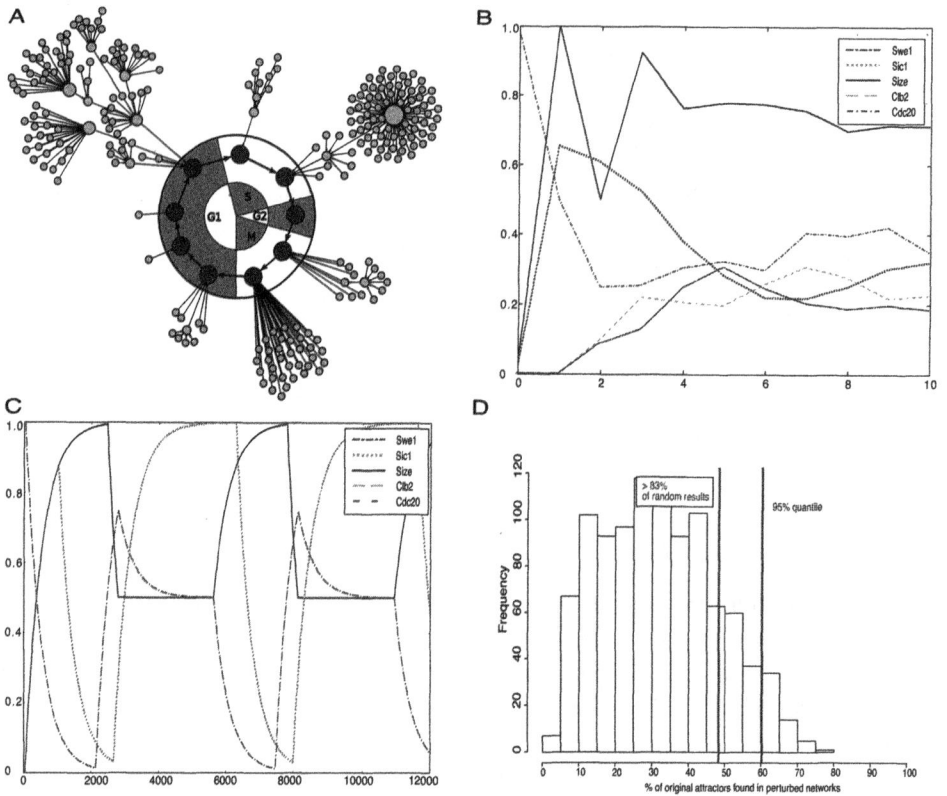

Fig. 3.   **A:** Core cell cycle attractor states with basins of attraction. All initial conditions with disabled osmotic stress and pheromone converge on this cyclic attractor. **B:** Average values of core cell cycle components over time out of 100 simulations, when the system is updated assynchronously. The fact that the species do not converge to one or zero suggests the absence of a singleton attractor. We confirm the existence of a complex cyclic attracor (Fig. 4). **C:** Partially linear differential equation simulation of selected species demonstrating the timing of core cell cycle components. **D:** Comparison with 1000 random Boolean models. Our model is more likely to exhibit the original attractor when perturbed than random networks. However, our significance threshold of $p < 0.05$ is not reached.

Additionally, we compared our network to 1000 randomly generated Boolean networks with the same number of nodes and edges and a similar structure as our cell cycle network. The random networks were subjected to 1000 random rule perturbations each and we determined in how many cases the attractors of the unperturbed network could be recovered. To ensure statistical significance, a statistical test was applied (see Materials and Methods), and the value at the 95% quantile of the obtained distribution was then compared with the mean value we derived from 1000 random rule perturbations of our boolean cell cycle model. We find that our cyclic attractor can be recovered more often than for the average of the random networks but did not reach our significance theshold of $p < 0.05$ (Fig. 3D). We conclude that the core network is more robust towards single mutations which would affect the way biochemical reactions or other events are executed than most random networks, however a high risk remains that mutations and severe perturbations lead to severe alterations of the cyclic attractor or cause it to collapse. Since we were at this stage exclusively interested in the robustness of the cyclic core cell cycle attractor, the "Hog1" and "Fus3" species were forced to remain off for all simulations and the random networks were generated using 8 nodes only.

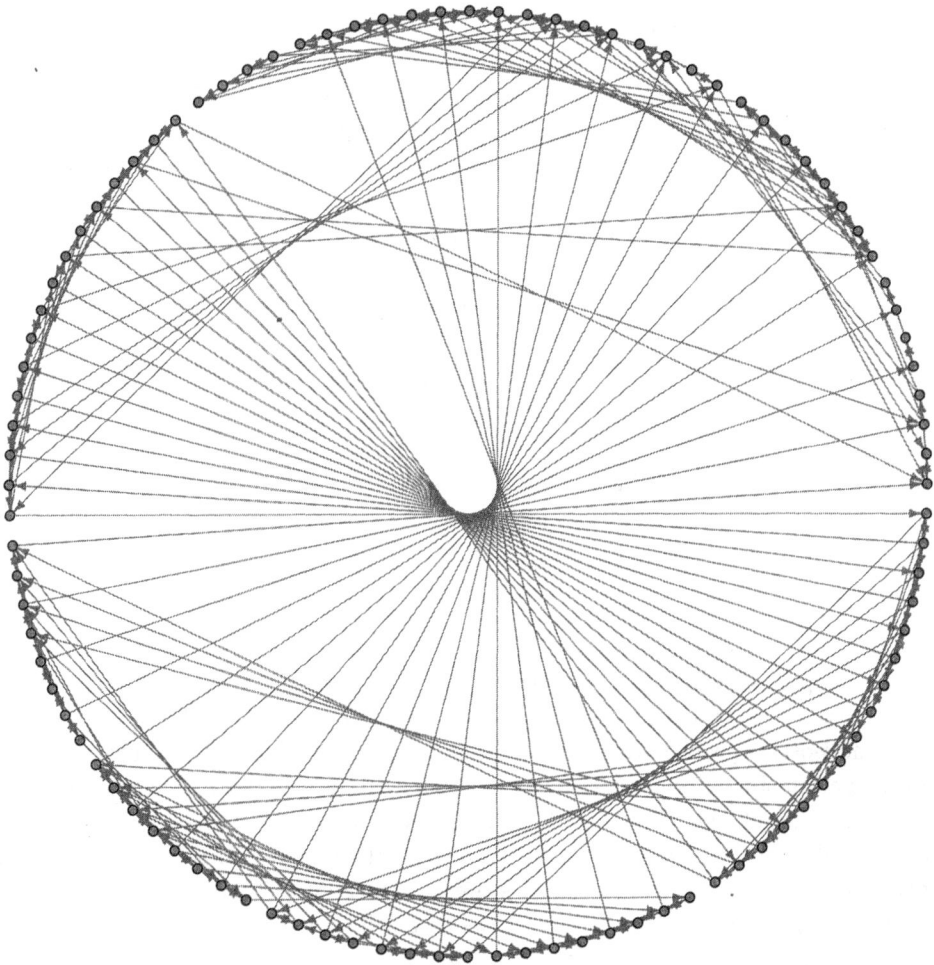

Fig. 4.   Complex attractor of the core cell cycle network without input from the pheromone or HOG pathway. We confirmed that all states of the cyclic attractor of the cell cycle model under synchronous updating are reached and that it is obtained independently of the initial configuration.

### 4.2. *Osmotic stress causes highly robust G1 and G2 arrests in silico*

To simulate the effect of osmotic stress on cell cycle progression, we switched "Hog1" on permanently. In the biological system, Hog1p is activated by osmotic stress and mediates cell cycle arrest in G1 and G2 by mechanisms described in detail in Sec. 2. Arrest singleton states in G2 were characterized by enabled "Swe1". All other point attractors could be assigned to early or late G1 depending on whether "Sic1" was enabled or not (Fig. 5, Tab. 3). However, also G2 arrest states exist, in which "Sic1" is switched on. These steady states might either be the result of initial conditions which would have little biological significance or of initial conditions, for which the G1 arrest would be too late. 256 states of the total 1024 states of our Boolean model are associated to exclusive exposure to osmotic stress. We confirmed that asynchronous updating does not affect the existence, number and composition of singleton attractors found during synchronous simulation of the network (data not shown) suggesting that also slight temporal variations of executing biochemical reactions and other events in the cell cycle network are unable to disrupt arrest.

Table 3.   Arrest steady states and their assignment to cell cycle phases under osmotic stress and pheromone exposure.

| Crit. Size | Cln3 | Sic1 | Cln2 | Clb5 | Swe1 | Clb2 | Cdc20,14 | Hog1 | Fus3 | Arrest |
|---|---|---|---|---|---|---|---|---|---|---|
| 1 | 1 | 0 | 0 | 0 | 0 | 0 | 0 | 1 | 0 | late G1 |
| 1 | 1 | 1 | 0 | 0 | 0 | 0 | 0 | 1 | 0 | early G1 |
| 1 | 1 | 0 | 0 | 0 | 1 | 0 | 0 | 1 | 0 | G2 |
| 1 | 1 | 1 | 0 | 0 | 1 | 0 | 0 | 1 | 0 | G2 |
| 1 | 0 | 0 | 0 | 0 | 0 | 0 | 0 | 0 | 1 | late G1 |
| 1 | 0 | 1 | 0 | 0 | 0 | 0 | 0 | 0 | 1 | early G1 |
| 1 | 0 | 0 | 0 | 0 | 0 | 0 | 0 | 1 | 1 | late G1 |
| 1 | 0 | 1 | 0 | 0 | 0 | 0 | 0 | 1 | 1 | early G1 |
| 1 | 0 | 0 | 0 | 0 | 1 | 0 | 0 | 1 | 1 | G2 |
| 1 | 0 | 1 | 0 | 0 | 1 | 0 | 0 | 1 | 1 | G2 |

In comparison with random networks, cell cycle arrest in G1 and G2 in response to osmotic stress is a highly robust property ($p < 0.04$); 97% of the perturbed cell cycle networks with osmotic stress input converge to the original attractor (Fig. 6). Perturbation simulations for the cases of Hog1p activation were performed using a network in which the "Hog1" node and all rules affected by it were completely left out. Combined with the results of the analysis of the core cell cycle attractor, we can also conclude that release of the stress - setting "Hog1" to "0" would result in convergence on the cyclic attractor of the core cell cycle network.

### 4.3. *Mating pheromone robustly mediates G1 cell cycle arrest*

Arrest to pheromone in G1 - simulated by switching the model species "Fus3" on permanently - leads to comparibly robust point attractors in our model. We only find early and late G1 arrest states and no G2 arrest states in accordance with biological observations (see Sec. 2). The G1 arrest states are reached independently

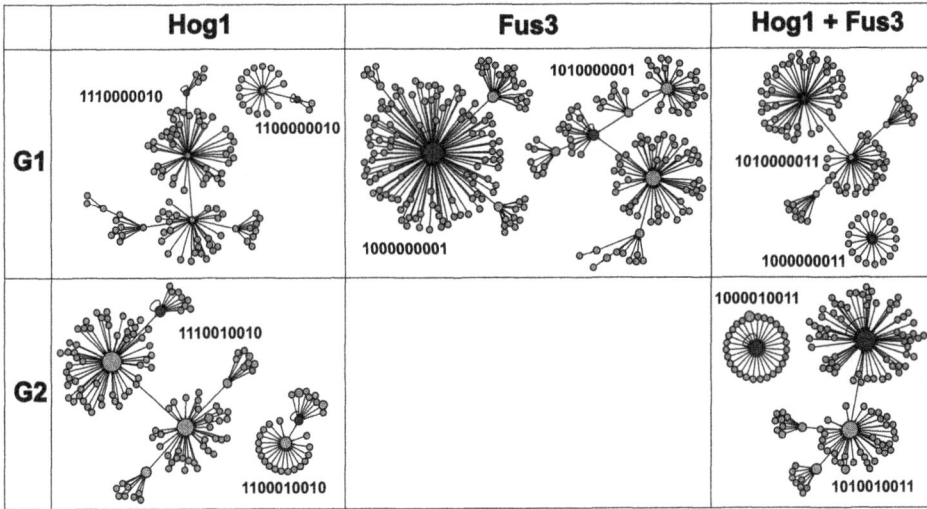

Fig. 5.   Point attractors under osmotic stress ("Hog1" permanently on) and pheromone exposure ("Fus3" permanently on) and their association to G1 and G2 cell cycle arrest states. For an explanation of the values of the singleton attractors refer to Tab. 3.

of the initial conditions and are robust towards asynchronous updating and hence towards slight variations of timing of reactions and events during the cell cycle phases. We demonstrate that the arrest due to the exposure to pheromone is also a more robust process than in random networks. Our comparison however yields that around 68% of the perturbed original networks converge to the singleton attractor of the unperturbed network (data not shown), better than the mean of the distribution of random networks which averages around 30%. However, our significance threshold of $p < 0.05$ is not met. Perturbation simulations for the cases of Fus3 activation were performed using a network in which the Hog1 node and all rules affected by it were completely left out. Release of pheromone exposure, by switching "Fus3" off, would cause the system to converge on the cyclic attractor of the core cell cycle.

## 4.4. *Simultaneous stimulation of both the HOG and the mating pathway yields highly robust cell cycle arrest*

Finally we simulated the simultaneous stimulation of the network with both osmotic stress and pheromone by switching both Hog1 and Fus3 permanently on. As shown in Fig. 5, G1 and G2 arrest states exclusively could be reached from the 256 initial conditions. Under asynchronous updating the states remain unchanged (data not shown). However, when compared to the random networks, the G1 and G2 arrest states under simultaneous stimulation are even more robust than in the osmotic stress case: 100% of the perturbed networks converge to a point attractor of the original network. Statistical testing yields a significance of $p < 0.001$.

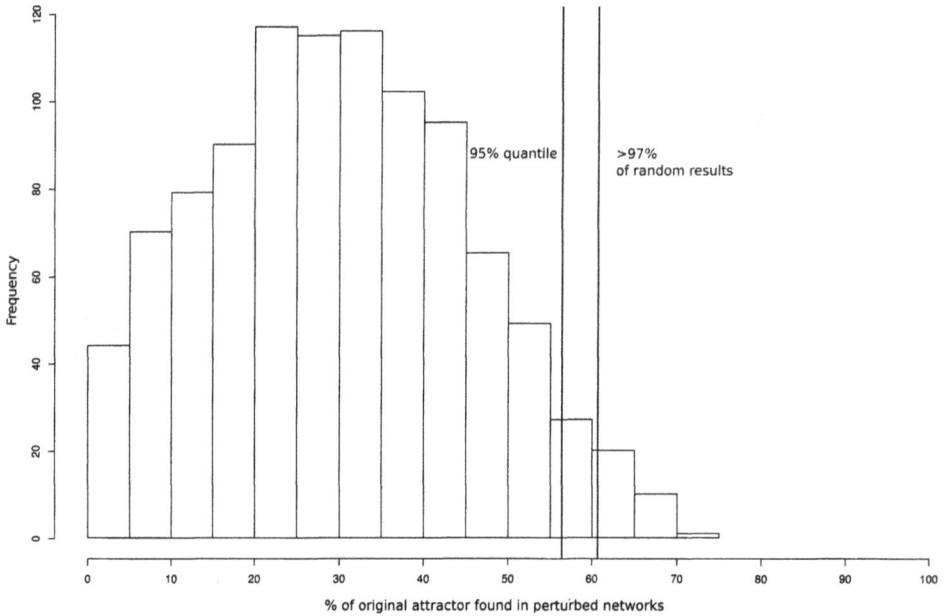

Fig. 6.   Comparison of the cell cycle model with "Hog1" input with random Boolean networks.

## 5. Discussion

We have presented a compact Boolean model for the budding yeast cell cycle which includes the interfaces to the HOG and pheromone pathways. We demonstrated that our model arrests in G1 and G2 depending on the initial conditions for osmotic stress and in G1 for pheromone exposure and that it is capable of converging back to the cyclic attractor if stress is lifted. A similar situation as for osmotic stress can be observed for the combined input of pheromone and osmotic stress. Similarly to previous studies [21, 23] we find that the attractor of the core cell cycle model is more robust to structural perturbations than attractors in average in random Boolean networks with compatible structure. Moreover, we find that upon asynchronous updating the rhythmic nature of the network might be preserved, suggesting that minor variations of the timing of reactions and events in the cell cycle have no major disruptive effects. However, the G1 and G2 arrest states in response to osmotic stress are significantly more robust than pheromone arrest in G1 and the core cell cycle itself. This finding could point to different physiological roles of the degree of robustness of the attractors: Non-hazardous situations such as pheromone exposure or mildly hazardous minor perturbations in the cell cycle network do not require such a high fidelity of response than the possibly life-threatening situation of osmotic shock.

## Acknowledgements

We would like to thank Falko Krause for his huge support by co-developing boolean-netGUI. This work was supported by grants of the German Research Council (DFG, International Research Training Group, GRK 1360), the European Commission (EU project Unicellsys HEALTH-2007-201142), the German Federal Ministry of Education and Research (BMBF, project Drug-iPS), to EK.

## References

[1] Albert, I., Thakar, J., Li, S., Zhang, R., and R. Albert, Boolean network simulations for life scientists, *Source Code Biol Med*, 3:16, 2008.

[2] Aldana, M., Coppersmith, S., and Kadanoff, L. P., *Perspectives and Problems in Nonlinear Science. A Celebratory Volume in Honor of Lawrence Sirovich*, chapter Boolean dynamics with random couplings, Springer Applied Mathematical Sciences Series., 2003.

[3] Amon, A., Tyers, M., Futcher, B., and Nasmyth, K., Mechanisms that help the yeast cell cycle clock tick: G2 cyclins transcriptionally activate G2 cyclins and repress G1 cyclins, *Cell*, 74(6):993–1007, 1993.

[4] Asano, S., Park, J.-E., Sakchaisri, K., Yu, L.-R., Song, S., Supavilai, P., Veenstra, T. D., and Lee, K. S, Concerted mechanism of Swe1/Wee1 regulation by multiple kinases in budding yeast, *EMBO J*, 24(12):2194–2204, 2005.

[5] Assmann, S. M. and Albert, R., Discrete dynamic modeling with asynchronous update, or how to model complex systems in the absence of quantitative information, *Methods Mol Biol*, 553:207–225, 2009.

[6] Belli, G., Gari, E., Aldea, M., and Herrero, E., Osmotic stress causes a G1 cell cycle delay and downregulation of Cln3/Cdc28 activity in Saccharomyces cerevisiae, *Mol Microbiol*, 39(4):1022–1035, 2001.

[7] Charvin, G., Oikonomou, C., Siggia, E. D., and Cross, F. R., Origin of irreversibility of cell cycle start in budding yeast, *PLoS Biol*, 8(1):e1000284, 2010.

[8] Clotet, J., Escote, X., Adrover, M. A., Yaakov, G., Gari, E., Aldea, M., de Nadal, E., and Posas, F., Phosphorylation of Hsl1 by Hog1 leads to a G2 arrest essential for cell survival at high osmolarity, *EMBO J*, 25(11):2338–2346, 2006.

[9] Cross, F. R., Schroeder, L., and Bean, J. M., Phosphorylation of the Sic1 inhibitor of B-type cyclins in Saccharomyces cerevisiae is not essential but contributes to cell cycle robustness, *Genetics*, 176(3):1541–1555, 2007.

[10] Cvrckova, F. and Nasmyth, K., Yeast G1 cyclins CLN1 and CLN2 and a GAP-like protein have a role in bud formation, *EMBO J*, 12(13):5277–5286, 1993.

[11] de Jong, H., Modeling and simulation of genetic regulatory systems: a literature review, *J Comput Biol*, 9(1):67–103, 2002.

[12] Escote, X., Zapater, M., Clotet, J., and Posas, F., Hog1 mediates cell-cycle arrest in G1 phase by the dual targeting of Sic1, *Nat Cell Biol*, 6(10):997–1002, 2004.

[13] Glass, L. and Kauffman, S. A., The logical analysis of continuous, non-linear biochemical control networks, *J Theor Biol*, 39(1):103–129, 1973.

[14] Harvey, S. L. and Kellogg, D. R., Conservation of mechanisms controlling entry into mitosis: budding yeast wee1 delays entry into mitosis and is required for cell size control, *Curr Biol*, 13(4):264–275, 2003.

[15] Hu, F., Gan, Y., and Aparicio, O. M., Identification of Clb2 residues required for Swe1 regulation of Clb2-Cdc28 in Saccharomyces cerevisiae, *Genetics*, 179(2):863–874, 2008.

[16] Jeoung, D. I., Oehlen, L. J., and Cross, F. R., Cln3-associated kinase activity in Saccharomyces cerevisiae is regulated by the mating factor pathway, *Mol Cell Biol*, 18(1):433–441, 1998.

[17] Kauffman, S. A., Metabolic stability and epigenesis in randomly constructed genetic nets, *J Theor Biol*, 22(3):437–467, 1969.

[18] Kitano, H., Systems biology: a brief overview, *Science*, 295(5560):1662–1664, 2002.

[19] Klipp, E., Liebermeister, W., Wierling, C., Kowald, A., Lehrach, H., and Herwig, R., *Systems Biology: A Textbook*, Wiley-Blackwell, 2009.

[20] Lew, D. J., Weinert, T., and Pringle, J. R., *The Molecular and Cellular Biology of the Yeast Saccharomyces: Cell Cycle and Cell Biology*, chapter Cell cycle control in Saccharomyces cerevisiae, 607–695. Cold Spring Harbor, NY: Cold Spring Harbor Laboratory Press, 1997.

[21] Li, F., Long, T., Lu, Y., Ouyang, Q., and Tang, C., The yeast cell-cycle network is robustly designed, *Proc Natl Acad Sci U S A*, 101(14):4781–4786, 2004.

[22] Lillacci, G. and Khammash, M., Parameter estimation and model selection in computational biology, *PLoS Comput Biol*, 6(3):e1000696, 2010.

[23] Mangla, K., Dill, D. L., and Horowitz, M. A., Timing robustness in the budding and fission yeast cell cycles, *PLoS One*, 5(2):e8906, 2010.

[24] Mendenhall M. D. and Hodge, A. E., Regulation of Cdc28 cyclin-dependent protein kinase activity during the cell cycle of the yeast Saccharomyces cerevisiae, *Microbiol Mol Biol Rev*, 62(4):1191–1243, 1998.

[25] Mizunuma, M., Miyamura, K., Hirata, D., Yokoyama, H., and Miyakawa, T., Involvement of S-adenosylmethionine in G1 cell-cycle regulation in Saccharomyces cerevisiae, *Proc Natl Acad Sci U S A*, 101(16):6086–6091, 2004.

[26] Muessel, C., Hopfensitz, M., and Kestler, H. A., BoolNet - an R package for generation, reconstruction, and analysis of Boolean networks, *Bioinformatics*, 2010.

[27] Nasmyth, K., Control of the yeast cell cycle by the Cdc28 protein kinase, *Curr Opin Cell Biol*, 5(2):166–179, 1993.

[28] Peter, M. and Herskowitz, I., Direct inhibition of the yeast cyclin-dependent kinase Cdc28-Cln by Far1, *Science*, 265(5176):1228–1231, 1994.

[29] Schwob, E., Boehm, T., Mendenhall, M. D., and Nasmyth, K., The B-type cyclin kinase inhibitor p40SIC1 controls the G1 to S transition in S. cerevisiae, *Cell*, 79(2):233–244, 1994.

[30] Schwob, E. and Nasmyth, K., CLB5 and CLB6, a new pair of B cyclins involved in DNA replication in Saccharomyces cerevisiae, *Genes Dev*, 7(7A):1160–1175, 1993.

[31] Shirayama, M., Toth, A., Galova, M., and Nasmyth, K., APC(Cdc20) promotes exit from mitosis by destroying the anaphase inhibitor Pds1 and cyclin Clb5, *Nature*, 402(6758):203–207, 1999.

[32] Sia, R. A., Bardes, E. S., and Lew, D. J., Control of Swe1p degradation by the morphogenesis checkpoint, *EMBO J*, 17(22):6678–6688, 1998.

[33] Talia, S. D., Skotheim, J. M., Bean, J. M., Siggia, E. D., and Cross, F. R., The effects of molecular noise and size control on variability in the budding yeast cell cycle, *Nature*, 448(7156):947–951, 2007.

[34] Verma, R., Annan, R. S., Huddleston, M. J., Carr, S. A., Reynard, G., and Deshaies, R. J., Phosphorylation of Sic1p by G1 Cdk required for its degradation and entry into S phase, *Science*, 278(5337):455–460, 1997.

[35] Visintin, R., Craig, K., Hwang, E. S., Prinz, S., Tyers, M., and Amon, A., The phosphatase Cdc14 triggers mitotic exit by reversal of Cdk-dependent phosphorylation, *Mol Cell*, 2(6):709–718, 1998.

[36] Visintin, R., Prinz, S., and Amon, A., CDC20 and CDH1: a family of substrate-specific

activators of APC-dependent proteolysis, *Science*, 278(5337):460–463, 1997.

[37] Waesch, R. and Cross, F. R., APC-dependent proteolysis of the mitotic cyclin Clb2 is essential for mitotic exit, *Nature*, 418(6897):556–562, 2002.

[38] Wijnen, H., Landman, A., and Futcher, B., The G(1) cyclin Cln3 promotes cell cycle entry via the transcription factor Swi6, *Mol Cell Biol*, 22(12):4402–4418, 2002.

# A DYNAMIC PROGRAMMING ALGORITHM TO PREDICT SYNTHESIS PROCESSES OF TREE-STRUCTURED COMPOUNDS WITH GRAPH GRAMMAR

YANG ZHAO[1]
tyoyo@kuicr.kyoto-u.ac.jp

TAKEYUKI TAMURA[1]
tamura@kuicr.kyoto-u.ac.jp

MORIHIRO HAYASHIDA[1]
morihiro@kuicr.kyoto-u.ac.jp

TATSUYA AKUTSU[1]
takutsu@kuicr.kyoto-u.ac.jp

[1] *Bioinformatics Center, Institute for Chemical Research, Kyoto University, Gokasho, Uji, Kyoto, 611-0011, Japan*

For several decades, many methods have been developed for predicting organic synthesis paths. However these methods have non-polynomial computational time. In this paper, we propose a bottom-up dynamic programming algorithm to predict synthesis paths of target tree-structured compounds. In this approach, we transform the synthesis problem of tree-structured compounds to the generation problem of unordered trees by regarding tree-structured compounds and chemical reactions as unordered trees and rules, respectively. In order to represent rules corresponding to chemical reactions, we employ a subclass of NLC (Node Label Controlled) grammars. We also give some computational results on this algorithm.

*Keywords*: organic syntheses; tree-structured compounds; dynamic programming algorithm; NLC grammars; unordered trees.

## 1. Introduction

Synthetic analysis of chemical compounds is one of important branches of organic chemistry. It is also important for design of new drugs, which is one of major goals of bioinformatics and systems biology. In order to synthesize target compounds, chemists have to consider a huge amount of possibilities on starting materials, intermediate compounds, and synthesis paths. For this reason, many computer-aided approaches have been proposed for synthesis design. Based on the notion of retrosynthetic analysis which was first proposed by Corey, programs such as LHASA [1], SYNCHEM [4], SECS [12], LILITH [11], TRESOR [9] were developed for synthesis planning. In order to construct a target compound, these programs analyze all possible retrosynthetic routes from some simpler available starting materials by applying reliable chemical transformations. However, none of these systems have been widely used because of its inefficiency due to a huge search space. Other programs such as EROS [3], WODCA [2], SYNGEN [5] have significant differences. These programs seek to how one could and should construct a target chemical structure from starting materials, without regard to the existence of synthetic methods and intermediate products. However, these programs have not yet been used widely. The reason might be that the prediction of synthetic routes could not be solved simply. Apart from practical aspects, much attention has not been paid to computational complexity so far, and most existing algorithms have exponential time complexities.

In this paper, we focus on the time complexity aspect of prediction of synthesis paths. As a first step towards development of a practical and polynomial time algorithm for the synthesis path problem, we define a kind of graph grammars that represent chemical reactions and propose a bottom-up dynamic programming algorithm for special classes of chemical reactions and chemical compounds: chemical reactions that do not decrease the sizes of molecules, and chemical compounds that have tree structures. It is to be noted that some restrictions are needed to develop a polynomial time algorithm. Otherwise, the problem would become intractable because it might be possible to simulate the universal Turing machine. Though the classes of chemical compounds and chemical reactions covered by our proposed method are quite limitted and far from practical, these might be extended to more general classes.

The organization of the paper is as follows. In Sec. 2, we define ECRGT (Elementary Chemical Reaction Grammar for Tree-structured molecules) by employing NLC grammars, which can be applied to unordered node labeled tree structures. Then, our main algorithm, called PGDP (Path Generation by Dynamic Programming), is introduced in Sec. 3. In Sec. 4, we conduct some computational experiments. Finally, we conclude with future work.

## 2. Chemical Reaction Grammar for Tree-Structured Molecules

In this section, we briefly review *node label controlled* (NLC) graph grammars, and introduce ECRGT which is a restricted variant of NLC grammars. We chose NLC graph grammars as our theoretical foundation because the complexity of NLC graph grammars has been well-studied [10].

### 2.1. *NLC graph grammars*

Graph grammars are often divided in two classes: node replacement grammars and hyperedge replacement grammars. As a simplest class of node replacement grammars, NLC grammars have been introduced by Janssens and Rozenberg in 1980 [7, 8] which only rewrite single nodes of undirected node labeled graphs. A simple form of an NLC grammar is a pair $(\alpha, S)$, written as $\alpha \to S$, where $\alpha$ denotes a label and $S$ denotes a graph. For applying such a form to a node $v$ in a graph $G$, we replace $v$ with $S$ if and only if $v$ is labeled by $\alpha$, and build the connection between $S$ and neighbors of $v$.

### 2.2. *ECRGT*

In order to represent rules corresponding to chemical reactions, we employ the concept of NLC grammars, and define a restricted variant, ECRGT (Elementary

Chemical Reaction Grammar for Tree-structured molecules). Though ECRGT may be regarded as a kind of *tree grammars* [13], we define it as a restricted variant of NLC grammars for possible future extensions of our proposed method. An ECRGT is defined as a system $(\Sigma, \mathcal{G}, \Delta)$, where $\Sigma$ is a finite nonempty set of labels, $\mathcal{G}$ is a finite set of undirected node labeled graphs over $\Sigma$, and $\Delta$ denotes a finite set of rewriting rules $\alpha \to S$, where $\alpha \in \Sigma$, $S \in \mathcal{G}$, and $\alpha \to S$ means that a single node labeled with $\alpha$ is replaced by graph $S$. We let the size of any graph $S$ in $\mathcal{G}$ be always smaller than 3 as a constraint of productions in an ECRGT. For such a constraint, a graph $S$ in $\mathcal{G}$ is a single node or an edge when $|S|$ is 1 or $|S|$ is 2, respectively, where $|S|$ denotes the number of nodes in $S$. Thus, productions in an ECRGT can be partitioned in two categories:

(i) when $|S| = 1$, $S$ is a single node labeled $\beta$, a production in ECRGT, called $ECRGT_1$, can be written as $\alpha \to \beta$, and this production rule rewrites label $\alpha$ of a node to label $\beta$ and keeps the original connection of this node;

(ii) when $|S| = 2$, $S$ is an edge with two nodes labeled $\beta$ and $\gamma$ respectively, a production in ECRGT, called $ECRGT_2$, can be written as $\alpha \to \beta\gamma$, and this production rule rewrites label $\alpha$ of a node to label $\beta$, adds a node labeled $\gamma$, and connects these nodes by an edge.

It is to be noted that we do not distinguish *terminal symbols* and *non-terminal symbols* because there is no such distinction in chemical reactions. It should also be noted that

(i) ECRGT only generates trees if the generation process starts from a tree,

(ii) ECRGT does not decrease the size of graphs.

In this paper, we only consider cases where the generation process starts from a tree. Therefore, we only consider trees and tree-structured chemical compounds. Figure 1 illustrates production rules of an ECRGT with $\Delta = \{c \to a, b \to ca\}$.

## 3. Dynamic Programming Algorithm for Generation of Synthesis Paths

In this section we present a dynamic programming algorithm, called PGDP (Path Generation by Dynamic Programming), for the synthesis path problem with ECRGT. We also show that this algorithm requires polynomial computational time even in the worst case.

We call compounds without rings *tree-structured compounds*. A tree-structured compound can be expressed as a tree if we treat atoms and kinds of atoms as nodes and their labels, respectively. When we predict synthesis paths from a given starting compound to a given target compound, we express these two compounds as a source tree and a target tree, respectively, and a synthesis path is modeled as a sequence of production rules from a given ECRGT. That is, we transform the synthesis path problem to a tree generation problem. An example of this transformed synthesis process is given in Fig. 2.

### 3.1. *PGDP algorithm*

Let $P$ be the source tree and $u$ be a node in $P$, $T$ be the target tree and $v$ be a node in $T$. $P$ and $T$ represent a start compound and a target compound, respectively. Let $\Sigma$ be a finite nonempty label set that consists of all kinds of labels included in $P$ and $T$. Let $\alpha$ be a label in $\Sigma$. Let $P(u)$ be the subtree of $P$ rooted at node $u$ and $T(v)$ be the subtree of $T$ rooted at node $v$. Define $\mathcal{L}[\alpha, v]$ be 1 if $T(v)$ can be generated from a single node labeled with $\alpha$. Otherwise, $\mathcal{L}[\alpha, v]$ is 0. Define $\mathcal{D}[u, v]$ be 1 if subtree $T(v)$ can be generated from $P(u)$. Otherwise, $\mathcal{D}[u, v]$ is 0. Here we partition the algorithm in two categories: (i) computation of $\mathcal{L}[\alpha, v]$ for each pair $(\alpha, T(v))$, and (ii) computation of $\mathcal{D}[u, v]$ for each node pair in $P$ and $T$. The main part of the algorithm is (ii). It computes $\mathcal{D}[u, v]$ from given source tree $P$, target tree $T$, and ECGRT system. Once $\mathcal{D}[u, v]$ are computed, a synthesis path can be obtained using the standard *traceback* technique.

#### 3.1.1. *A procedure for computing* $\mathcal{L}[\alpha, v]$

For computing $\mathcal{L}[\alpha, v]$, the pseudocode is given below, where during the computation, we produce a tree set, called *tree*[$x$] for each $x$ ($x \in \Sigma$) by applying ECRGT productions, and a tree set $T(v)$ dynamically for each $v$ ($v \in T$). For a label $x$ ($x \in \Sigma$), *tree*[$x$] is initialized as $\{x\}$ (i.e., tree consisting of a node with label $x$). During the generation of *tree*[$x$], as constraints, we restrict $ECRGT_2$ to be only used in roots of trees included in *tree*[$x$]; and $ECRGT_1$ is applied to any nodes

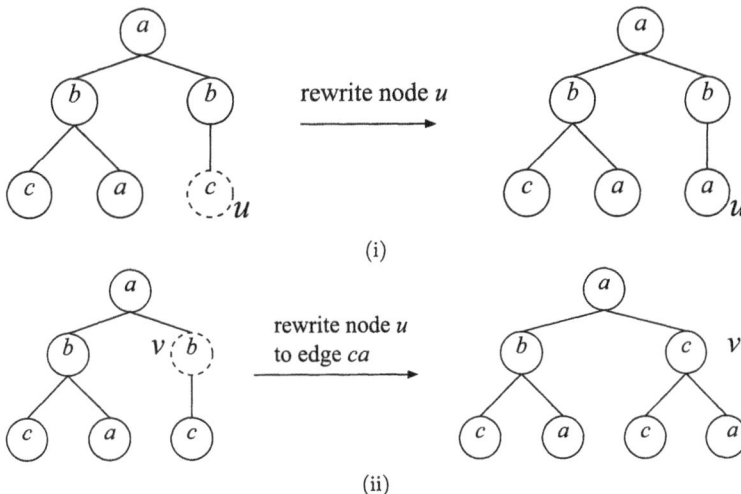

Fig. 1. Node rewriting processes of production in ECRGT. $P$ is a source tree, label($v$) = b, label($u$) = c. (i) illustrates application of production rule $c \to a$ to node $u$. (ii) illustrates application of production rule $b \to ca$ to node $v$.

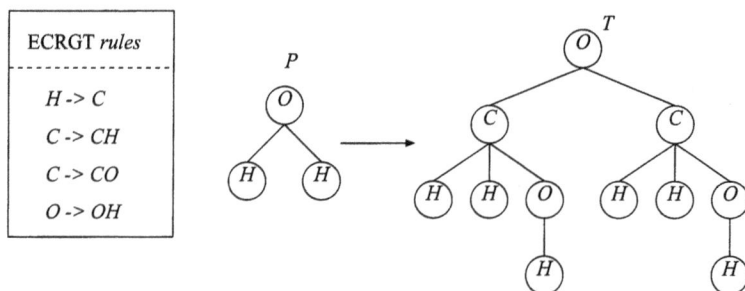

Fig. 2.   Illustration of an example of synthesis process with starting compound $H_2O$ and target compound $CH_2OHOCH_2OH$, which can be computer structured as a source tree and a target tree respectively. And ECRGT defined here is system $(\Sigma, \mathcal{G}, \Delta)$, where $\Sigma = \{C, H, O\}$, $\mathcal{G} = \{\{C\}, \{(C, H)\}, \{(C, O)\}, \{(O, H)\}\}$, $\Delta = \{H \rightarrow C, C \rightarrow CH, C \rightarrow CO, O \rightarrow OH\})$.

of trees in $tree[x]$. Let $h$ be the maximum height of trees in $tree[x]$, and $d$ be the maximum outdegree of target tree $T$ (i.e., the maximum number of children). We produce trees to generate $tree[x]$ which have $h \leq 1$ and maximum degree $\leq d$. For a node $v$ in $T$, if $v$ is a leaf, we let $T(v)$ be a set consisting of a tree with a single node labeled $x$. Let $child(v)$ denote the set of children of $v$. If $|child(v)| \neq 0$, assuming that $v_1, \ldots, v_l$ are the children of $v$, we let $T(v)$ be a set of trees rooted at $v$ with leaves labeled $\{x_1, \ldots, x_l\}$ such that $\mathcal{L}[x_j, v_j]=1$. Notice that, in this solution, if we fail to produce $T(v)$ for a node $v$, $\mathcal{L}[x, v]$ is always 0 for each $x$ ($x \in \Sigma$). In the following, a tree consisting of a single node labeled $x$ is denoted by $x$, and a tree of height 1 consisting of a root labeled $x$ and leaves labeled $y_1, y_2, \ldots, y_l$ is denoted by $(x, y_1, \ldots, y_l)$.

**Input:** $T$, ECRGT system
**Output:** $\mathcal{L}[\alpha, v]$
**Method:**
$d :=$ maximum degree of $T$
**for** each $x \in \Sigma$ **do**
   $tree[x] := \{x\}$
   $tree[x] := tree[x] \cup ECRGT_1(tree[x])$
   **for** $k := 1$ to $d$ **do**
      $tree[x] := tree[x] \cup ECRGT_2(tree[x])$
      $tree[x] := tree[x] \cup ECRGT_1(tree[x])$
**for** each $v$ of $T$ in postorder **do**
   **if** $|child(v)| = 0$
   **then** $T(v) := \{label(v)\}$
   **else** $T(v) := \{(label(v), y_1, y_2, \cdots, y_l) | \mathcal{L}[y_j, v_j] = 1 \text{ for all } v_j \in child(v)\}$
   **for** each $x \in \Sigma$ **do**
      **if** $T(v) \cap tree[x] \neq \emptyset$
      **then** $\mathcal{L}[x, v] := 1$
      **else** $\mathcal{L}[x, v] := 0$
**end**

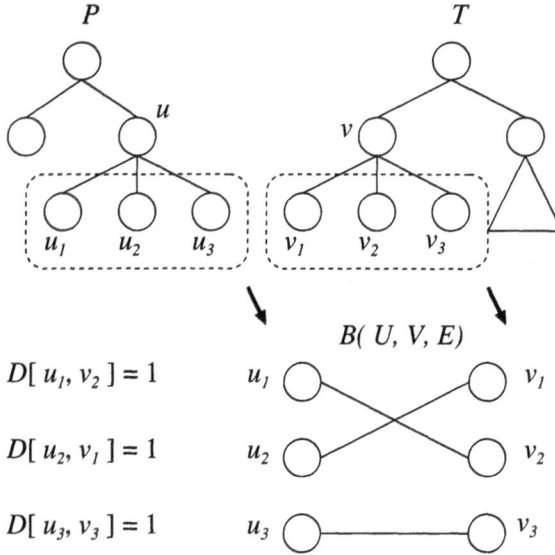

Fig. 3. Construction of a bipartite graph $B(U, V, E)$ for pair$(u, v)$, where $u \in P, v \in T$, $U = \{child(u)\}$, $V = \{child(v)\}$, $E = \{(u_i, v_j) | \mathcal{D}[u_i, v_j] = 1, u_i \in child(u), v_j \in child(v)\}$.

### 3.1.2. *PGDP for $\mathcal{D}[u, v]$*

As mentioned above, we gave a dynamic programming solution to compute each $\mathcal{L}[\alpha, v](\alpha \in \Sigma, v \in T)$ by applying a defined ECRGT system. Here in this PGDP algorithm, we let the results of $\mathcal{L}[\alpha, v]$ be the results of $\mathcal{D}[u, v]$, if and only if $u$ is a leaf node of $P$, and $label(u) = \alpha$. And we also construct a bipartite graph $B(U, V, E)$ for each pair$(u, v)(u \in P, v \in T)$, when $u$ and $v$ have the same size of nonempty child set. Let $U$ be $child(u)$, and $V$ be $child(v)$. For any $u_i \in U, v_j \in V$, if $\mathcal{D}[u_i, v_j]$ is 1, we let $(u_i, v_j)$ be an edge in $E$ (see also Fig. 3). We compute $\mathcal{D}[u, v]$ to be 1, if and only if the graph $B(U, V, E)$ has a complete bipartite matching. The pseudocode for computing $\mathcal{D}[u, v]$ is given below.

**Input:** $P$, $T$, $\mathcal{L}[\alpha, v]$
**Output:** $\mathcal{D}[u, v]$
**Method:**
for each $u$ in $P$ **do**
  for each $v$ in $T$ **do**
    **if** $|child(u)| = 0$
    **then** $\mathcal{D}[u, v] := \mathcal{L}[label(u), v]$
    **else if** $|child(u)| = |child(v)|$
        **then** Construct bipartite graph $B(U, V, E)$
            $U := \{u_1, u_2, \cdots, u_h\}$
            $V := \{v_1, v_2, \cdots, v_l\}$
            $E := \{(u_i, v_j) | \mathcal{D}[u_i, v_j] = 1\}$
            **if** $B$ has a complete bipartite matching
            **then** $\mathcal{D}[u, v] := 1$
            **else** $\mathcal{D}[u, v] := 0$

       **else** $\mathcal{D}[u,v] := 0$

**end**

## 3.2. *Time complexity*

In Sec. 3.1.1, for computing $\mathcal{L}(\alpha, v)$, we produce a tree set $tree[x]$ for each $x$ ($x \in \Sigma$) in advance. In this production, the number of possible trees is bounded by $O(|\Sigma|^d)$. If the degree of $T$ is decided and the number of labels in $P$ and $T$ is limited, which equivalently mean that $d$ and $|\Sigma|$ are constants, we get that the part of $|\Sigma|^d$ is a constant as well. Therefore, construction of $tree[x]$ can be done in constant time. In order to compute $\mathcal{L}[\alpha, v]$, we dynamically generate a tree set $T(v)$ for each $v$ ($v \in T$), which need time $O(n)$, where $n = |T|$ and $m = |P|$. Thus, the former part of this algorithm runs in $O(n)$ time.

In Sec. 3.1.2, for each leaf in $P$, we employ the result of $\mathcal{L}[\alpha, v]$ to be the result of $\mathcal{D}[u, v]$, if and only if the label of $u$ is $\alpha$. This procedure needs computational time $O(m \cdot n)$. For nodes which are not leaves in $P$, in order to compute $\mathcal{D}[u, v]$, we construct bipartite graphs and find complete bipartite matchings of these graphs. We know that, the algorithms for bipartite matching problem have time complexity $O(d^{2.5})$ [6]. Thus, this part of the algorithm runs in $O(m \cdot n \cdot d^{2.5})$ time. If $d$ is a constant, the part of $d^{2.5}$ is a constant as well. Thus, the latter part of the algorithm has time complexity $O(m \cdot n)$.

Overall, integrating the time complexity of these two sections mentioned above, our algorithm consumes time $O(m \cdot n + n) = O(m \cdot n)$. Thus, we see that this algorithm for the synthesis problem of tree-structured compounds has polynomial time complexity.

## 4. Computational Experiments

We implemented the algorithm introduced above and performed some computational experiments. In our implementation, we did not employ the $O(d^{2.5})$ time bipartite matching algorithm, instead used a simple exhaustive method because $d$ is very small in practice (usually, $d \leq 8$).

## 4.1. *Experiment 1*

We chose a biochemical pathway from 2-oxo-glutarate to N-acetyl-glutamyl-P through L-glutamate and N-acetyl-glutamate. Chemical structures of these compounds are shown in Fig. 4. Considering 2-oxo-glutarate as the starting compound and N-acetyl-glutamyl-P as the target compound, Fig. 5 gives the tree structures of these compounds without hydrogen atoms. It is to be noted that we excluded hydrogen atoms and did not distinguish between single bonds and double bonds so that we can only use production rules that do not delete nodes. In this case, the label set $\Sigma$ is defined as $\{C, O, N, P\}$. The productions of ECRGT are defined as $\{O \rightarrow N, N \rightarrow NC, C \rightarrow CO, C \rightarrow CC, O \rightarrow OP\}$ (see also Fig. 5). After the initialization about starting tree $P$, target tree $T$, $\Sigma$ and productions of ECRGT, the algorithm dynamically computes $\mathcal{L}[\alpha, v]$ for each $\alpha \in \Sigma$ and $v \in T$, and $\mathcal{D}[u, v]$ for each $u \in P$ and $v \in T$. The results are given in Fig. 6. Finally, the algorithm traces back the results of $\mathcal{D}[u, v]$ to get the generation path from the starting tree to the target tree (see also Fig. 7). Since we only used simple production rules, this path is not the same as that in Fig. 4. However, it is intrinsically the same as the one if we decompose single reactions in Fig. 4 to multiple reactions. For computing this example, the algorithm took CPU time 0.003 sec. on a PC with Core2 Quad CPU 3.0GHz and 2GB memory.

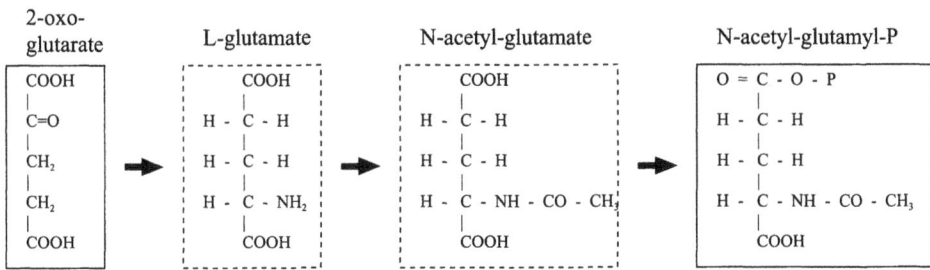

Fig. 4.   A pathway from 2-oxo-glutarate to N-acetyl-glutamyl-P through intermediate compounds L-glutamate and N-acetyl-glutamate.

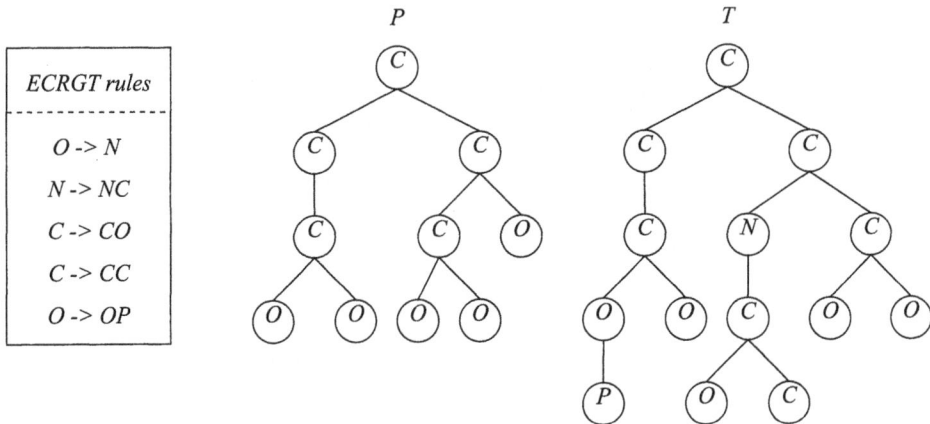

Fig. 5.   Tree structures of 2-oxo-glutarate and N-acetyl-glutamyl-P without hydrogen atoms which are regarded as the starting tree $P$ and target tree $T$.

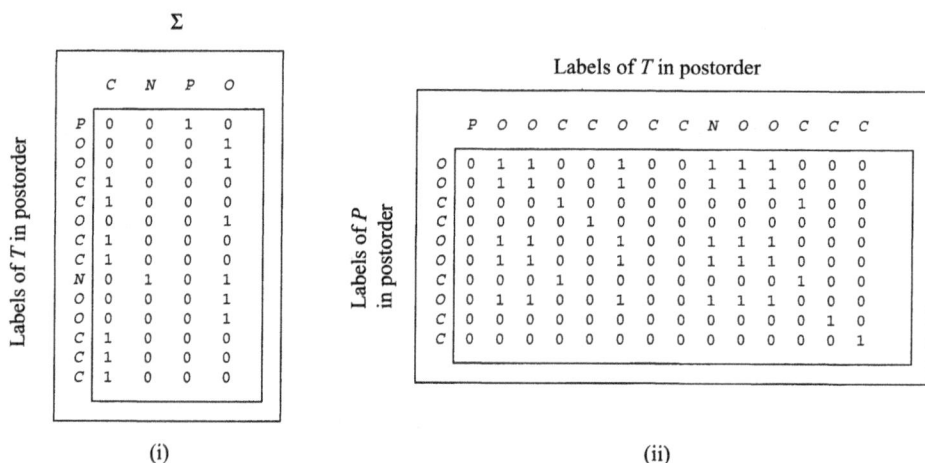

Fig. 6.   Dynamic programming tables (i) $\mathcal{L}[\alpha, v]$ and (ii) $\mathcal{D}[u, v]$ for 2-oxo-glutarate and N-acetyl-glutamyl-P.

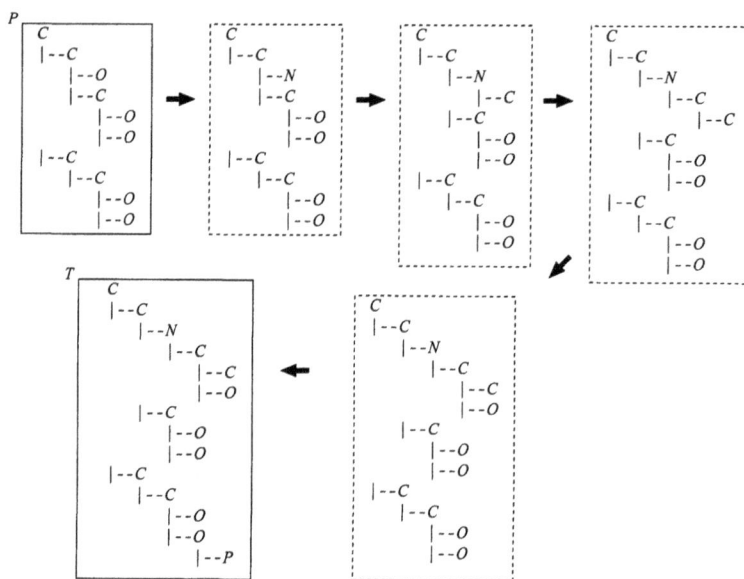

Fig. 7.   Tracing back the results shown in Fig. 6(ii), the algorithm produces the generation path from 2-oxo-glutarate to N-acetyl-glutamyl-P.

## 4.2. Experiment 2

Another example is to compute a pathway from 2-oxobutyrate to 4-aspartyl-P through L-homoserine. Chemical structures of these compounds are shown in Fig. 8. Considering 2-oxobutyrate as the starting compound and 4-aspartyl-P as the target compound, Fig. 9 gives the tree structures of these compounds without hydrogen atoms. As in the previous case, the label set $\Sigma$ is defined as $\{C, O, N, P\}$. The productions of ECRGT are defined as $\{O \rightarrow N, C \rightarrow CO, O \rightarrow OP\}$ (see also Fig. 9). The results are given in Fig. 10 and Fig. 11. For computing this example, the algorithm took CPU time 0.002 sec.

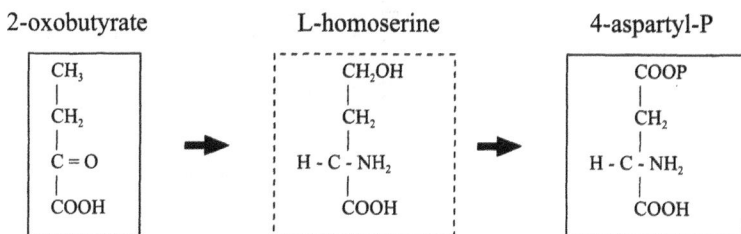

Fig. 8. A pathway from 2-oxobutyrate to 4-aspartyl-P through intermediate compound L-homoserine.

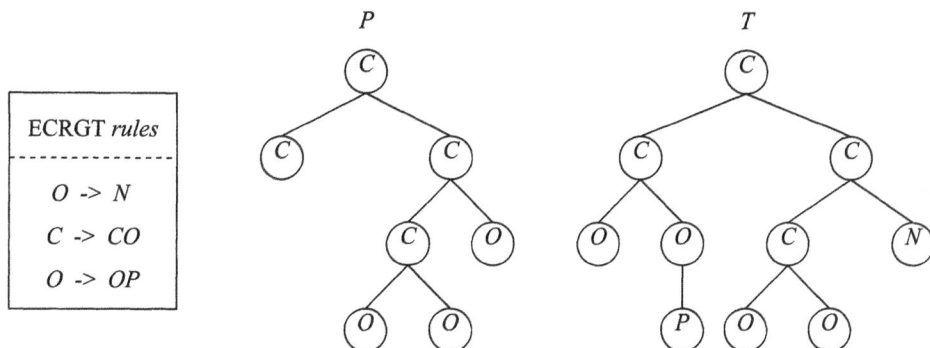

Fig. 9. Tree structures of 2-oxobutyrate and 4-aspartyl-P without hydrogen atoms which are regarded as the starting tree $P$ and target tree $T$.

## 5. Conclusion

In this paper, we proposed a dynamic programming algorithm to predict synthesis paths of target tree-structured compounds. We defined ECRGT graph grammars to represent chemical reactions. Regarding the tree-structured compounds as undirected node labeled trees, the algorithm computes generation paths from starting trees to target trees which denote starting materials and target compounds, respectively. Different from many existing methods, our proposed algorithm can compute a generation path between two given compounds even if a pool of possible intermediate compounds is not given. The algorithm was shown to have polynomial time

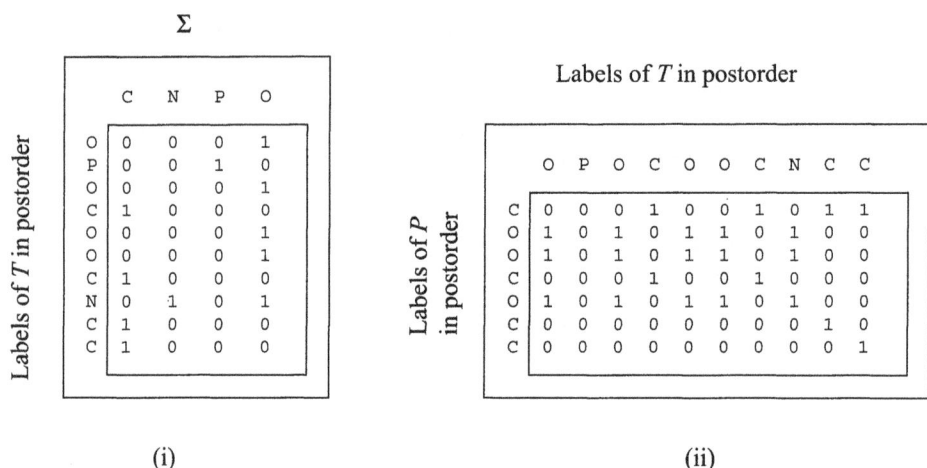

Fig. 10.   Dynamic programming tables (i) $\mathcal{L}[\alpha, v]$ and (ii) $\mathcal{D}[u, v]$ for 2-oxobutyrate and 4-aspartyl-P.

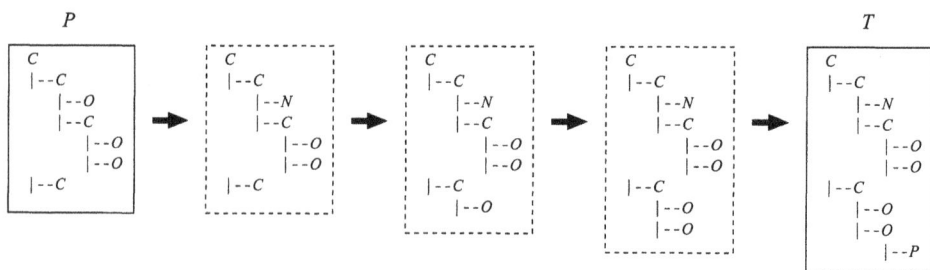

Fig. 11.   Tracing back the results shown in Fig. 10(ii), the algorithm produces the generation path from 2-oxobutyrate to 4-aspartyl-P.

complexity. The results of computational experiments suggest that our proposed algorithm is very fast in practice.

Further studies should be done to handle more complex compounds. Although our algorithm can predict synthesis paths of tree-structured compounds, it cannot predict synthesis paths of compounds which have rings. Another inadequacy is that we restricted graph grammars so that the productions can only add nodes but cannot delete nodes mainly because of time complexity issue. However, there are a plenty of chemical reactions that make target compounds have fewer elements than starting compounds. Therefore, improvement of the algorithm should be done to cover wider classes of chemical reactions and chemical compounds. We restricted synthesis paths so that each path starts from a single compound. However, there exist many synthesis paths starting from multiple compounds. Therefore, improvement of the algorithm should also be done to predict synthesis paths starting from multiple compounds.

## References

[1] Corey, E. J., Cramer, R. D., and Howe, W. J., Computer-assisted synthetic analysis for complex molecules, *J. Am. Chem. Soc.*, 94(2):440–459, 1972.

[2] Gasteiger, J., Ihlenfeldt, W.-D., Fick, R., and Rose, J. R., Similarity concepts for the planning of organic reactions and syntheses, *J. Chem. Inf. Comput. Sci.*, 32(6):700–712, 1992.

[3] Gasteiger, J., Ihlenfeldt, W.-D., Rose, P., and Wanke, R., Computer-assisted reaction prediction and synthesis design, *Anal. Chim. Acta*, 235:65–75, 1990.

[4] Gelernter, H. L., Sanders, A. F., Larsen, D. L., Agarwal, K. K., Boivie, R. H., Spritzer, G. A., and Searleman, J. E., Empirical explorations of SYNCHEM, *Science*, 197(4308):1041–1049, 1977.

[5] Hendrickson, J. B., Organic synthesis in the age of computers, *Angew. Chem. Int. Ed. Engl.*, 29(11):1286–1295, 1997.

[6] Hopcroft, J. E. and Karp, R. M., An $n^{5/2}$ algorithm for maximum matchings in bipartite graphs, *SIAM. J. Comput.*, 2(4):225–231, 1973.

[7] Janssens, D. and Rozenberg, G., On the structure of node-label-controlled graph languages, *Info. Sci.*, 20(3):191–216, 1980.

[8] Janssens, D. and Rozenberg, G., Restriction, extensions and variations of NLC grammars, *Info. Sci.*, 20(3):217–244, 1980.

[9] Moll, R., Context description in synthesis planning, *J. Chem. Inf. Comput. Sci.*, 34(1):117–119, 1994.

[10] Rozenberg, G. (ed.), *Handbook of Graph Grammars and Computing by Graph Transformation*, World Scientific, 1997.

[11] Sello, G., From childhood to adolescence, *J. Chem. Inf. Comput. Sci.*, 34(1):120–129, 1994.

[12] Wipke, W. T., Ouchi, G. I., and Krishnan, S., Simulation and evaluation of chemical synthesis-SECS, *Artif. Intell.*, 11(1-2):173–193, 1978.

[13] http://tata.gforge.inria.fr/

# AUTHOR INDEX

www.ingramcontent.com/pod-product-compliance
Lightning Source LLC
Chambersburg PA
CBHW081537190326
41458CB00015B/5570